Principles of Membrane Processes in Liquid Separation

Principles of Membrane Processes in Liquid Separation fills a gap in membrane science by focusing on fundamental principles rather than materials or applications. It covers three key areas: the distinction between concentration polarization and cake formation in membrane processes, state equation modeling for reverse osmosis desalination optimization, and mechanisms of water and ion transport across membranes. The text provides theoretical foundations while offering practical insights for improved membrane process design and operation, serving both as an educational resource for students and a reference for engineering professionals.

This book:

- Addresses fundamental principles of membrane processes in liquid separation rather than materials or fabrication methods.
- Provides advanced modeling approaches and optimization techniques for reverse osmosis desalination that enable more efficient process design and operation.
- Resolves longstanding conceptual challenges in membrane science, including proper characterization of concentration polarization and improved boundary conditions for ion transport equations.

Principles of Membrane Processes in Liquid Separation

Theories, Models, and Quantitative Analysis

Lianfa Song

CRC Press
Taylor & Francis Group
Boca Raton London New York

CRC Press is an imprint of the
Taylor & Francis Group, an informa business

Designed cover image: Shutterstock

First edition published 2027
by CRC Press
2385 NW Executive Center Drive, Suite 320, Boca Raton FL 33431

and by CRC Press
4 Park Square, Milton Park, Abingdon, Oxon, OX14 4RN

CRC Press is an imprint of Taylor & Francis Group, LLC

ISBN: 978-1-041-24790-6 (hbk)
ISBN: 978-1-041-24791-3 (pbk)
ISBN: 978-1-003-74356-9 (ebk)

DOI: 10.1201/9781003743569

Typeset in Palatino
by KnowledgeWorks Global Ltd.

To my wife Fuxia and daughters Lisa and Julie.

Contents

About the Author

Dr. Lianfa Song received his BS and MS degrees from Peking University, China, in 1982 and 1984, respectively. He went to the United States to study for his PhD degree in UCLA in 1990 and obtained his degree in March 1993. After two and a half years as a postdoctoral research associate at UCLA and the Oak Ridge National Laboratory, he worked as an assistant professor at Hong Kong University of Science and Technology (HKUST) in 1996 and as an associate professor in 2002 at National University of Singapore (NUS). He became an associate professor in the Department of Civil and Environmental Engineering at Texas Tech University in September 2007 and was promoted to full professor in 2013. Dr. Song's main research interests include membrane processes, colloidal phenomena, and modeling and optimization of environmental engineering systems. In the last few years, his research has focused on reverse osmosis (RO) processes on water treatment and reclamation. He has developed new concepts and models for the performance of full-scale RO processes in order to improve process efficiency holistically. His work in the membrane field has been well received and recognized by the membrane community. He is a member of the North American Membrane Society and the International Water Association.

Symbols

The majority of symbols used in this book are listed below. The first appearance of the symbols is indicated by equation number or section number.

$\mathbf{a}$	a constant ratio of time-dependent thickness to equilibrium thickness of cake layer, Eq. (4.23)
$\mathbf{A}$	membrane permeability, Eq. (1.11)
$\mathbf{a_1}, \mathbf{a_2}, \mathbf{a_2}$	coefficients to be determined, Section 5.3.2
$\mathbf{A_{os}}$	osmotic membrane permeability
$\mathbf{a_p}$	radius of particles, Eq. (1.3)
$\mathbf{A_s}$	correction factor, Eq. (2.8)
$\mathbf{A_s}^*$	membrane correction factor, Eq. (2.12)
$\mathbf{B}$	benefit, Eq. (10.10)
$\hat{\mathbf{B}}$	specific benefit, Eq. (10.12)
$\hat{\mathbf{B}}_0$	ultimate specific benefit, Eq. (10.14)
$\mathbf{C}$	concentration of retained particles, Eq. (2.5)
$\mathbf{c}$	concentration, Eq. (1.1)
$\mathbf{C}$	dimensionless concentration, Eq. (5.16)
$\mathbf{c_+}, \mathbf{c_-}$	cation concentration and anion concentration, Eq. (11.32)
$\mathbf{c^0}$	concentration at standard condition, Eq. (9.1)
$\mathbf{c_0}$	concentration of feed water, Eq. (7.13)
$\mathbf{C_1}$	cost for feed water pretreatment, Eq. (6.1)
$\mathbf{c_1}$	feed water pretreatment price, Eq. (6.5)
$\mathbf{C_2}$	cost for membrane process, Eq. (6.1)
$\mathbf{c_3}$	concentrate handling price, Eq. (6.6)
$\mathbf{C_3}$	cost for concentration handling, Eq. (6.1)
$\mathbf{c_a}$	average concentration, Eq. (7.17)
$\mathbf{c_d}$	price for draw solution pretreatment, Eq. (10.8)
$\mathbf{c_e}$	energy price, Eq. (6.3)
$\mathbf{c_e}$	concentration at the exit of membrane channel, Section 5.3.2
$\mathbf{C_e}$	dimensionless concentration at the exit of membrane channel, Eq. (5.20)
$\mathbf{C_e}$	energy cost, Eq. (6.3)
$\mathbf{c_f}$	feed concentration, Eq. (1.4)
$\mathbf{c_f}$	foulant concentration, Eq. (8.5)
$\mathbf{C_g}$	cake concentration of retained particles, Eq. (4.12)
$\mathbf{c_g}$	concentration of gel (cake) layer, Eq. (2.19)
$\mathbf{c_i}$	concentration at the interface of skin layer and porous layer, Eq. (9.19)
$\mathbf{c_{i0}}$	ion concentration in bulk solution on the side of membrane surface at 0, Eq. (11.62)

c_i^0	ion concentration on membrane surface at 0, Eq. (11.62)
c_{iL}	ion concentration in bulk solution on the side of membrane surface at L, Eq. (11.63)
c_i^L	ion concentration on membrane surface at L, Eq. (11.63)
c_m	molar concentration, Eq. (1.1)
c_m	membrane price, Eq. (6.2)
c_m	concentration inside membrane, Section 11.1.1
C_m	membrane cost, Section 6.1.2
c_{m0}	concentration on membrane surface on feed side, Eq. (11.1)
c_{ml}	concentration on membrane surface on permeate side, Eq. (11.5)
C_n	number concentration of retained particles, Section 2.3.3
c_n	number concentration, Eq. (1.2)
c_p	permeate concentration, Eq. (1.4)
c_s	specific cost of permeate, Eq. (6.8)
C_t	total cost of desalination, Eq. (6.1)
c_v	solid volume fraction, Eq. (1.3)
c_w	concentration on channel wall, Eq. (7.15)
c_w	water concentration, Eq. (9.11)
C_w	wall concentration of retained particles, Eq. (2.6)
D	diffusion coefficient, Eq. (2.3)
D_+, D_-	cation diffusion coefficient and anion diffusion coefficient, Eq. (11.44)
E	energy, Section 1.5.4
E	electric field, Section 11.2.1
e	elementary charge, Section 11.2.2
$\hat{E}$	specific energy, Eq. (1.12)
F	filtration coefficient, Eq. (8.18)
F	Faraday constant, Eq. (9.9)
f	feed water fraction, Eq. (10.4)
F_c	convective force, Eq. (9.10)
F_d	diffusive force, Eq. (2.1)
F_e	electrostatic force, Section 11.2.1
F_f	drag force, Eq. (2.2)
f_{os}	osmotic coefficient, Eq. (1.6)
G	Gibbs free energy, Section 2.2.1
G^0	Gibbs free energy at standard condition, Section 2.2.1
H	channel height of membrane channel, Eq. (5.1)
I	cake resistivity in MFI measurement, Eq. (8.2)
I_{CC}	cost corrected index, Eq. (10.8)
J	flux, Eq. (1.10)
J_-	anion flux, Eq. (11.42)
J_+	cation flux, Eq. (11.41)
J_c	convective flux, Eq. (9.11)
J_d	diffusive flux, Section 9.2.3
J_s	solute flux, Section 11.1.1

k	Boltzmann constant, Eq. (1.5)
k	a constant relating fouling strength to foulant concentration, Eq. (8.5)
K	friction coefficient, Eq. (5.4)
K	partition coefficient, Eq. (11.1)
$\mathbf{k_d}$	diffusive friction coefficient, Eq. (9.3)
$\mathbf{k_f}$	convective friction coefficient, Eq. (9.10)
L	channel length of membrane channel, Eq. (5.1)
l	membrane thickness, Eq. (9.12)
M	amount of retained particle, Eq. (2.5)
$\mathbf{M_0}$	cost to run a reference pretreatment process of flowrate Q_0, Eq. (6.5)
$\mathbf{M_1}$	cost to run a reference concentrate handling facility of flowrate Q_1, Eq. (6.6)
MFI	modified fouling index, Eq. (8.2)
n	number of membrane elements in a pressure vessel, Eq. (6.2)
$\mathbf{N_1}$	correction rate of retained particles, Eq. (3.9)
$\mathbf{N_2}$	carrying capacity, Eq. (3.10)
$\mathbf{N_{avo}}$	Avogadro number, Eq. (1.2)
$\mathbf{N_F}$	filtration number, Eq. (2.11)
$\mathbf{p^{(m)}}$	pressure inside membrane at the interface, Eq. (9.20)
$\mathbf{p_d}$	pressure on draw solution, Eq. (10.24)
Pe	Peclet number, Eq. (7.26)
$\mathbf{p_e}$, $\mathbf{p_v}$	prices for membrane element and pressure vessel, Eq. (6.2)
$\mathbf{p_f}$	pressure on feed water, Eq. (10.24)
P	solute passage coefficient, Eq. (11.7)
PN	polarization number, Eq. (7.18)
q	electrical charge, Section 11.2.1
Q	flowrate, Eq. (1.14)
$\mathbf{q_1}$, $\mathbf{q_2}$	electrical charges, Section 11.2.1
$\mathbf{Q_c}$	concentrate flowrate, Section 6.1
$\mathbf{Q_f}$	feed flowrate, Eq. (1.7)
$\mathbf{Q_p}$	permeate flowrate, Eq. (1.7)
r	rejection rate, Eq. (1.4)
r	distance between two point charges, Section 11.2.1
R	gas constant, Eq. (1.5)
R	permeate recovery, Eq. (1.7)
$\mathbf{R_{cl}}$	resistance of cake layer, Eq. (2.21)
$\mathbf{r_{cl}}$	specific resistance of cake layer, Eq. (2.21)
$\mathbf{R_m}$	membrane resistance, Eq. (1.11)
$\mathbf{R_{m0}}$	resistance of clean membrane, Eq. (8.3)
$\mathbf{R_{mf}}$	resistance of fouled membrane, Eq. (8.3)
S	membrane area, Eq. (1.10)
S	structural parameter, Eq. (10.22)
$\mathbf{S_1}$, $\mathbf{S_2}$	scale effect coefficients, Section 6.1.3
SDI	silt density index, Eq. (8.1)

$\mathbf{S_e}$	membrane area in a membrane element, Eq. (6.2)
$\mathbf{S_f}$	fouling strength, Eq. (8.3)
$\mathbf{T}$	absolute temperature, Eq. (1.5)
$\mathbf{T}$	total elapsed test time for SDI test, Eq. (8.1)
$\mathbf{t'}$	dummy integration variable, Section 4.2.2
$\mathbf{t}$	operation time, Eq. (3.4)
$\mathbf{t_1}$	time to collect a given volume filtration in clean water filtration in fouling strength measurement, Eq. (8.6)
$\mathbf{t_2}$	time to collect a given volume filtration in feed water filtration in fouling strength measurement, Eq. (8.6)
$\mathbf{T_e}$, $\mathbf{T_v}$	service times of membrane element and pressure vessel, Eq. (6.2)
$\mathbf{t_f}$	final time to collect 500 mL filtrate, Eq. (8.1)
$\mathbf{t_i}$	initial time to collect 500 mL filtrate, Eq. (8.1)
$\mathbf{t_{ss}}$	time to reach steady state
$\mathbf{U}$	mobility coefficient, Eq. (11.46)
$\mathbf{u}$	tangential velocity, Eq. (3.8)
$\mathbf{u_0}$	tangential velocity at the entrance of membrane channel, Eq. (5.1)
$\mathbf{u_a}$	average tangential velocity
$\mathbf{u_L}$	tangential velocity at the exit of membrane channel, Eq. (5.1)
$\mathbf{v}$	transverse velocity, Eq. (7.7)
$\mathbf{V}$	volume, Eq. (1.12)
$\mathbf{V_0}$	initial volume
$\mathbf{v_1}$	initial permeate velocity at the entrance with the presence of a cake layer, Eq. (4.15)
$\mathbf{V_c}$	volume of concentrate, Section 1.4.3
$\mathbf{V_d}$	volume of draw solution, Eq. (10.10)
$\mathbf{V_f}$	volume of feed water, Eq. (1.7)
$\mathbf{V_i}$	molar volume, Eq. (9.8)
$\mathbf{v_m}$	electromigration velocity, Eq. (11.13)
$\mathbf{V_p}$	volume of permeate, Eq. (1.7)
$\mathbf{V_w}$	molar volume of water, Eq. (9.10)
$\mathbf{v_w}$	transverse velocity on channel wall, Eq. (7.10)
$\langle \mathbf{v_i} \rangle$	average diffusive velocity, Eq. (9.4)
$\bar{\mathbf{v}}$	average permeate flux, Eq. (3.18)
$\mathbf{w}$	a variable defined by Eq. (3.16)
$\mathbf{W}$	membrane power density, Eq. (10.17)
$\mathbf{x}$	coordinate along membrane surface, Eq. (3.9)
$\mathbf{X}$	location of division line between equilibrium section and nonequilibrium section of cake layer, Eq. (4.22)
$\mathbf{X}$	dimensionless coordinate along membrane surface, Eq. (5.16)
$\mathbf{X}$	density of fixed charge of membrane, Eq. (11.34)
$\mathbf{x'}$	dummy integration variable, Eq. (3.9)
$\mathbf{y}$	coordinate perpendicular to membrane surface, Eq. (2.6)
$\mathbf{y_1}$, $\mathbf{y_2}$	functions in graphical method defined by Eq. (5.29)
$\mathbf{z_i}$	valence of ion, Eq. (9.9)

Greek symbols

α	crossflow filtration number defined by Eq. (3.23)
β	crossflow filtration number with presence of cake layer, Eq. (4.16)
γ	shear rate, Eq. (3.8)
δ'	dummy integration variable, Section 4.2.2
δ	thickness of cake layer, Eq. (2.21)
δ_{eq}	equilibrium thickness of cake layer, Eq. (4.21)
ε	porosity, Eq. (10.23)
ε	relative permittivity of the medium, Section 11.2.1
ε_0	vacuum permittivity, Section 11.2.1
θ	angle between electric field and normal direction of surface, Section 11.2.1
θ	conversion factor from Joule to kWh, Section 1.5.3
θ	dimensionless parameter for solid volume, Eq. (2.6)
λ	Debye length, Eq. (11.21)
μ	viscosity, Eq. (2.2)
μ	chemical potential Eq. (9.1)
μ^0	chemical potential at standard condition, Eq. (9.1)
$\mu_0^{(m)}$	standard chemical potential inside membrane at the interface, Section 9.4.1
π_c	circumference ratio, Eq. (1.3)
π	osmotic pressure, Eq. (1.5)
π_d	osmotic pressure of draw solution, Eq. (9.25)
π_{eff}	effective osmotic pressure, Eq. (10.22)
π_f	osmotic pressure of feed water, Eq. (9.25)
σ	surface charge density, Section 11.2.1
σ_0	charge in the transition layer adjacent to membrane surface at 0, Eq. (11.58)
σ_L	charge in the transition layer adjacent to membrane surface at L, Eq. (11.61)
σ_m	charge on membrane, Eq. (11.66)
τ	shear stress, Eq. (7.1)
τ	tortuosity, Eq. (10.23)
φ	electrical potential, Eq. (9.9)
φ_0	potential on membrane surface at 0, Eq. (11.58)
ω	compound cost coefficient, Eq. (6.15)
Δp	pressure difference, Section 1.5.4
ΔP	dimensionless pressure, Eq. (5.18)
Δp_{cl}	pressure difference on cake layer, Eq. (2.19)
Δp_{cp}	pressure difference on concentration polarization layer, Eq. (2.10)
Δp_{cp}^{Max}	maximum pressure difference on concentration polarization layer, Eq. (2.18)
Δp_m	pressure difference across membrane, Eq. (2.17)

$\Delta \mathbf{p}^*$	characteristic pressure, Eq. (5.26)
$\Delta \mathbf{R}_m$	increment in membrane resistance, Eq. (8.3)
$\Delta \mathbf{t}$	operation period, Eq. (1.14)
$\Delta \mathbf{x}$	step size in space, Section 5.2.2
$\Delta \pi$	osmotic pressure difference, Eq. (1.5)
$\Delta \pi_{mix}$	osmotic pressure of mixture, Eq. (10.31)
$\Delta \Pi_0$	dimensionless osmotic pressure, Eq. (5.18)
$\Delta \varphi_D$	Donnan difference, Eq. (11.32)
$\Delta \varphi_{dif}$	diffusion potential, Eq. (11.45)
$\Delta \varphi_L$	potential difference across the transition layer adjacent to on membrane surface at L, Eq. (11.58)
$\Delta \varphi_N$	Nernst potential, Eq. (11.30)
$\Delta \varphi_{TMS}$	Teorell–Meyer–Sievers (TMS) potential, Eq. (11.47)

Subscripts and superscripts

f	feed water
b	bulk solution
c	concentrate
m	membrane
g	gel
p	permeate
w	water
v	volume
a	average
s	solute
d	diffusion
0	location of membrane surface
L	location of membrane surface

Preface

Membranes emerged in the middle of last century as a means to separate species that are carried in various gaseous and liquid media. The applications of membrane processes keep increasing as new membrane materials, new membrane processes, and new membrane applications are developed over the years. Membrane processes have now become a vital separation technique with various applications in many fields, including chemical separations, biomedical purifications, and water and water treatments. As a student and a professor in this field for over 30 years, I personally experienced the difficulties and challenges to master the fundamental principles of membrane processes. It usually requires decades for a professional in this field to develop a relatively complete understanding of membrane processes from piecewise knowledge scattered in various publications and information from different sources. Therefore, there is a great need for the well-prepared textbooks, which can be adapted by universities into regular curricula, to bring students up quickly to the required level to meet the needs in this field.

The objective of this book is to present systematically the fundamental principles of membrane processes in liquid separation. This kind of book is currently unavailable because most books on membranes are about membrane materials, fabrication methods, and practical applications. The book will be suitable for college students to learn membrane processes and conduct research in this field. It can also be a good reference for practical engineering to design better membrane processes or run them more efficiently.

The book starts with a general introduction in Chapter 1 about membrane processes in liquid separation. It covers the basic background and applications of membranes and membrane processes. It also provides definitions and calculations of the parameters used to describe or assess the performance of membrane processes, such as hydraulic pressure, concentration of species rejected by membranes, water fluxes, and energy consumptions. The knowledge included in this chapter are essential to follow the more advanced topics and discussions in the book. The rest of the book is divided into three distinct parts with detailed descriptions provided below.

Part I is devoted to the roles and impact of retained species on permeate fluxes of membrane processes. Performance and efficiency of a membrane system are affected by the accumulations of species (particulates, solutes, and ions, …) rejected by the membrane. Furthermore, the different states of the retained species (concentration polarization or cake layer) can affect a membrane process in different ways and to different extents. Therefore, it is compulsory to track the amount of retained rejected species and knowledge of their existence for a thorough understanding of the behaviors of membrane processes. Creative thinkings and delicate deviations are made in this textbook/monograph to provide a sound understanding of the roles of retained species in the membrane separation processes. By introducing a model binary membrane system, fundamental differences between the two important phenomena, namely concentration polarization and cake formations, are revealed, and much deeper insights of membrane processes are obtained.

Part II focuses on reverse osmosis processes in seawater and brackish water desalination, which is by far the largest application of membranes in liquid separations. The typical configuration of reverse osmosis processes in desalination is typically represented by a long narrow membrane channel. Such a reverse osmosis process can be well described with a state equation of the process. The state equation intrinsically links four state variables of the membrane process together, which are recovery (performance), osmosis pressure (feed water property), characteristic pressure (membrane device characteristics), and driving pressure (operating/manipulating strategy). With the state equation, simulation and optimization of the reverse osmosis desalination processes can be carried out efficiently with great elegance and ease. In addition, a more accurate and convenient characterization method for fouling strength of feed water is systematically introduced in the book, which is superior to the existing silt density index (SDI) and modified fouling index (MFI).

Part III discusses the mechanisms and principles of water and ion transport across membranes, which form the fundamental basis for membrane in liquid separation. For water transport, the role of hydraulic pressure is well understood and described adequately with Darcy's law. However, the role of chemical potential (osmotic pressure) on water transport remains elusive and represents a long-lasting serious challenge in this field. An attempt is made in this book to explain the low effectiveness of osmotic pressure for water transport that is different from the traditional one based on internal concentration polarization. As for ion transport, although Nernst–Planck–Poisson (NPP) equation is a well-accepted governing equation, the boundary conditions have not been appropriately handled traditionally. Any solutions (either analytical or numerical) of the NPP equation cannot represent the problems appropriately without the well-imposed boundary conditions. A boundary-updating scheme is introduced to ensure the right solutions to the problems under consideration.

My first encounter with membrane science and technology was in 1994 when I was a postdoctoral researcher at the University of California, Los Angeles. My advisor, Professor Menachem Elimelech, then led me into this fascinating field, and I became fond of it immediately. Thereafter, I have remained in this field as a major research area for my entire academic career. I am so lucky to enter this field at the time with abundant accumulation of findings and knowledge that can trigger the transition of membrane technology from laboratory wonders to a solid branch of science. It is my sincere hope that this book makes its contributions to this transition. At this moment,

I would like to express my deep gratitude to Professor Elimelech. It is impossible for me to write this book without his guidance and encouragement.

Last, I would like to say that mathematics is inevitable to accurately and precisely describe many phenomena in membrane processes. However, the textbook/monograph emphasizes understanding of the vivid physical process over immersing in the cold mathematical deviations. The book is prepared for college students and practical engineers with college degrees. Mathematics is kept at a level that most college students in science and engineering majors should master. Basic knowledge of calculus and ordinary differential equations is sufficient for the majority of materials in this book. Most calculations in this book were done in a Microsoft Excel spreadsheet. The basic knowledge of partial differential equations and numerical solution methods is required only for the last two chapters about ion transport across the membrane. This part can be treated optionally by those who have not been sufficiently prepared.

1

Introduction

1.1 Background

Membrane technology that initially emerged in the 1920s as a laboratory curiosity has become an important separation technique of huge commercial value in various industrial applications. Nowadays membranes are widely used in separation, concentration, and purification of chemicals in streams and desalination of seawater and brackish water [1–3]. Membrane separations are physical processes that are usually completed without chemical additions and elevated temperature. These features make them especially suitable for separations of delicate products as often encountered in pharmaceutical and biological industries that can be otherwise damaged or altered by chemicals or heat [4–6].

1.1.1 Membrane and Membrane Separation Processes

Functionally, a membrane can be defined as a thin film of certain material, because of its physical or chemical properties, that can serve as a selective barrier to different components in streams, under an appropriate driving force. Membranes can be made of many different materials, including organic polymers and inorganic ceramics.

Membranes are widely used in separation, concentration, and purification of various species in liquids. Usually, a stream of mixture is split by the membrane into two streams of permeate and concentrate (Figure 1.1). The permeate stream contains the species that can pass through the membrane, while the concentrate stream contains higher concentrations of the species that are rejected by the membrane.

Separation in pressure-driven membrane processes are achieved by size exclusion. Figure 1.2(a) shows that a binary stream of particles and water molecules is separated by a membrane. The small water molecules pass through the pores of membrane from the feed side to permeate side in the pressure-driven membrane processes. The particles are rejected by the membrane because of their bigger size than pore size. As a result, the feed stream is split into two streams: the permeate stream is particle free and the remaining stream has a higher particle concentration in it.

DOI: 10.1201/9781003743569-1

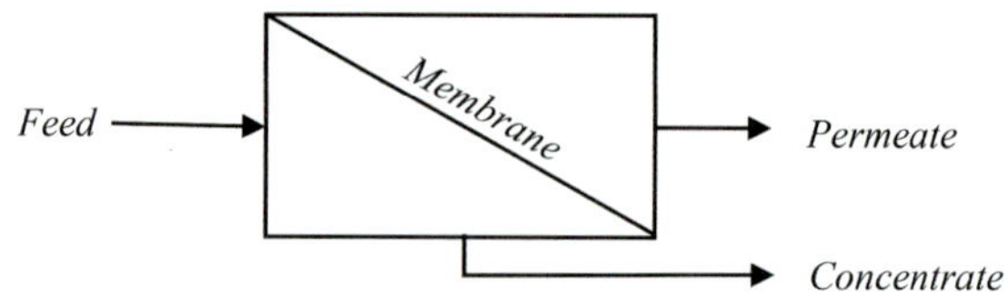

FIGURE 1.1
Schematic of a membrane separation process. A feed stream is separated by a membrane into permeate and concentrate streams.

Separation can also be achieved when water molecules remain still while particles transport through the membrane. A membrane becomes a selective barrier of charged particles (i.e., ions) when it carries fixed electrical charge. As shown in Figure 1.2(b), the membrane has positive fixed charge in it. As a result, anions can pass through the membrane, but cations cannot. With alternative placement of oppositely charged membranes, the feed stream can be split into two streams of high and low salt concentrations. This is the principle of desalination by electrodialysis.

Two major driving forces in liquid separations are mechanical force (pressure) and electromotive force. They can separate a salt concentration into two with different concentrations by either moving water or ions across the membranes. Membrane separation is a process of decreasing entropy. According to the second law of thermodynamics, energy is required to complete the splitting. This is the reason that mechanical energy and electrical energy are consumed to produce purified water of low salinity in pressure-driven membrane process or electrodialysis process, respectively. In principle, energy can be produced by reversing the separation process, which is the controlled mixing of two solutions of different salt concentrations. Counterparties of pressure-driving membrane processes and electrodialysis processes are pressure-retarded osmosis and inverse electrodialysis.

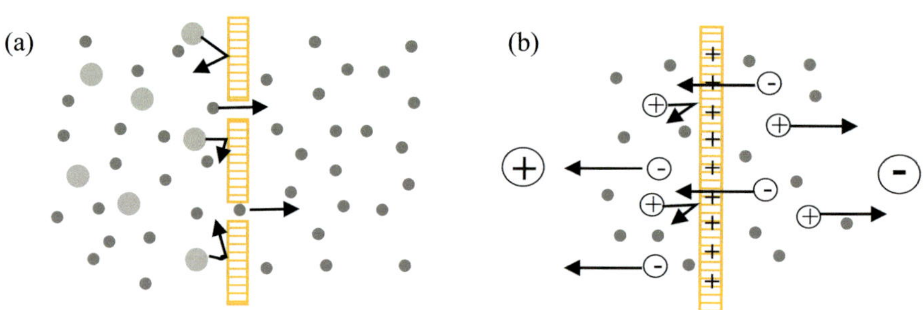

FIGURE 1.2
Two principal mechanisms in liquid separation. (a) Size exclusion: Small water molecules pass through the pores of membrane, but large particles cannot. (b) Electromigration: Counterions pass through the charged membrane, but co-ions cannot.

Concentration gradient, or in another term, chemical potential gradient, of a species is not a driving force for separation or purification though it can drive species across membranes. Diffusion of species across a membrane under concentration gradient is a mixing process that virtually leads to equalized concentrations across membrane. Diffusive transport of species across membrane occurs spontaneously because entropy increases during the process.

1.1.2 Classification of Synthetic Membranes

There are various different ways to classify membranes. Pressure-driven membranes in liquid separation can be classified into four categories according to pore size and the targeted species: microfiltration (MF), ultrafiltration (UF), nanofiltration (NF), and reverse osmosis (RO) [1,2,4]. The targeted species that can be rejected by different membranes are presented in Figure 1.3.

MF membrane is the coarsest with pore size greater than 0.1 μm. It can remove suspended solids, silt, and other particulate matters. Bacteria and most other microorganisms in water can also be removed by MF membranes. MF membranes are commonly used in drinking water treatment to reduce turbidity and pathogens. Because of large membrane pore sizes, the driving pressures in MF are relatively small and usually equal to or less than 2 bars.

The next category of membranes is UF. Pore sizes of UF membranes are usually in the range of 10–100 nm. UF membranes reject colloids and a large variety of organic compounds. Therefore, they are widely used in the separation, concentration, and purification of many chemical, biological, and biochemical products, such as protein and whey. Viruses in water that can pass through MF membranes can also be removed by UF membranes. The applied pressure in UF is in the range of 2–5 bars.

Lastly, RO membranes have the smallest pore sizes that only allow water to pass through. RO membranes usually have excellent salt rejection rates. Therefore, they are mainly used to desalt seawater and brackish water to

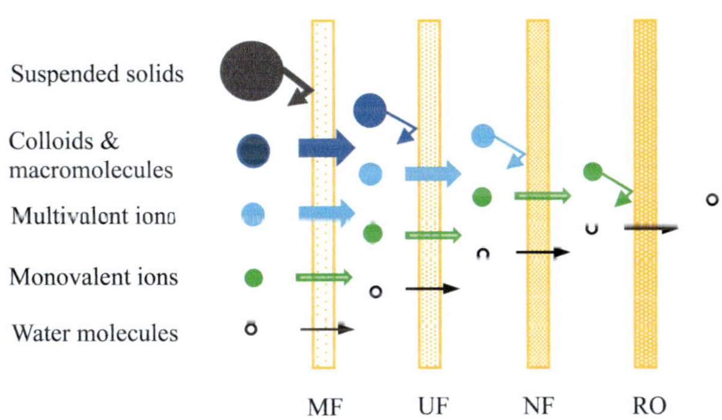

FIGURE 1.3
Classification of pressure-driven membranes and their targeted species to separate.

TABLE 1.1

Characteristics of Pressure-Driven Membranes

Membrane	Pore Size (nm)	Pressure (Bar)	Species to Reject
Microfiltration	>100	0–2	Particulates, silt, microorganisms
Ultrafiltration	5.0–100	2–5	Colloids, macromolecules
Nanofiltration	0.5–5.0	5–10	Multivalent ions, organic molecules
Reverse osmosis	<0.5	7–80	Monovalent ions

produce potable drinking water. The membranes can also be used to remove trace amount of organic matter in the treated wastewater for non-potable and potable reuse. Because of the tightness of the membranes, high driving pressures are required to push water through the membranes. Depending on the salinity of feed water, the applied pressure varies from as low as 7 bars for brackish water desalination to over 70 bars for seawater desalination. NF membranes can be regarded as a subcategory of RO membranes on the loose end. NF membranes have a relatively poor rejection rate to monovalent ions but can be a good barrier to divalent and multivalent ions. The membranes are commonly used to reduce hardness in drinking water with applied pressure that can be lower than 5 bars.

Some characteristics of the pressure-driving membranes are briefly summarized in Table 1.1.

Membranes can be classified into symmetric membranes, asymmetric membranes, and composite membranes according to their structure [2,3,7], as portrayed in Figure 1.4. Symmetric membranes have uniform pore size and porosity across membrane thickness. Most MF membranes have symmetric structure. Because of the large pore size, the resistance due to membrane thickness is not a concern in the operation. Some UF membranes with large pore size may be made with symmetric structures when the membrane resistance is not seriously too high.

Asymmetric membranes, on the other hand, are characterized by a reduced pore size and porosity from one side to the other side of a membrane with a dense skin layer formed on the denser end of the membrane thickness. Separation is achieved by the thin skin layer because of its small pore size, while the rest of the thick and porous part of the membrane provides physical

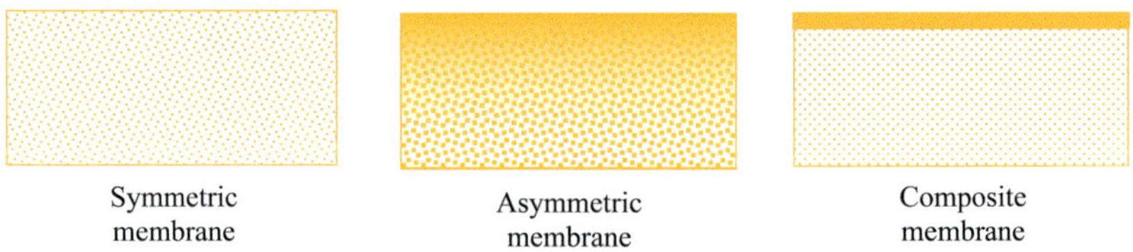

FIGURE 1.4
Structures of symmetric membrane, asymmetric membrane, and composite membrane.

support and strength. The technology to form such asymmetric structure with the same polymers and chemicals is a milestone in membrane fabrication history. A big advantage of the asymmetric membranes is that their resistance is significantly reduced over the symmetric membranes without sacrificing rejection rate. Therefore, it is widely used in UF and RO membranes where high permeate flux under relatively low pressure is essential to the economy of separation processes.

Composite membranes, compared with the asymmetric membranes, consist of two or more layers of different materials. For example, the thin-film composite (TFC) polyamide RO membranes typically contain a nonwoven fabric layer on the bottom, a polysulfone (PS) layer in the middle, and a polyamide layer on top. The fabric and PS layers provide the needed mechanical strength and the basis for the skin layer to form. The separation function is primarily achieved by the top polyamide layer. Similar to the asymmetric membranes, the dense skin layer for separation in the composite membranes can be made very thin to reduce membrane resistance while the required physical strength of the membranes are maintained.

1.1.3 Historical Development

The phenomenon of osmosis, which is the spontaneous transport of water through a semipermeable membrane from a diluted solution to a more concentrated one, was noticed as early as in the middle of eighteenth century. The first synthetic membrane was developed by A. Fick in 1855 with nitrocellulose [4,8]. Nitrocellulose remained as a major membrane material in the early days of membrane fabrications and studies. At that time, the main interest in membranes was to conduct fundamental studies of osmosis phenomena and diffusion processes.

Membrane fabrication and application progressed steadily in the following century. The noted milestones in MF membranes, such as the methods of pore size control, the application to the removal of particles from liquid and gaseous streams, and cultivation of microorganisms on membranes, were adopted later by U.S. government agency as a standard procedure for drinking water analysis. Meanwhile, there were attempts to develop RO membranes with the purpose to desalt seawater and brackish water to produce potable water. Sourirajan [9] at University of California, Los Angeles and Reid and Breton [10] at University of Florida independently demonstrated that the homogeneous cellulose acetate membrane could produce freshwater from seawater. However, the membranes in the initial stages were not of any practical uses because of the extremely low water fluxes under considerable high pressures. By 1960, membranes were used only in a limited number of laboratories and small-scale industrial applications.

A breakthrough as far as industrial membrane applications were concerned was achieved by the development of asymmetric membranes by Loeb and Sourirajan in 1962 [11]. These membranes consist of a top layer of thickness less than 1 μm supported by a porous sublayer of thickness about 50–200 μm. The top layer or skin determines the salt rejection rate, while the porous sublayer provides necessary physical strength and support for the skin layer. This type of asymmetric membrane can produce more than ten times permeate flux of the homogeneous membranes under similar operating conditions. This landmark invention made RO a potential desalination method of practical commercial value.

The Loeb and Sourirajan method was later used in fabrication of UF membranes. It significantly reduced resistance of the membranes and expanded applications of the membranes in chemical, biological, and biochemical fields and in water and wastewater treatment. The development of polyamide RO membranes in the 1980s [12–14] was another miracle in the history of membrane technology that made the current widespread and diversified applications of membrane processes possible.

Brief summaries of historical developments of MF membranes, UF membranes, and RO membranes are provided in the following sections.

1.1.3.1 Microfiltration

MF refers to filtration processes that use porous membranes to separate suspended particles with diameter between 0.1 and 10 μm. MF membranes were first developed with collodion (nitrocellulose) by Bechold, Zsigmondy, and Bachmann in the 1920s and 1930s [2]. For a long period, a blend of cellulose acetate and nitrocellulose remained as the major materials for MF membranes. However, other materials have been gradually used for MF membranes, including polyvinylidene fluoride, polyamide, polyolefins, and polytetrafluoroethylene.

The practical use of MF membranes could be traced back to World War II when a rapid method was developed in Germany to monitor for bacterial contamination in drinking water, which remains a significant application nowadays. The first commercial application of MF membranes was in biological and pharmaceutical manufacturing in the 1960s. Since then, the applications of MF membranes have been used as an effective means to remove microorganisms in pharmaceutical industry, to eliminate particles in rinsing water in semiconductor industry. It was also widely used in various processes in food processing industry, such as sterilization of beer and wine and clarification of various fruit juices.

In 1990s, because of more stringent regulations and control for pathogens in drinking water, MF membrane was increasingly used as an alternative to chlorine disinfection. The disinfection with MF is especially effective for the removal of cryptosporidium and giardia because these microorganisms are hardly killed by chlorine. Another important application of MF membrane in water industry is the treatment of municipal wastewater. MF membranes are used to replace the secondary clarifier in the form of membrane bioreactor for better effluent quality, which can be further reclaimed for reuse.

1.1.3.2 Ultrafiltration

UF uses a fine porous membrane to separate water and micro solutes from macromolecules and colloids. The average pore diameter of UF membranes is in 1–100 nm range. The first synthetic UF membranes were prepared by Bechhold [2,4] from collodion (nitrocellulose) at about the same time he fabricated MF membrane with the same polymer. By the mid-1920s, UF membranes were available for laboratory use. The crucial breakthrough for UF membranes was as a spin-off of the development of anisotropic cellulose acetate membranes by Loeb and Sourirajan in 1962. Their goal was to produce high-flux RO membranes, but it was realized that the technique could be applied to fabrication of other membranes. Michaels and his coworkers [15] produced UF membranes from the same cellulose acetate shortly after the invention of Loeb–Sourirajan method.

Most UF membranes are prepared with Loeb–Sourirajan method and therefore have an asymmetric structure. Cellulose acetate and other cellulose polymers were the primary membrane materials. Many other polymers were used later to fabricate UF membranes, including polyacrylonitrile copolymers, aromatic polyamides, PS, and polyvinylidene fluoride. As a result, UF membranes with higher physical strength and chemical stability were developed so that the membranes can be extended to more applications. These materials are still widely used to fabricate UF membranes today.

UF membranes are usually used as a barrier to separate colloids and macromolecules from water and small molecular weight species. Separation is primarily achieved by size exclusion depending on the relative size of membrane pores and the sizes of species to be separated though charges and hydraulic condition can also affect separation in certain extent. The applied pressure is in the range of 2–10 bars. Osmotic pressure does not play an important role in UF because the membranes do not reject ions.

Nowadays, UF membranes are widely used in various industrial processes. One early successful application of UF membranes was to recover electrocoat paint from rinse water in automobile industry in 1970. UF membranes are widely used in food industry, such as recovering proteins from whey, dietary milk processing, and beverage clarifying. Another important application of UF membrane is water industry. UF membranes can be used independently as an effective barrier to remove bacteria and viruses from drinking water or as a pretreatment for RO desalination processes.

1.1.3.3 Reverse Osmosis

RO membranes are permeable to water but essentially impermeable to salt. The ability of some membranes to separate small solutes from water has been known for a very long time. Modern interest dates from the work of Reid and Breton, who in 1958 showed that cellulose acetate film could perform this type of separation [10]. Their membrane was basically a homogeneous film of 5–20 μm thick. The membrane could remove 98% of salt in the permeate water, though its flux was very low to be of any practical use.

The breakthrough invention that made RO a practical process was the development of the asymmetric cellulose acetate membranes by Loeb–Sourirajan in 1962 [11]. Instead of a symmetric or homogeneous film, the Loeb–Sourirajan membranes had a very thin skin layer with a thick layer of much more porous structure. The rejection of salt was completed with the thin skin layer, while the porous layer provided support to the skin layer and physical strength of the membrane. The asymmetric membranes could produce permeate fluxes that were ten times higher than the best asymmetric membranes of Reid and Breton type with essentially equivalent salt rejection rates.

Asymmetric cellulose acetate membranes were the industry standard through the 1960s to the mid-1970s, until Cadotte invented the interfacial polymerization method for the fabrication of composite membranes in the late 1970s [12–14]. The composite membranes were made first by casting a polymer (usually PS) on nonwoven fabric to form a film of microporous structure. Then a thin dense layer was formed on top of the microporous layer with polyamide by the method of interfacial polymerization. The composite membranes showed extremely high salt rejections combined with good water fluxes. Similarly to the asymmetric membranes, the top layer of composite membranes serves as the functional layer to reject salt, while other layers provide physical strength and support.

Due to the high permeate flux and superior chemical stability of polyamide, it rapidly replaced cellulose acetate as the primary material for RO membranes. The most recent development, beginning in the mid-1980s, was the introduction of low-pressure NF membranes. The composite membrane made of polyamide, mainly in the form of a spiral-wound module, has replaced anisotropic cellulose acetate membranes. The invention of composite RO membranes is the key breakthrough to make large-scale membrane desalination possible, which currently produces millions of cubic meters of freshwater daily.

1.2 Membrane Materials and Preparation

1.2.1 Membrane Materials

The basic requirements for membrane materials are that they can be fabricated with the structures that can serve as a selective barrier to different components in liquid and gas streams. High permeability and selectivity are desirable features of membrane materials. At the same time, physical strength and chemical stability of the materials under the intended working are other preferred properties because the membranes made of these materials tend to have long service times. Membranes should be able to maintain their integrity under pressure and resistance to the scratching and wearing

TABLE 1.2

Common Materials in Liquid Separation Membranes

Membrane	Materials
Microfiltration	Cellulose nitrate, cellulose acetate, polyacrylonitrile, polyamide, polycarbonate, polyethylene, polypropylene, polysulfone, polytetrafluoroethylene, polyvinylidene fluoride
Ultrafiltration	Cellulose acetate, polysulfone, polycarbonate, polyacrylonitrile, polyethersulfone, polyvinylidene fluoride
Nanofiltration	Polyamide, polysulfone
Reverse osmosis	Cellulose acetate, polyamide

by the suspended particulate materials in the liquid. Membranes should also have the capability to resist damage from chemical reactions caused by low or high pH, oxidants, and other chemicals in the streams. Furthermore, the materials should be easily fabricated into the flat sheet or hollow fiber membranes for practical applications.

Though membranes can be made virtually with organic polymers, inorganic ceramics, metals, and other materials, most membranes are made of various organic polymers. Polymers are formed by polymerization of a large number of monomers. The monomers can be of the same type or a combination of different types. The properties of polymers depend on the properties of monomers and also on the total number of monomers, usually measured by molecular weight of the polymers. Because there are large numbers of monomers available and polymers can have a large range of molecular weight, the polymers may be made with various desired properties to fabricate membranes of specific purposes. Over the years, a long list of polymers has been used to fabricate various membranes [16–21]. Table 1.2 is a partial list of polymers commonly used to make commercial membranes. A large number of polymers can be used to make MF membranes, but only a couple of polymers can be used for RO and NF membranes.

1.2.1.1 Cellulose Acetate

Cellulose acetate is a classic membrane material used by the pioneers of modern membrane technology to create various membranes. Cellulose and its derivatives are generally linear, rodlike, and rather inflexible molecules, which are considered desired important properties for RO and UF membranes. The major sources of cellulose are wood pulp, cotton fibers, and other plants and fruits.

Cellulose acetate is produced in the acetylation process, in which some of the hydroxyl groups on the cellulose molecule are substituted by acetate groups. Cellulose acetate is a versatile plastic with considerable mechanical strength and chemical stability. It can be used to fabricate all membranes, from the coarsest to the finest. Especially, it is one of a few polymers for RO membranes because of its excellent water permeability and high rejection rate to salt. The degree of polymerization of cellulose is an important property that affects membrane characteristics and performances. The cellulose acetate polymers with a molecule weight of 25,000–80,000 are preferred for membrane fabrication.

Cellulose acetate membranes are used in RO desalination processes due to their low cost, hydrophilicity, and resistance to fouling, but they have limitations such as susceptibility to chlorine degradation and a narrow pH operating range. These membranes work by selectively allowing water to pass through while blocking dissolved salts and other solutes under high pressure. They are often blended with triacetate to improve performance, though they generally have lower flux and salt rejection than newer polyamide membranes and are being replaced in many commercial RO applications.

1.2.1.2 Polysulfone

PSs are a family of high-performance thermoplastics, including PS, polyethersulfone (PES), and poly-aryl-ether-sulfone (PAES). PSs are well-known for their toughness and stability at high temperatures. Membranes can be easily made of PS with reproducible properties and controllable size of pores down to 40 nm. These properties make it a preferred material for MF and UF membranes. It is also used to form the support layer on the nonwoven fabric of RO membranes.

PS membranes can withstand high temperatures, making them resistant to degradation from steam and hot water because of their thermal stability. Their hydrolytic and thermal stability allows them to be sterilized using methods like steam and autoclaving. They are also chemically inert across a wide pH range of 0.5–13, which simplifies cleaning and allows for their use in various chemical environments. They also have good resistance to oxidative attack. PS membranes possess high compressive strength, which is beneficial for pressure-driven processes.

There are some disadvantages of PS membranes. The raw material and processing of PS are relatively expensive. They may require specialized equipment and expertise for processing. Their naturally hydrophobic nature can lead to significant membrane fouling, which reduces permeate flux.

1.2.1.3 Polyamide

Polyamides are semicrystalline thermoplastics with low density and high thermal stability. It is one of the most important construction plastics due to its excellent wear resistance, good coefficient of friction, and very good temperature resistance and impact strength. Polyamides all have main polymer chains of amide groups (–NHCO–), which include aliphatic and aromatic polyamides. The basic aliphatic polyamides are referred to as nylon, which are semicrystalline polymers. They are thermoplastics and possess good thermal stability and mechanical strength.

Polyamides are the key polymers in the fabrication of the TFC RO membrane. The thin dense polyamide layer forms on the top of a PS porous support layer by interfacial polymerization of diamines and diacyl chloride. This thin polyamide layer has a much better salt rejection rate and higher water flux than the cellulose acetate counterpart. Today, most RO and NF membranes use a polyamide active layer with the TFC structure. However, polyamide membranes have lower tolerance to free chlorine than cellulose acetate membranes.

1.2.2 Membrane Fabrications

Membranes can be manufactured with various methods depending on membrane materials, membrane types, and applications. The general requirements for a membrane to serve the intended applications are size and uniformity of pore, permeability for the solvent and small species to pass the membranes, and the shapes of membranes (flat sheet or hollow fiber) for convenient practical applications. The materials for MF membranes are the most diversified, and so are the manufacturing methods. Table 1.3 lists the commonly used methods to fabricate MF membranes with brief descriptions.

UF membranes and cellulose acetate RO are commonly manufactured with the phase inversion method. The TFC polyamide RO membranes are

TABLE 1.3

Manufacturing Methods of Microfiltration Membranes

Method	Materials	Description
Phase inversion	Cellulose acetate, polypropylene, polyamide, polyimide, acrylonitrile	A polymer is dissolved in a solvent to form a dope, which is then cast into a film and solidified by a non-solvent or heat. The pore structure is formed during this solidification process.
Stretching	Polypropylene, polytetrafluoroethylene	A thermoplastic polymer is extruded into a sheet and then stretched perpendicularly to the extrusion direction to induce pore formation. This method creates a structure where small ruptures form in the material.
Track-etching	Polyethylene terephthalate, polycarbonate, polyester	A polymer film is bombarded with high-energy ions, which creates ion tracks. These tracks are then chemically etched to form pores. This method creates a straight-through pore structure (capillary-pore).
Sintering	Polytetrafluoroethylene powders, polyethylene powders	A support is coated with a mixture of particles (e.g., ceramic or polymer) and then compressed and heated. The spaces between the compressed particles form the pores.
Electrospinning	Polyacrylonitrile, cellulose	A high electric field is applied to a polymer solution, causing it to form nanofibers and create a nonwoven membrane.

mainly made with interfacial polymerization. These two important methods are described in more detail below.

1.2.2.1 Phase Inversion Method

Phase inversion is a process in which an initially homogeneous polymer solution is transformed from a liquid state to a solid state in a controlled manner. This method is the most commonly used, and it helps to fabricate polymeric porous membranes. This method is widely used in manufacturing symmetric and asymmetric membranes, from coarse MF membranes to fine RO membranes. The preparation of asymmetric RO membranes with cellulose acetate was initially developed by Loeb and Sourirajan in 1962 and later was used to manufacture UF membranes extensively.

In the membrane manufacturing process, a polymer (e.g., cellulose acetate) is dissolved in a suitable solvent (e.g., acetone). Magnesium perchlorate or formamide is added as a swelling agent. After the solution is cast, the evaporation of solvent results in an elevated concentration of polymer at the solution/air interface. The polymer precipitates out of solution to form a skin layer of asymmetric membranes. The remaining solvent in the bulk of the solution evaporates more slowly. Eventually, the swelling agent starts to separate out as a different phase in the mixture. Now there are two liquid phases in the substructure: the polymer-solvent as the concentrated phase and the swelling agent as the disposed phase. As the solvent evaporates further, the polymer solidifies into a solid matrix to form the porous support layer of membranes. By managing the process of phase transition, the membrane morphology can be controlled [4].

There are different ways for carrying out the phase inversion process, such as immersion precipitation, thermal phase separation, evaporation phase separation, and vapor-induced phase separation [3,7,20]. Most commercial membranes are prepared with immersion precipitation. In the wet immersion precipitation method, a relatively concentrated solution is cast on a plate at low temperature. The solvent is allowed to evaporate only partially in open air before the plate is immersed into water to complete the phase inversion. The skin layer is formed by gelation. The asymmetric structures of membranes can be formed by controlled evaporation of solvent in the phase inversion process.

1.2.2.2 Interfacial Polymerization

Currently, the RO membranes are predominantly the TFC type, which are made of polyamide with the interfacial polymerization method. This method was initially developed by Cadotte in the late 1970s [12–14]. In this method, the thin film is not made directly with polymer but formed in the manufacturing process with two very active monomers at the interface of two immiscible solutions. The support layer, which is usually a UF membrane of PS, is immersed in an aqueous solution of a reactive monomer of amine type. The membrane is then immersed in another bath containing a water-immiscible solvent containing an acid chloride reactive monomer (e.g., trimesoyl chloride). These two monomers react with each other to form a polymer layer on the top of the support layer. Heat treatment is usually required to complete the interfacial polymerization reaction.

The advantage of interfacial polymerization is the reaction is a self-inhibiting process, in which the supply of reactant for further reaction is hindered by the already formed polymerization layer. Therefore, the polymerization layer can be extremely thin so that the resistance to water can be really small. For example, the separation barrier is only 20 nm in the interfacial polymerization composite membrane, FT-30, developed by FilmTec in 1978 [14]. The polyamide RO membranes developed with the interfacial polymerization method are greatly superior to cellulose acetate membranes in bother salt rejection rate and permeate flux under similar conditions [20]. Especially, a significant amount of energy can be saved in seawater and brackish water desalination because much lower pressures are required for the same flux or recovery with the polyamide composite membranes than the cellulose acetate membranes.

1.3 Liquid Separation with Membranes

1.3.1 Impurities in Liquids and Concentrations

A liquid usually contains impurities in many forms, such as suspended particulate matter, dissolved neutral molecules, and charged ions. Though the liquid in this book refers to water most of the time, the discussions and conclusions are generally applicable to most other liquids. The impurities in a liquid can be products of industrial manufacturing processes or indigenous materials in natural environments. The useful products from chemical reactions usually need to be separated and/or concentrated for their intended purposes. The detrimental species must be removed to purify the liquid for further use. The choice of separation methods depends on the amount and physical and chemical properties of impurities in the liquid.

The amount of a species in a mixture is usually measured and expressed in *concentration*, which is defined as the ratio of measurements of the species and the mixture. Different measurements can be used for the species and the mixture. According to the measurements, concentration can be expressed in the units of mass per volume, moles per volume, mass or molar ratios, and many others. Water compositions are indicated by the concentrations of various impurity species in water. The commonly used concentration units in water and wastewater and conversion factors between units are briefly discussed below.

Mass concentration is the most used measurement for an impurity species in water and wastewater. It specifies the mass of the species in a unit volume of solution, usually expressed in the unit of milligram per liter (mg/L). For example, when the masses of total suspended solids (TSS) and total dissolved solids (TDS) in 1 L of water are found to be 150 and 2000 mg, respectively, the corresponding mass concentrations are noted as 150 and 2000 mg/L. Another commonly used concentration unit is part per million (ppm). This unit is fundamentally different from the unit of mg/L because it is the weight ratio of impurity to total solution. However, the ppm unit is often used interchangeably with mg/L unit without conversion. The reason is that 1 L of water weighs roughly 1 million milligrams most of the time. The two units cannot be used exchangeably for other liquids or waters with high contents of impurities, whose densities deviate significantly from 1 million milligrams per liter.

Molar concentration is another commonly used concentration unit, which is defined as the number of moles of species in a liter of liquid (mol/L or M). One mole is the number 6.023×10^{23}, which is the Avogadro's number. Molar concentration is convenient in calculations of the physical and chemical properties of a liquid, such as vapor pressure and osmotic pressure. A related concentration unit to molar concentration is *the number concentration* (#/L). The number of species (particulates, molecules, or ions, …) per unit volume is directly used as concentration. Number concentration is fundamentally the same as molar concentration and therefore shares the same advantages and disadvantages.

The conversion between mass concentration and molar concentration is

$$c_m = \frac{c}{MW} \tag{1.1}$$

where c_m is the molar concentration, c is the mass concentration, and MW is the mass of a mole of particles or molar weight of dissolved species. The molar concentration can be further converted to number concentration by

$$c_n = c_m N_{avo} \tag{1.2}$$

where c_n is the number concentration, and N_{avo} is the Avogadro's number.

Solid volume fraction is one convenient concentration unit in membrane processes when a cake layer is involved, which is defined as the volume of particles to the total volume of suspension. When identical spheres are assumed for the particles in water, solid volume fraction is related to number concentration with

$$c_v = \frac{4}{3}\pi_c a_p{}^3 c_n \tag{1.3}$$

where c_v is the solid volume fraction, π_c is the circumference ratio, and a_p is the particle radius. The symbol π_c is used for circumference ratio in this book because the symbol π is reserved exclusively for osmotic pressure.

Example 1.1

A groundwater sample contains 2500 mg/L of sodium chloride (NaCl). What are the molar concentration and number concentration of sodium chloride in the water sample?

SOLUTION

The molar weight of NaCl is 58.5 g/mol. Therefore, the molar concentration is

$$c_m = \frac{c}{MW} = \frac{2500\ mg/L}{58.5\ g/mol} \times \frac{1\ g}{1000\ mg} = 0.0427\ M = 42.7\ mM\ \left(mM = 10^{-3}\ M\right)$$

The number concentration is then

$$c_n = c_m N_{avo} = 0.0427\ mol/L \times 6.023 \times 10^{23}/mol = 2.57 \times 10^{22}/L$$

The molar concentration and number concentration of sodium chloride in the water sample are 42.7 mM and $2.57 \times 10^{22}/L$, respectively.

Example 1.2

A suspension contains 0.5% volume fraction of particles. The particles are assumed to be spherical in shape with a radius of 2 nm. Find the number concentration and molar concentration of the suspension.

SOLUTION

The number concentration and molar concentration can be calculated from the volume fraction with Eqs. (1.2) and (1.3), respectively.

The number concentration

$$c_n = \frac{c_v}{\frac{4}{3}\pi_c a_p^{\ 3}} = \frac{0.005}{\frac{4}{3} \times 3.14 \times \left(2 \times 10^{-9}\ m\right)^3} = 1.49 \times 10^{23}/m^3$$

The molar concentration

$$c_m = \frac{c_n}{N_{avo}} = \frac{1.49 \times 10^{23}/m^3}{6.023 \times 10^{23}/mol} = 0.248\ mol/m^3.$$

The number concentration and molar concentration of particles in the suspension are $1.49 \times 10^{23}/m^3$ and $0.248\ mol/m^3$, respectively.

Different symbols are used here to indicate different concentration units for the sake of clarity in the discussions. The symbol "c" without a subscript is generally used in the rest of the book when the concentration unit is clear or specified in context. Subscripts are only used occasionally in cases where unnecessary confusion may be caused.

The effectiveness of membrane separation is conveniently indicated and measured by concentration changes of species in liquids. When a feed stream is split into a concentrated stream and a diluted stream, a large concentration ratio of the two streams represents a better separation. The concentration of the rejected species in the permeate stream is a more commonly used indicator for membrane separation. The lower concentration in the permeate stream indicates the better separation of the membrane. Of course, zero concentration in the permeate stream indicates a perfect separation.

1.3.2 Driving Pressure

Separation of impurities from a liquid can be achieved by various methods. Suspended particulates can be removed by a series of operations of coagulation, flocculation, and sedimentation. Dissolved species, such as hardness and heavy metals, can be removed by chemical precipitation. On the other hand, some liquid can be partially removed from the mixture with the impurity left in the remaining more concentrated liquid. For example, in the thermal distillation process, potable water is produced by first evaporating water into vapor and then condensing it into liquid again, leaving the impurities behind in the concentrate. Membrane technology, as a relatively new separation method, has grown into a major separation method over the last half century. It is now the preferred method for potable water production through desalination of seawater and brackish water and reclamation of treated wastewater. Membrane technology has many advantages over other separation methods, such as no chemical addition, no phase change, simple operation, and many others.

One principal driving force for water to pass through the membrane is the difference of hydraulic pressures applied on both sides of the membrane. Pressure is generally defined as the force per unit area perpendicular to the surface. Specifically, pressure in a liquid is the force exerted by the liquid on a unit area of a surface in contact with it. The basic metric unit of pressure is pascal (Pa), which equals 1 Newton force per square meter, as shown in Figure 1.5.

Because the basic pressure unit is relatively small, some bigger units are commonly used for pressures in membrane processes, which are the standard atmosphere (atm), bar, and pound per square inch (psi). Conversions between these units are given in Table 1.4.

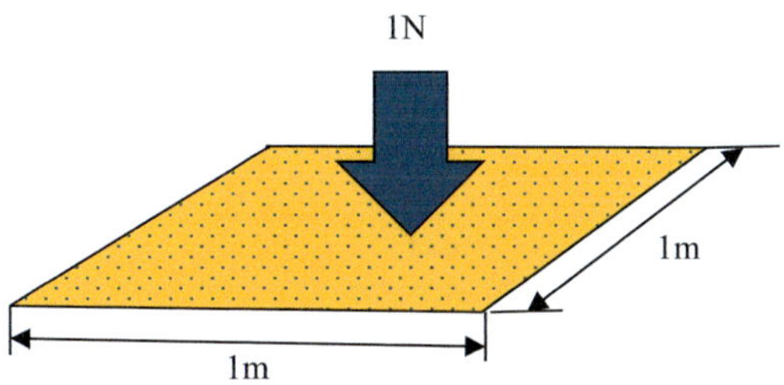

FIGURE 1.5
A pascal pressure is defined as 1 Newton normal force on an area of 1 m^2.

1.3.3 Rejection Rate

As a permselective barrier, the most important property of a membrane is how good it is at separating the target species from the liquid. One membrane can remove one species completely but other species only partially. It depends both on the size or other relevant properties of the target species and the membrane characteristics. The efficiency of a membrane to separate a particular species is commonly expressed in rejection rate, which is defined as

$$r = 1 - \frac{c_p}{c_f} \tag{1.4}$$

where r is the rejection rate, and c_p and c_f are the concentrations in permeate and feed streams, respectively. Rejection rate is usually expressed in percentage. A rejection rate of 0% means the membrane does not reject the target species at all, while a rejection rate of 100% indicates complete removal of the target species from the liquid.

Nowadays, membranes with various pore sizes are available in the market. It is possible to find suitable membranes that have adequate rejection rates to any target species in a liquid. Even for the smallest sodium and chloride ions in water, the best available RO membranes can have a rejection rate up to 99.7%.

TABLE 1.4
Conversion Between Pressure Units

Unit	Pa	atm	bar	psi
1 Pa	1	9.87×10^{-6}	10^{-5}	1.45×10^{-4}
1 atm	101,325	1	1.01	14.70
1 bar	100,000	0.99	1	14.50
1 psi	6895	0.068	0.069	1

1.3.4 Osmotic Pressure

Osmosis is a phenomenon that is intrinsically associated with semipermeable membranes. A semipermeable membrane allows water to pass through it but rejects the solutes in the solution. When a semipermeable membrane separates a solution from pure water in two chambers without a hydraulic pressure difference, a water flow spontaneously moves across the membrane from pure water to the solution, as shown in Figure 1.6(a). This water flow is called the osmosis flow.

The spontaneous water flow can be slowed down by applying a hydraulic pressure on the solution side of the membrane. The water flow can be completely ceased by choosing an appropriate hydraulic pressure. If the membrane is completely semipermeable and it separates a solution and pure water, the hydraulic pressure that stops the water flow is termed "osmotic pressure" of the solution (see Figure 1.6(b)). Osmotic pressure is not real pressure because there is not a physical force pressed on the solution. It is a physicochemical property of the solution that is associated with a perfect semipermeable membrane. Osmotic pressure is named a kind of "pressure" only because it is defined and measured with a hydraulic pressure difference across the membrane.

Osmotic pressure is a colligative property that depends on the number of particles, molecules, or impurities in any other form in a solution. Osmotic pressure of a solution/suspension can be calculated with the van't Hoff equation:

$$\Delta\pi = kT\sum_i c_{n,i} = RT\sum_i c_{m,i} \tag{1.5}$$

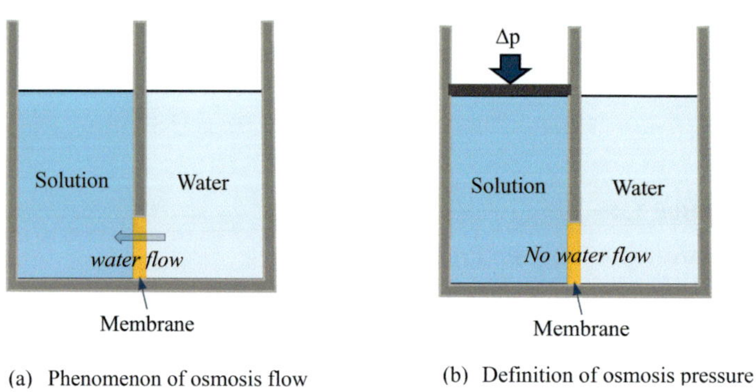

FIGURE 1.6
(a) Phenomenon of osmosis flow and **(b)** definition of osmotic pressure.

where $\Delta\pi$ is the osmotic pressure, k is the Boltzmann constant (=1.38 × 10^{-20} Pa·L/K), and R is the gas constant (=8314 Pa·L/(mol·K)), T is the absolute temperature, and $c_{n,i}$ and $c_{m,i}$ are the number concentration and molar concentration of i-component (particulate, molecule, or ion) in the solution or suspension.

Example 1.3

Determine the osmotic pressures of solutions with one particle and a mole of particles in 1 L of water, respectively.

SOLUTION

Consider the room temperature of 20°C, i.e., $T = 20 + 273 = 293\ K$. The number concentration and molar concentration are, respectively, $n = 1 / L$ and $m = 1\ mol/L$ in the two cases.

Therefore, with the van't Hoff equation, the osmotic pressure of one particle in 1 L of water is

$$\pi = kT\sum_i c_{n,i} = 1.38 \times 10^{-20}\ \text{Pa} \cdot \text{L/K} \times 293\ K \times 1/L = 4.04 \times 10^{-18}\ Pa$$

and the osmotic pressure of 1 molar of particles in 1 L of water is

$$\pi = RT\sum_i c_{m,i} = 8314\ \text{Pa} \cdot \text{L/(mol} \cdot \text{K)} \times 293\ \text{K} \times 1\ mol/L = 2.44 \times 10^{6}\ Pa$$

$$= 24.4\ bar$$

The osmotic pressures are $4.04 \times 10^{-18}\ Pa$ and $2.44 \times 10^{6}\ Pa$ in the solutions with particle concentrations of 1/L and 1 mol/L, respectively.

When the composition of a solution/suspension is known, the osmotic pressure can be calculated with the van't Hoff equation. Otherwise, an empirical equation of the form below is commonly used to link the osmotic pressure to the concentration of TDS in feed water:

$$\Delta\pi = f_{os}c \tag{1.6}$$

where f_{os} is the osmotic coefficient, and c is the TDS concentration. Usually, TDS concentration is given in the unit of milligram per liter (mg/L). The units of bar/(mg/L) or Pa/(mg/L) can be used for f_{os}. Compared with the van't Hoff equation, it can be easily found that the osmotic coefficient f_{os} = 8.33 × 10^{-4} bar/(mg/L) for a solution of sodium chloride at a temperature of 20°C. The value of 7.35 × 10^{-4} bar/(mg/L) can be generally used for seawater [22]. Eq. (1.6) is more convenient to use than Eq. (1.4) because a thorough water analysis is not required. However, the coefficient f_{os} is a function of temperature and composition of the water and calibration for different water samples is required for the sake of accuracy.

Example 1.4

The Dead Sea has a salt concentration of about 340 g/L, while the average seawater salt concentration is about 35 g/L. Calculate the osmotic pressures of water in Dead Sea and average seawater with the linear correlation equation.

SOLUTION

The seawater has an osmotic pressure coefficient f_{os} = 73.5 Pa/(mg/L).
The osmotic pressure of water in Dead Sea is

$$\Delta\pi = f_{os}c = 73.5\ Pa/\left(mg\ /\ L\right)\times 340{,}000\ mg/L = 2.5\times 10^{7}\ Pa = 250\ bar$$

The osmotic pressure of average seawater is

$$\Delta\pi = f_{os}c = 73.5\ Pa/\left(mg/L\right)\times 35{,}000\ mg/L = 2.57\times 10^{6}\ Pa = 25.7\ bar$$

The osmotic pressures are 250 *bar* and 25.7 *bar*, respectively.

1.4 Membrane Processes

A membrane process is an engineering setup to use membranes to achieve the desired separation. Membrane modules and process configurations are developed to suit various applications efficiently and economically.

1.4.1 Membrane Modules

Due to the relatively small permeate production rate per unit membrane area, a large membrane area is usually required in most engineering applications. There is a critical need for a membrane process to utilize a large membrane area efficiently in a small space. Various membrane modules are developed to serve this purpose. A membrane module is an engineering unit into which a large area of membrane is packed. To utilize the membrane effectively, the membrane modules should be able to supply feed water to all membrane areas and remove the rejected species from the membrane surface readily. At the same time, intermixing of the permeate, the concentrate, and the feed streams should be prevented. The commonly used membrane modules are plate-and-frame, hollow fiber, and spiral-wound. A brief description of each membrane module is provided here [2–4].

1.4.1.1 Plate-and-Frame Module

A plate-and-frame module consists of a flat sheet (plate) of rigid material, on which a membrane is placed. The edges of the membrane are sealed as an envelope with the plate in it as a physical support. A pressure difference is applied on the membrane to drive water through the membrane from outside in. Permeate is collected by a layer of netlike material (spacers) that is sandwiched between plate and membrane and flows out of the modules through an exit. The plate-and-frame model is featured with a relatively high cost and small surface-to-volume ratio, and therefore, with relatively limited applications. An advantage of the plate-and-frame module is that the rejected species can be effectively removed from the membrane surface. Therefore, this module is often used with MF and UF membranes to treat surface water with high concentration of suspended solids.

1.4.1.2 Hollow Fiber Module

This module characteristically consists of a huge number of hollow fibers. One hollow fiber is a tiny tube with a membrane as the wall that separates feed flow and permeate flow. Because diameters of the hollow fibers are usually small, varying from a few millimeters to less than 1 mm, the hollow fiber modules can achieve a very high surface-to-volume ratio. Hollow fiber is usually run with feed water on the shell side (outside) of the fiber. Hollow fibers can be used with or without housing dependent on applications. The unhoused hollow fiber modules are increasingly being used in water and wastewater treatments. Permeate passes through the membrane by applying negative pressure (suction) to the inside or "lumen" of the fiber. Hollow fibers in the housed modules are placed in the sealed vessels on which positive pressure can be added to drive water from the shell side to the lumen of fibers. Although hollow fiber modules were once used with RO membranes, nowadays they are most associated with MF/UF membranes.

1.4.1.3 Spiral-Wound Module

This module consists of several membrane leaves wrapped around a central pipe. Each membrane leaf is sealed as an envelope with a permeate spacer inside. Feed spacers are placed between membrane leaves. Feed water enters one end of the module under the required pressure. Permeate passes through the membrane and flows along the permeate spacer into the central pipe for collection. Both permeate and concentrate streams exit from the other end of the module. Spiral-wound module is popular for RO in water desalination. When placed in pressure vessels, it can sustain high pressures that are usually required for water desalination. This module is of relatively large surface-to-volume ratio with relatively low costs. A standard industrial spiral-wound module of 8 in. in diameter and 40 in. in length can host a membrane area up to 40 m^2.

Characteristics of the three membrane modules are given in Table 1.5. More detailed descriptions of these three membrane modules can be found in some other published books. One good example is the one authored by M. Cheryan [4].

TABLE 1.5

Characteristics of Membrane Modules

	Plate-and-Frame	Hollow Fiber	Spiral-Wound
Packing density, m^2/m^3	30–500	500–9000	200–800
Relative cost	High	Low	Low
Main applications	UF, MF	MF, UF, RO	RO
Resistance to fouling	Good	Poor	Moderate

1.4.2 Operation Modes

A membrane module can be operated in dead-end mode or crossflow mode, as schematically shown in Figure 1.7. In dead-end mode, the feed stream flows vertically toward the membrane, and the permeate stream flows out of the module from the other side of the membrane. Because there is no exit for the rejected species, they accumulate inside the membrane module during operation. When accumulation is up to a certain level, operation is ceased, and the accumulated species are backwashed out of the membrane module. Then the module is back to work again. Therefore, membrane modules work intermittently in the dead-end operation mode.

In crossflow operation mode, the feed stream enters one end of the membrane module and flows parallelly on one side of the membrane. There are two outflows of the membrane module operated in crossflow mode. At the exit end, permeate comes out from the one side, while the concentrate stream leaves the module from the other side of the membrane. Because the rejected species are carried out of the membrane module by the concentrate stream, the separation operation can be carried out continuously in crossflow mode.

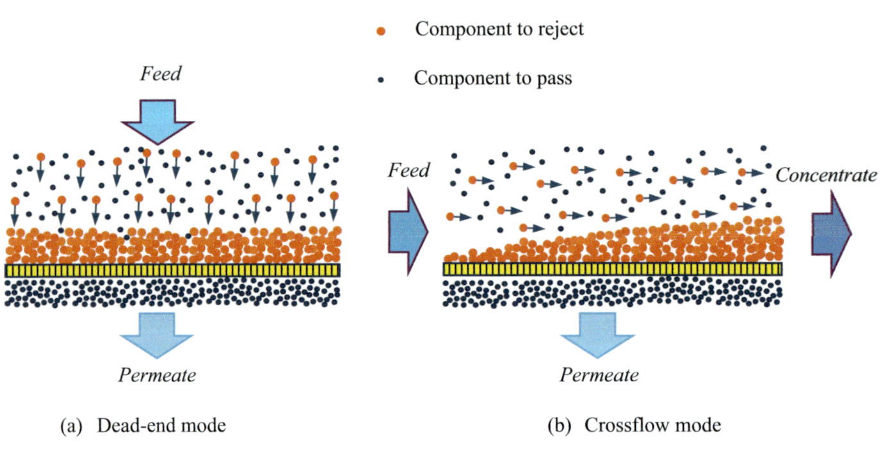

FIGURE 1.7
(a) Dead-end and **(b)** crossflow operation modes of membrane processes.

This is a desirable advantage of crossflow mode over dead-end mode in many applications.

1.4.3 Process Configurations

Membrane modules are used in various configurations to achieve the required separation goals. An important design parameter for a membrane process is the recovery, which is defined as the ratio of permeate volume to feed volume, i.e.,

$$R = \frac{V_p}{V_f} \times 100\% = \frac{Q_p}{Q_f} \times 100\% \tag{1.7}$$

where R is the recovery, V_p and V_f are the volumes of permeate and feed, respectively, and Q_p and Q_f are the flowrates of permeate and feed, respectively. Usually, the recovery is given in percentage.

A membrane process is preferred to work at as high as possible a recovery for better economic benefits. It is obvious for membrane processes in solvent recovery and product concentration because the solvent can be of high cost, and the product is usually more valuable in high concentration. In water desalination, because of the cost involved with pumping and pretreatment of feed water, it is economically desirable to turn as much as feed water into permeate. A brackish desalination process can work most economically at a recovery above 80%.

The required recovery in most cases cannot be achieved with a single pass of feed water through a single membrane module. A certain flowrate (or velocity) is usually required for the concentrate stream in a membrane module to carry out the species rejected by the membrane. Otherwise, the accumulation of the rejected species inside the membrane module can have significant adverse impacts on performance. For this reason, the recovery of a membrane model is usually managed at 10–15%. Certain arrays of membrane modules are required to achieve the high recovery goal of the whole membrane system and, at the same time, to satisfy the low recovery requirement for individual membrane modules.

Furthermore, membrane modules for industrial applications are provided in a limited number of specifications, for which the feed flowrates are specified for satisfactory performance. On the other side, the flowrates of feed water from practical applications can vary in a wide range, from several liters per hour in chemical separation to thousands of cubic meters per day in water desalination. Process configurations are selected to serve two purposes: (1) to accommodate various flowrates and (2) to achieve the required system recovery while satisfying the requirements specified for membrane modules. Four configurations of batch, feed-and-bleed, modules in series, and concentrate staging will be introduced below.

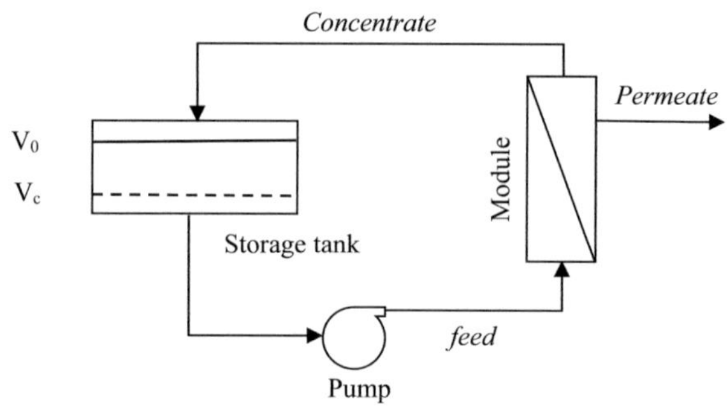

FIGURE 1.8
Batch membrane system.

1.4.3.1 Batch

A batch membrane system is used for the feed water flowrate that is smaller than the required feed flowrate to a membrane module. The batch configuration is schematically shown in Figure 1.8. Feed water is first accumulated to a certain amount in a storage tank before separation. Then the feed water in the storage tank is pumped into the membrane module at the required pressure and flowrate. The permeate flows out of the module as product, while the concentrate (retentate) is returned to the storage tank. There is a mixing mechanism in the storage tank that mixes the returned concentrate flow with the water in the tank. As permeate comes out of the membrane system, water in the storage tank reduces in volume but increases in concentration with time. The separation operation stops when the required volume reduction (or recovery) is reached. The (initial) volume of feed water and the (final) volume of concentrate (retentate) in the tank are indicated by V_0 and V_c in the figure.

Recovery of this batch operation is calculated by

$$R = \frac{V_0 - V_c}{V_0} \times 100\%$$

The operation of a batch membrane system is therefore intermittent. The storage tank is emptied by the end of operation. A new batch of feed water is then collected for the next run.

1.4.3.2 Feed-and-Bleed

In cases where the flowrate of feed water is in the level of about one-tenth of the required feed flowrate to a membrane module, continuous separation can be managed in a membrane process with the feed-and-bleed configuration. As shown in Figure 1.9, feed water of flowrate Q_f is pumped into the

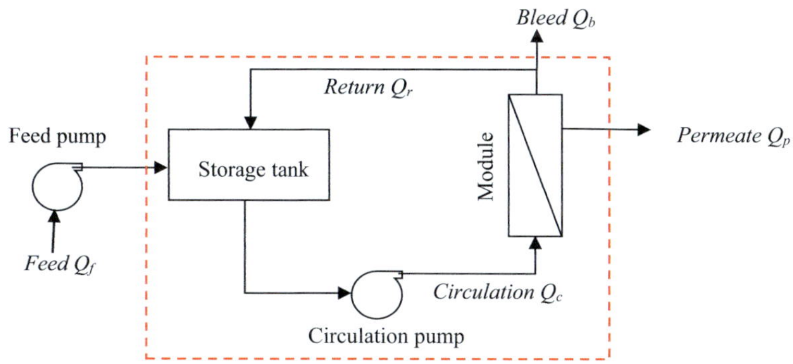

FIGURE 1.9
Feed-and-bleed membrane system.

storage tank by a feed pump at the required pressure for membrane separation. Feed water in the storage tank is then circulated at flowrate Q_c to the membrane module with a circulation pump. The membrane module produces permeate at flowrate Q_p. The concentrate out of the membrane module is split into two streams. A small fraction of concentrate flow Q_b bleeds out of the system, while the majority of concentrate flow Q_r returns to the storage tank.

With this configuration, the circulation flowrate satisfies the required feed flowrate to the membrane module because it can be several times higher than the feed water flowrate, which is not sufficient to feed directly to the membrane module. In Figure 1.9, the components included in the frame of break line can be treated as a membrane system. Continuous operation of the membrane system is possible if mass balance can be maintained. It means that the feed flowrate is equal to the sum of the permeate flowrate plus the bleed flowrate. There are two different recoveries with the feed-and-bleed configuration. They are the module recovery and the system recovery that are defined, respectively, by

$$R_{module} = \frac{Q_p}{Q_c} \times 100\%$$

and

$$R_{system} = \frac{Q_p}{Q_f} \times 100\%$$

Because the circulation flowrate Q_c is much higher than the feed flowrate Q_f of the feed water, the feed-and-bleed configuration can satisfy the required low module recovery (<10–15%) and high system recovery (>50%) at the same time.

Example 1.5

A feed-and-bleed membrane system is operated at a module recovery of 10% and a system recovery of 80%, respectively. If salt concentration in feed water is 2000 mg/L and the membrane has a perfect rejection, determine the salt concentrations in the bleed stream and in the storage tank, respectively.

SOLUTION

Both salt concentrations can be readily calculated with mass balance relationships and the definition of recovery. The salt concentration in the bleeding stream is

$$c_b = \frac{c_f}{1 - R_{system}} = \frac{2000 \text{ mg/L}}{1 - 0.8} = 10{,}000 \; mg/L$$

Then the concentration in the storage tank that is pumped into the membrane module is

$$\begin{aligned} c_{Tank} &= c_b \left(1 - R_{module}\right) = 10{,}000 \; mg/L \times (1 - 0.1) \\ &= 9000 \; mg/L \end{aligned}$$

The concentration fed to the membrane module is much higher (nine times) than that in the original feed water. Therefore, one cost that this feed-and-bleed configuration has to pay for continuous operation is the elevated concentration in the storage tank, which results in higher pressure requirement and energy consumption in separation.

1.4.3.3 Modules (Elements) in Series

When the flowrate of feed water is comparable to the required feed flowrate of a membrane module, it can be directly fed to the membrane module without the need for concentrate circulation. The storage tank and the circulation pump used in the feed-and-bleed configuration are no longer needed. Due to the limited recovery of a membrane module, the required system recovery can be achieved by using multiple modules in series. The membrane modules are often called membrane elements in this configuration. As shown in Figure 1.10, multiple membrane elements are arranged in series in a pressure vessel. The central pipes of the membrane elements that collect permeate are connected to

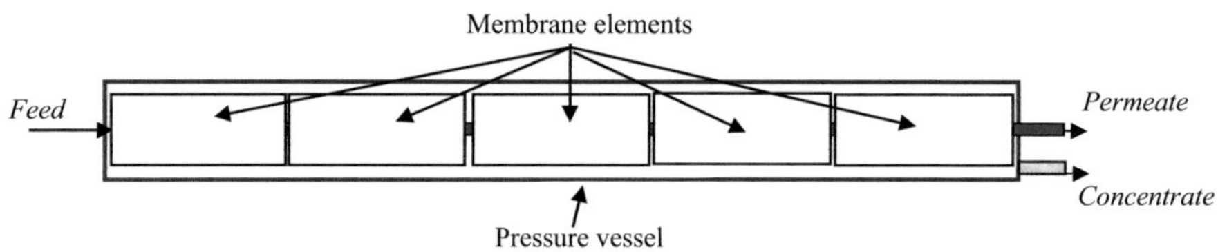

FIGURE 1.10
Multiple membrane elements in series in a pressure vessel.

each other. By the way, pressure vessels are needed to house the spiral-wound membrane elements to resist high pressure used in desalination applications.

Feed water enters the first element on the left end of the pressure vessel. Permeate is collected into the permeate pipe in the center of the element. The concentrated stream that comes out of an element immediately enters the next one as a feed stream. This process repeats itself until the last element. Finally, two streams of permeate and concentrate come out of the pressure vessel. This configuration of elements in series is commonly used in membrane processes for water desalination. A pressure vessel with six to eight elements in series usually works at a typical system recovery around 50%, while the individual element recovery in the series is kept below 10–15% as required.

1.4.3.4 Concentrate Staging

The recovery of about 50% of a pressure vessel with elements in series is usually satisfactory for seawater desalination. Higher recoveries are generally not recommended for a pressure vessel. One consideration is that the flowrate in the membrane element by the exit might be too small to take the rejected species out of the membrane process. The accumulation of the rejected species inside of a membrane process can block the passage of the water through the membrane and the flow along the channel.

When a higher than 50% recovery is pursued such as in brackish water desalination, the method of concentrate staging is often used with pressure vessels containing membrane elements in series. A schematic of concentrate staging configuration in three stages is shown in Figure 1.11. The boxes represent membrane units with a feed flow in and a permeate flow and a concentrate flow out of each box. The capital letters "C" and "P" indicate the concentrate side and permeate side of the membrane, respectively. Concentrate staging means that the concentrate flow from a membrane unit serves as the feed flow to the next membrane unit. In this way, a larger fraction of feed water can be recovered as permeate. Figure 1.11 shows that for a

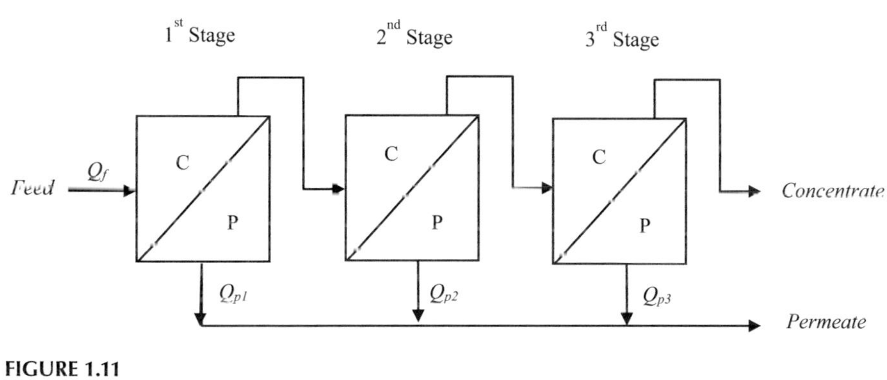

FIGURE 1.11
Schematic of concentrate staging in three stages.

feed stream of flowrate Q_f to the membrane system, three permeate streams of flowrates, Q_{p1}, Q_{p2}, and Q_{p3}, are produced, respectively.

The recovery of a concentrate staging system can be calculated from the recoveries of individual membrane units. Let R_1, R_2, and R_3 be the recoveries of membrane units at the first, second, and third stages, respectively. The permeate flowrates in the three stages can be calculated with

$$Q_{p1} = R_1 Q_f$$

$$Q_{p2} = R_2 \left(Q_f - Q_{p1}\right)$$

$$Q_{p3} = R_3 \left(Q_f - Q_{p1} - Q_{p2}\right)$$

In the last two equations, the terms in the brackets are the feed flowrates to the membrane units in the second and third stages, respectively. The recovery of the concentrate staging membrane system is

$$R = \frac{Q_{p1} + Q_{p2} + Q_{p3}}{Q_f}$$

Substituting the three equations for the permeate flowrates into the equation for system recovery, we obtain

$$R = R_1 + R_2 + R_3 - R_1 R_2 - R_2 R_3 - R_3 R_1 + R_1 R_2 R_3 \tag{1.8}$$

When there are only two stages in a membrane process, let $R_3 = 0$ in Eq. (1.8); the system recovery reduces to

$$R = R_1 + R_2 - R_1 R_2 \tag{1.9}$$

In practical desalination applications, Christmas tree structure is often used for concentrate staging. The membrane units employed in each stage are pressure vessels holding multiple membrane elements in series. Such a pressure vessel usually has a recovery of around 50%. Therefore, the concentrate flowrate from the pressure vessels is about half of the feed flowrate. The concentrate streams of two pressure vessels are usually combined to feed one pressure vessel in the next stage (see Figure 1.12). In this way, relatively constant feed flowrates to each pressure vessel in

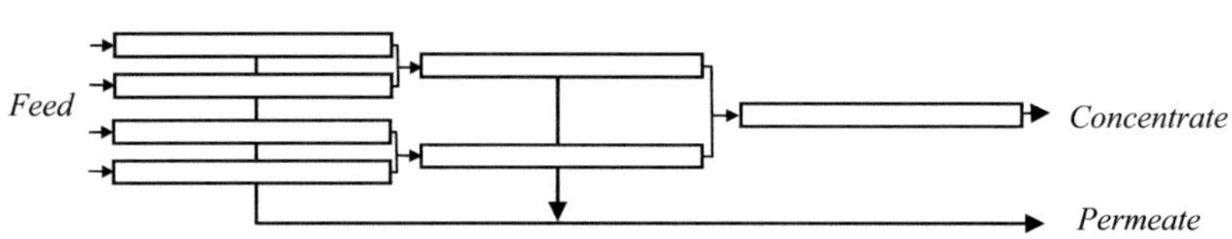

FIGURE 1.12
Arrangement of pressure vessels with Christmas tree structure.

the system are roughly maintained. This is beneficial to the membrane process.

Example 1.6

If the individual recovery of each membrane unit is assumed to be 50%, calculate the system recoveries of a membrane process with a concentrate staging configuration of two stages and three stages, respectively.

SOLUTION

For two stages,

$$R = R_1 + R_2 - R_1R_2 = 0.5 + 0.5 - 0.5 \times 0.5 = 0.75 = 75\%$$

For three stages,

$$\begin{aligned} R &= R_1 + R_2 + R_3 - R_1R_2 - R_2R_3 - R_3R_1 + R_1R_2R_3 \\ &= 0.5 + 0.5 + 0.5 - 0.5 \times 0.5 - 0.5 \times 0.5 - 0.5 \times 0.5 + 0.5 \times 0.5 \times 0.5 \\ &= 0.875 = 87.5\% \end{aligned}$$

The system recoveries are 75 and 87.5% for membrane processes with concentrate staging of two stages and three stages, respectively.

1.5 Performance

Performance of a membrane process is usually assessed by the quality and quantity of permeate. Nowadays, permeate quality is usually not a serious concern because the proper membrane to produce the required quality can be selected from the available membranes of a big variety of rejection rates. More attention is turned to the permeate quantity that can be produced by a membrane process. The economy of a membrane process is tightly related to its productivity of the products of desired quality.

1.5.1 Permeate Flux

One primary indicator for the performance of a membrane process is the permeate flux, which is defined as the permeate flowrate per unit membrane area:

$$J = \frac{Q_p}{S} \tag{1.10}$$

where J is the permeate flux, Q_p is the permeate flowrate, and S is the membrane area. When permeate flux is determined, the total membrane area needed to produce the required permeate flowrate is readily calculated from Eq. (1.10).

TABLE 1.6

Units of Permeate Flux and Conversion

Unit	$m^3/(m^2 \cdot s)$	LMH	GFD
$m^3/(m^2 \cdot s)$ or m/s	1	3.600×10^6	2.121×10^6
LMH	2.778×10^{-7}	1	0.589
GFD	4.715×10^{-7}	1.698	1

The units for permeate flux can be directly derived from the units used for flowrate and area. The basic metric unit of flux is $m^3/(m^2 \cdot s)$ when the basic length unit of meter and time unit of second are used. The permeate velocity (m/s) is completely exchangeable for permeate flux. In fact, permeate velocity sometimes is more convenient to use than permeate flux. For instance, the velocity makes more sense than the flux when the transport phenomenon is being studied in the flow adjacent to the membrane surface. Because the basic flux unit of $m^3/(m^2 \cdot s)$ is very large for practical membrane processes, some smaller flux units are also used for convenience. The two commonly used ones are the liter per square meter per hour (LMH) and gallon per square feet per day (GFD). The conversions between these units are provided in Table 1.6.

1.5.2 Basic Filtration Equation

Of fundamental importance to pressure-driven membrane processes is *the basic filtration equation* [15,23–27]. The equation relates the permeate flux across a piece of membrane to the feed water property, membrane characteristics, and applied pressure by

$$J = A\left(\Delta p - \Delta\pi\right) = \frac{\Delta p - \Delta\pi}{R_m} \tag{1.11}$$

where A is the membrane permeability, Δp is the pressure difference across the membrane, $\Delta\pi$ is the osmotic pressure difference across the membrane, and R_m is the membrane resistance. Though permeability is more commonly used as a symbol of membrane property, resistance sometimes is more convenient to use in certain practical circumstances, such as predicting permeate flux with retained species involved.

Pressure difference is the variable that can be manipulated in a membrane process to achieve the operation objective. Feed water property affects permeate flux through its osmotic pressure, which is proportional to the concentration of species to be separated by the membrane. Osmotic pressure of feed water reduces permeate flow by reducing the effective driving force from the applied pressure. In some applications, such as seawater desalination, a large fraction of applied pressure is spent to overcome the osmotic pressure, and the net driving pressure $\left(\Delta p - \Delta\pi\right)$ can be

TABLE 1.7

Units of Permeability and Conversion Factors

Unit	m/(s·Pa)	LMH/bar	GFD/psi
m/(s·Pa)	1	3.600×10^{11}	1.463×10^{10}
LMH/bar	2.778×10^{-12}	1	0.041
GFD/psi	6.837×10^{-11}	24.615	1

only a small fraction of the applied pressure Δp. In some other applications such as MF and UF, the term of osmotic pressure can be dropped out of Eq. (1.11) because it is usually negligibly small compared to the applied pressure.

It can be seen from Eq. (1.11) that the permeability of a membrane is the permeate flux produced under a unit net driving pressure. Therefore, the unit for membrane permeability can be derived from the units of pressure and permeate flux. The basic metric unit for permeability is m/(s·Pa) when the basic metric units are used for pressure and flux, respectively. The metric unit LMH/bar and the imperial unit GFD/psi are the other units of permeability commonly used in practices. The conversion factors among the permeability units are listed in Table 1.7. The units for membrane resistance R_m are the reciprocals of those for membrane permeability.

Example 1.7

A salt solution that contains 0.5 mol/L sodium chloride (NaCl) is filtrated through a piece of RO membrane. Calculate the permeate flux of the membrane, assuming the membrane to have a perfect salt rejection. The feed water temperature is 20°C, the membrane permeability is 3.0×10^{-12} m/(s·Pa), and the applied pressure is 50 bar. Express the result in units of m/s, LMH, and GFD.

SOLUTION

One NaCl molecule dissociates into a sodium ion Na^+ and a chloride ion Cl^-; the osmotic pressure of the feed water can be determined with the van't Hoff equation:

$$\pi = RT\left(c_{Na^+} + c_{Cl^-}\right) = 8314\ L\ Pa/(K\ mol) \times 293\ K \times \left(0.5\ mol/L + 0.5\ mol/L\right)$$
$$= 2.44 \times 10^6\ Pa = 24.4\ bar$$

With the basic membrane filtration equation, the permeate flux is calculated as

$$J = A\left(\Delta p - \Delta\pi\right) = 3.0 \times 10^{-12}\ m/(s\ Pa) \times \left(5.0 \times 10^6\ Pa - 2.44 \times 10^6\ Pa\right)$$
$$= 7.69 \times 10^{-6}\ m/s$$

In LMH

$$J = 7.69 \times 10^{-6}\, m/s \times 3.600 \times 10^{6} = 27.7\ LMH$$

In GFD

$$J = 27.7\ LMH \times 0.5889 = 16.3\ GFD$$

The permeate flux is $7.69 \times 10^{-6}\ m/s$, or $27.7\ LMH$, or $16.3\ GFD$.

1.5.3 Energy Requirement for Separation

Energy is needed to pump water into the membrane modules and drive it through the membrane in pressure-driven membrane processes. Especially, desalination of high salt content streams, such as seawater, is an intensive operation. The incurred cost can account for more than half of the total cost in seawater desalination [28–32]. Specific energy is commonly used to measure energy consumption in water desalination, which is defined as the amount of energy for per unit volume of permeate produced.

The basic metric unit for specific energy is J/m^3. It is interesting to point out that the unit of specific energy is in the same form as the basic metric unit of pressure. The basic metric unit for pressure is

$$Pressure\ unit = Pa = \frac{Newton}{m^2}$$

Multiplying the length dimension m to both the numerator and denominator results in the unit for specific energy:

$$\frac{Newton}{m^2} \times \frac{m}{m} = \frac{Joule}{m^3} = Specific\ energy\ unit$$

Joule is a very small energy unit in desalination applications. A more appropriate energy unit is the kilowatt-hour (kWh). The conversion between these two units is

$$\theta = \frac{Joule}{kWh} = \frac{Joule}{1000 \times \frac{Joule}{s} \times 1h \times \frac{3600\,s}{1h}} = \frac{1}{3.6 \times 10^{6}}$$

where θ is the conversion factor. In the following discussions, specific energy is first calculated in the unit of Joule per cubic meter (i.e., Pa). Then the result is converted to kilowatt-hour (kWh) by dividing by a factor of 3.6×10^{6}.

Theoretical minimal energy requirement for liquid separation is first analyzed with a reversible membrane process below. Then the energy consumption in practical membrane processes is discussed.

1.5.4 Theoretical Minimal Energy

Energy consumption rate in liquid separation largely depends on the technologies and operation strategies used. However, there is a thermodynamic minimal energy requirement for liquid separation regardless of technologies. This minimal energy is often employed as a benchmark to compare the energy efficiency of different technologies and operations in liquid separation.

The minimal specific energy to produce water out of solutions is the specific energy consumed in a reversible membrane process. For this purpose, a virtual reversible membrane process is constructed, as shown in Figure 1.13. The membrane device consists of a cylinder with a perfect membrane on the bottom and a piston to seal it on the other end. Feed water is initially placed in the cylinder between the piston and the membrane. Permeate is produced when the piston moves downward in the cylinder. To render the process reversible, the driving pressure is maintained to equal the osmotic pressure inside the cylinder. From the thermodynamic principle, the energy spent in this reversible process is the minimum energy for liquid separation.

Let V_0 and c_0 be the initial volume and salt concentration of feed water, respectively. As permeate comes out of the membrane, salt concentration inside the cylinder increases according to

$$c = \frac{c_0 V_0}{V}$$

where V and c are the variables for volume and salt concentration in the cylinder. The osmotic pressure inside the cylinder is linearly related to salt concentration, i.e.,

$$\Delta\pi = f_{os}\frac{c_0 V_0}{V} = \Delta\pi_0 \frac{V_0}{V}$$

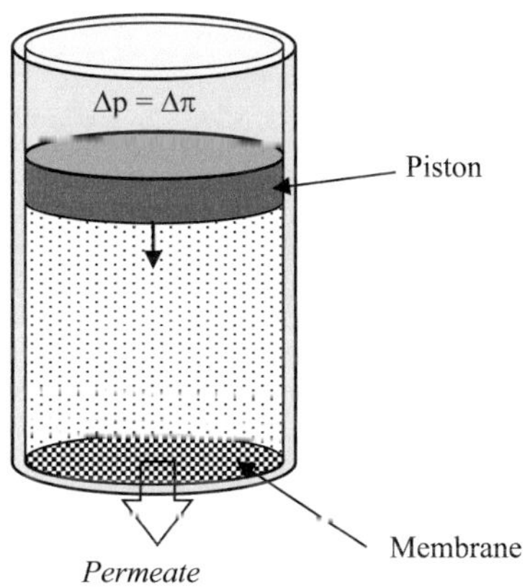

FIGURE 1.13
A reversible membrane process. Permeate is produced by a pressure that is always equal to osmotic pressure.

where $\Delta\pi$ is the variable for osmotic pressure, f_{os} is the osmotic pressure coefficient, and $\Delta\pi_0$ $(= f_{os}c_0)$ is the initial osmotic pressure of feed water. The energy spent on the piston to push water out of the membrane is

$$E = -\int_{V_0}^{V} \Delta p \, dV$$

A negative sign "−" appears because the volume change dV is negative and the energy is always positive. Since the applied pressure is always equal to the osmotic pressure in the reversible process, the applied pressure in the above integration operation can be replaced by the osmotic pressure. Therefore,

$$E = -\int_{V_0}^{V} \Delta\pi_0 \frac{V_0}{V} dV = -\Delta\pi_0 V_0 \ln\frac{V}{V_0}$$

The volume of permeate produced during this process is $V_0 - V$; therefore, the specific energy is

$$\hat{E} = \frac{E}{V_0 - V} = -\Delta\pi_0 \frac{V_0}{V_0 - V} \ln\frac{V}{V_0} \qquad (1.12)$$

In this reversible membrane process, the recovery can be defined as

$$R = \frac{V_0 - V}{V_0}$$

Then we have

$$\frac{V_0}{V_0 - V} = \frac{1}{R}$$

and

$$\frac{V}{V_0} = \left(\frac{V}{V_0} - 1\right) + 1 = -\frac{V_0 - V}{V_0} + 1 = 1 - R$$

Substituting the above two expressions into Eq. (1.12) results in

$$\hat{E} = \Delta\pi_0 \frac{1}{R} \ln\frac{1}{1-R} \qquad (1.13)$$

Eq. (1.13) shows that specific energy is a function of recovery. Let the recovery approach zero, i.e., $R \to 0$, the smallest specific energy for liquid separation can be obtained as the initial osmotic pressure of feed water $\Delta\pi_0$. It should be reminded that the unit of specific energy in Eq. (1.13) is J/m^3 when the Pascal is used as the unit of pressure. A factor of 3.6×10^6 is needed to convert it to the energy unit of kilowatt-hour.

Example 1.8

Calculate the specific energies for seawater desalination at the recovery of zero and subsequently for every 10% increase in recovery up to 70%. Present the result graphically.

SOLUTION

The osmotic pressure of average seawater is 2.57 × 10⁶ Pa (Pa = J/m³). Then the specific energy at the recovery approaching zero is

$$\begin{aligned}\hat{E} &= \frac{1\,kWh}{3.6\times10^{6}\,\mathrm{J}}\Delta\pi_0 \\ &= \frac{1\,kWh}{3.6\times10^{6}\,\mathrm{J}}\times5.57\times10^{6}\ J/m^3 \\ &= 0.71\ kWh/m^3\end{aligned}$$

Eq. (1.11) (with the conversion factor) is used to calculate the specific energy for nonzero recoveries. At R = 10%

$$\begin{aligned}\hat{E} &= \frac{1\,kWh}{3.6\times10^{6}\,\mathrm{J}}\Delta\pi_0\frac{1}{R}\ln\frac{1}{1-R} \\ &= \frac{1\,kWh}{3.6\times10^{6}\,\mathrm{J}}\times5.57\times10^{6}\ J/m^3\times\frac{1}{0.1}\ln\frac{1}{1-0.1} \\ &= 0.75\,kWh/m^3\end{aligned}$$

The specific energies are calculated similarly for other recoveries. The results are graphically shown in the figure below.

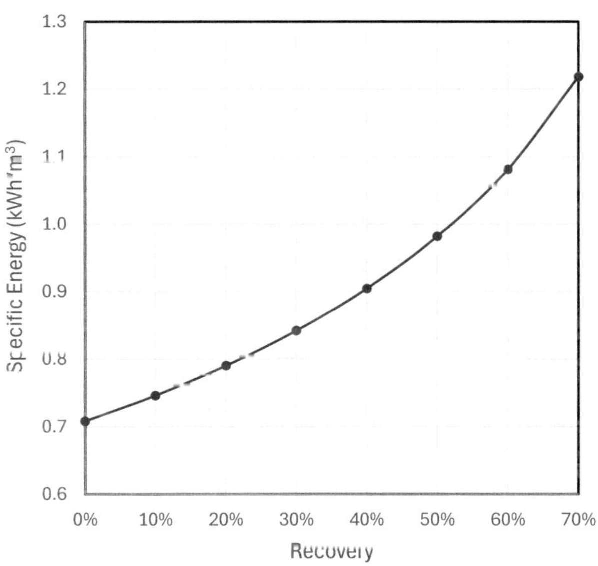

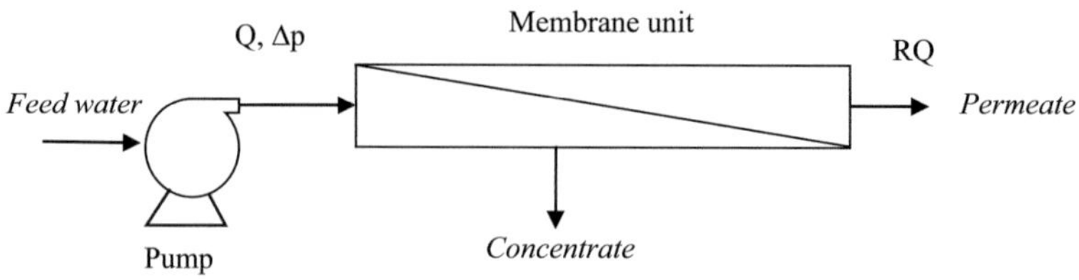

FIGURE 1.14
Schematic of a practical pressure-driven membrane unit.

1.5.5 Energy Consumption in Practical Membrane Processes

A schematic of a practical membrane unit is shown in Figure 1.14. Feed water is pumped into the membrane unit at pressure ΔP and flowrate Q. The permeate stream comes out of the membrane unit at flowrate RQ, with R being the recovery of the membrane unit.

The specific energy in such a membrane unit can be determined straightforwardly. In a time interval Δt, the energy spent on pumping water into the membrane unit is

$$E = \Delta p Q \Delta t$$

Because the volume of permeate produced in the same time interval is $RQ\Delta t$, the specific energy is then calculated with

$$\hat{E} = \frac{\Delta p Q \Delta t}{RQ\Delta t} = \frac{\Delta p}{R} \tag{1.14}$$

Again, the pressure unit of Pascal should be used in Eq. (1.14) to calculate the specific energy in the energy unit of Joule. The energy unit is then converted to kilowatt-hour by dividing by a factor of 3.6×10^6 to present the result in a more convenient unit of kilowatt-hour.

Example 1.9

The pressure required in reverse osmotic membrane processes for seawater desalination is usually between 800 and 1100 psi. What is the specific energy in seawater desalination at the typical recovery of 50%?

SOLUTION

When $\Delta P = 800\ psi$, the specific energy is

$$\hat{E} = \frac{\Delta p}{R} = \frac{800\ psi}{0.5} \times \frac{6895\ Pa}{1\ psi} \times \frac{1}{3.6 \times 10^6\ Pa/kWh} = 3.06\ kWh/m^3$$

When $\Delta P = 1100\ psi$, the specific energy is

$$\hat{E} = \frac{\Delta p}{R} = \frac{1100\ psi}{0.5} \times \frac{6895\ Pa}{1\ psi} \times \frac{1}{3.6 \times 10^6\ Pa/kWh} = 4.21\ kWh/m^3$$

Therefore, the specific energy for seawater desalination with RO membrane is typically in the range of 3.06–4.21 kWh/m^3.

1.6 Coverage

Background information and basic concepts about the membrane process are introduced in this chapter. Such knowledge is essential to understand the more advanced topics of membrane processes, which are the primary goals of this textbook. The rest of the book is divided into three parts on the roles of retained species in membrane processes, principles of membrane desalination, and transport phenomena across membranes.

Part I: *Fundamentals of Retained Particles.* The behaviors of the species retained by the membrane are among the most fascinating phenomena in membrane separation processes. The performance of the membrane process is significantly affected or even largely controlled by the amount and states of retained species accumulated. The states and roles of retained species in membrane processes are analyzed rigorously with the fundamental thermodynamic and hydraulic principles. Especially with the introduction of a model membrane system of binary solution, the two states of retained species, namely concentration polarization and cake formation, can be clearly separated and delineated. Consequently, the equation for permeate flux can be rigorously developed for membrane devices with or without a cake layer. Important contents in this part include the critical roles of the amount of retained species in concentration polarization, the occurrence of limiting flux, and an explanation of the flux paradox. It is also demonstrated that the cake formation is a mechanism to bring the operation of a crossflow filtration from nonequilibrium operation to equilibrium operation.

Part II: *Reverse Osmosis Desalination.* This part focuses on the quantitative prediction of the performance of the RO membrane in desalination applications. The practical application of RO membrane in water desalination is characterized by a very large aspect ratio of length to height of the feed channel. The performance of such a channel can be rigorously predicted with a concise expression – the state equation. The long narrow channel can be characterized by the characteristic pressure for the RO membrane channel. When the applied pressure is smaller than the characteristic pressure, the recovery of the membrane channel is governed by mass transfer that is a strong function of membrane resistance and geometry of the membrane channel. However, when the applied pressure is greater than the characteristic pressure, the recovery of the membrane channel is completely controlled by thermodynamic parameters and independent from membrane resistance and channel geometry. A much more effective characterization method for the fouling strength of feed water is introduced. It can be measured easily with a simple device as for SDI measurement. On the other side, it bears clearly defined physical meaning and can be used straightforwardly in simulation and assessment of fouling development in full-scale RO desalination processes.

Part III: *Membrane Transport Phenomena*. This part focuses on understanding the fundamental principles and mechanisms for the transport of various species through membranes. The nature and mechanism of water transport through the membrane are critically examined, which is vital for the understanding of the phenomena associated with membrane processes, especially under net osmotic driving force. Transport of noncharged solutes across membranes is rigorously examined. Analytical equations are derived from the first principles for flux and rejection ratio of solutes when both convection and diffusion are involved for solute transport. It reveals that the partition of solutes unfavorably to the membrane is essential for membrane separation. The transport of charged species, the ions, across the membrane is much more complicated than that of the uncharged species because of the involvement of electromigration. The development of membrane potential and its interplay with ion transport make the delineation of ion transport a serious challenge. Rigorous formulation and numerical simulation of ion transport across the membrane are also presented.

1.7 Problems

1. What is a membrane? What are the two major driving forces for liquid separation by membranes?
2. Is concentration gradient a driving force for membrane separation? Give your reasons.
3. Briefly describe membrane categories in pressure-driven membrane processes.
4. What is the significance of the Loeb–Sourirajan method in membrane technology?
5. What are the principal materials and manufacturing methods for RO membranes nowadays?
6. A solution is made by dissolving 1000 mg of $CaSO_4$ in a liter of water. What is the molar concentration and number concentration of the solution?

7. Using van't Hoff equation calculate the osmotic pressure of the water sample in Question 6 at temperature of 20 and 35°C, respectively.
8. The particle concentration of a suspension is 1000 mg/L. The specific weight of the particles is 1.2. Assume the particles are uniform in size with a perfect spherical shape of 0.5 μm in radius, and calculate (1) the volume fraction of the solids in water and (2) the number concentration of particles.
9. An analysis of a water sample of 250 mL shows that the weights of four major ions are 632.5 mg Na^+, 887.5 mg Cl^-, 200 mg Ca^{2+}, and 720 mg SO_4^{2-}, respectively. Assume other ions and solutes in the water sample to be negligible, and determine the mass concentration, molar concentration, and number concentration of TDS in the water sample.
10. The average TDS concentration of seawater is 35,000 mg/L. What is its molar concentration, assuming the TDS is sodium chloride only?
11. The concentration of suspended solids in a water sample is 500 mg/L. The specific weight of particulates is 1.05. It is known that the particulates are divided equally in weight into two radii of 0.1 and 0.2 μm, respectively. Assuming a uniform perfect spherical shape of the particles, what are the number concentration and molar concentration of particulates in the water sample?
12. The pressure of 30 psi is used in the SDI test. What is the pressure in units of bar and Pascal?
13. A piece of RO membrane of 10 cm^2 is under the pressure difference of 100 psi. What is the total force applied on the piece of membrane in the unit of Newton?
14. What is the nature of osmotic pressure? How to define the osmotic pressure of a solution?
15. Two solutions of different salts with the same mass concentration are separated by a semipermeable membrane as shown below. What is the flow direction of water across the membrane?

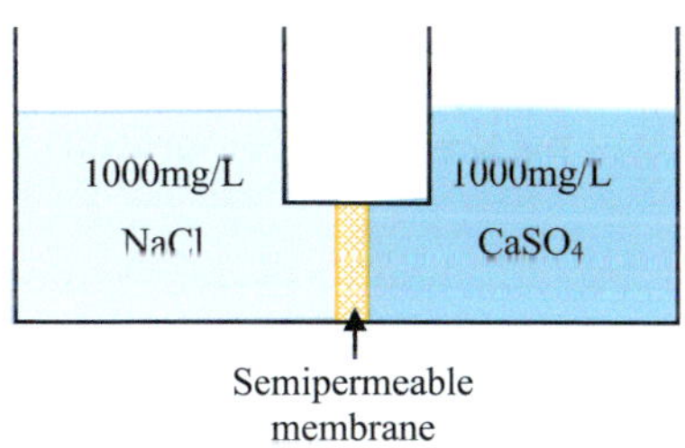

16. Estimate the osmotic pressure of seawater with an average TDS of 35,000 mg/L.
17. Seawater contains on average 4.7 mg/L boron, and the recommended boron concentration for drinking water by the World Health Organization (WHO) is 0.5 mg/L. The boron rejection rate of a seawater RO membrane is 65%. What is the boron concentration in the permeate of the membrane? Does it satisfy the WHO standard for boron in drinking water?
18. Brackish water has TDS (salinity) of 2500 mg/L. An RO desalination plant is planned to produce drinking water with TDS <500 mg/L. What is the minimum rejection rate of the membrane required?
19. An NF membrane has rejection rates of 95 and 60% for cations Ca^{2+} and Na^+, respectively. The concentrations of Ca^{2+} and Na^+ in a brackish water are 500 mg/L and 2000 mg/L. Calculate Ca^{2+} and Na^+ concentration in the permeate of the membrane.
20. A membrane process with feed-and-bleed configuration is shown below. The flowrates are indicated in the figure. Calculate the concentration in each flow and express it in terms of the feed concentration c_f.

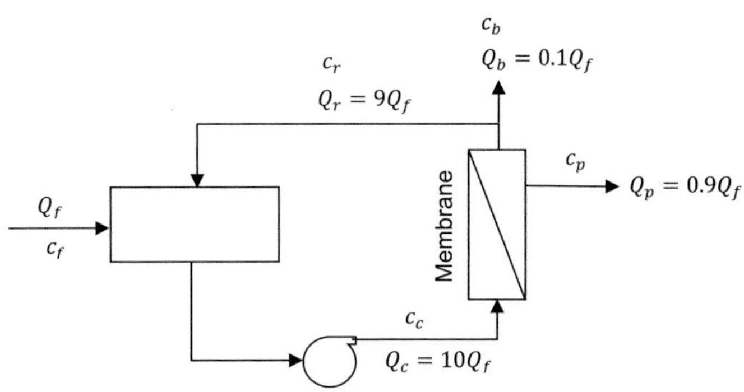

21. A membrane process is run in batch configuration. Feed salt concentration is 2000 mg/L, and the salt rejection rate by the membrane is 95%. Derive the expression of salt concentration in the permeate as a function of recovery. What is the salt concentration in the retentate by the end of operation with a recovery up to 80%?
22. In seawater desalination, sometimes the permeate of the first stage is fed to the second stage (permeate staging) for higher salt removal. When the salt rejection rates are r_1 and r_2 for the two membrane

processes, derive the expression for the overall rejection rate of the membrane process.

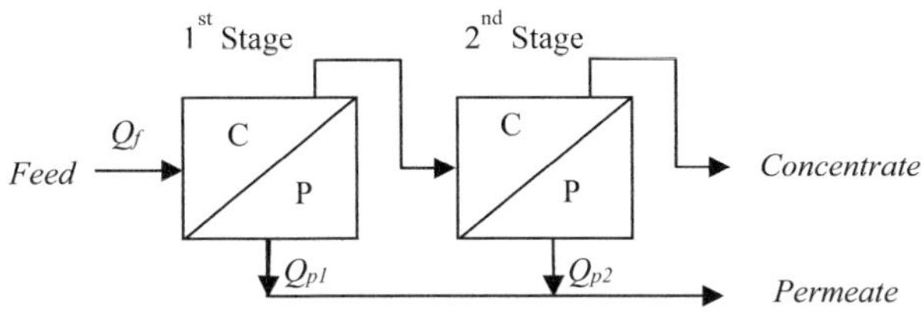

23. A membrane process of two stages on the concentrate stream is shown below (concentrate staging, i.e., concentrate out of the first stage is used as feed to the second stage). If the recoveries of the first and second stages are 60 and 40%, respectively, what is the overall recovery of the membrane process?

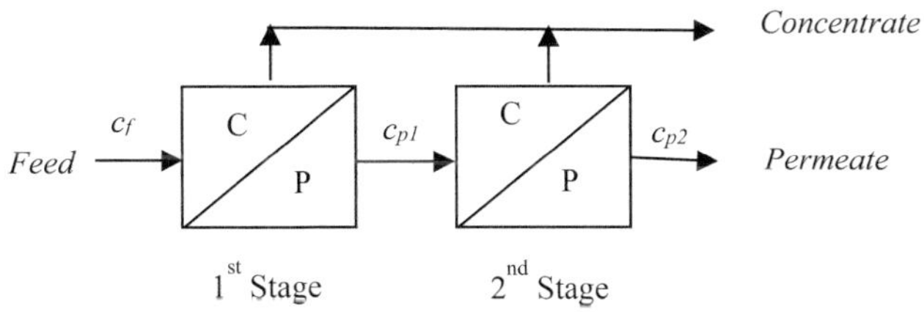

24. The permeate passing through two stages of membrane is usually needed for adequate boron removal due to the low rejection rate of boron by RO membranes. Seawater has a boron concentration of about 5 mg/L, and drinking water is considered safe to have a boron concentration below 1 mg/L. If the same membrane is used in both stages, what is the minimal rejection rate of the membrane for boron that is required?
25. The effective area of a piece of MF membrane is 47 mm in diameter. The filtrate collected over 5 minutes is 2150 mL in a test. (1) Calculate permeate flux in the test with pure water. (2) Determine the membrane resistance and permeability of the membrane when the pressure used in the test is 30 psi.
26. A desalination plant of capacity of 1000 m^3/day is designed to run at an average permeate flux of 5×10^{-6} m/s. (1) Convert the flux to the units of LMH and GFD. (2) Calculate the area of membrane needed.
27. The typical resistance of brackish water RO membranes is 1.0×10^{11} Pa·s/m. Convert the resistance to the unit of bar·h·m^2/L and psi·day·ft^2/gal, respectively.
28. Determine the permeability of the membrane in Q22 and express it in units of m/(s·Pa), L/(h·m^2·bar), and gal/(day·ft^2·psi), respectively.

29. In a brackish water desalination membrane process, the salt concentration is 2000 mg/L. The osmotic pressure can be estimated with the rule of thumb $\Delta\pi = 0.01c$ (with psi and mg/L as the units of pressure and salt concentration, respectively). The membrane resistance is 10^{11} Pa·s/m. Calculate permeate flux at the pressure of 10 bar (pay attention to the units for pressures in different parameters).
30. A seawater desalination process with RO membrane uses a driving pressure of 60 bar. The salt concentration is 34,500 mg/L, and the membrane resistance is 3×10^{11} Pa·s/m. Assuming single salt NaCl in the seawater, the osmotic pressure of seawater can be calculated with the van't Hoff equation. Calculate permeate flux at a temperature of 20°C.
31. Demonstrate that the unit for specific energy in water desalination is the same unit for pressure.
32. How many Joules is 1 kWh?
33. A pump station is needed to lift feed water up 50 m from the source to the site of desalination plant. Calculate the theoretical minimum energy needed per cubic meter of feed water.
34. A desalination plant of capacity 1000 m^3/day works at a pressure of 10 bar with a recovery of 75%. Calculate the total energy consumed daily and the specific energy of this membrane process.
35. Determine the theoretical minimum specific energy of a membrane process in desalination of brackish water of salt concentration of 5000 mg/L at a recovery of 75%.

References

1. Fane, A.G., Wang, R., & Jia, Y. (2011). Membrane Technology: Past, Present and Future. In: Wang, L.K., Chen, J.P., Hung, Y.T., Shammas, N.K. (eds) Membrane and Desalination Technologies. Handbook of Environmental Engineering, vol 13. Humana Press, Totowa, NJ.
2. Baker, R.W. (2004). Membrane Technology and Applications (2nd ed.). John Wiley & Sons, Ltd, Chichester, West Sussex.
3. Mulder, M. (2012). Basic Principles of Membrane Technology (2nd ed.). Springer Science & Business Media, Dordrecht, The Netherlands.
4. Cheryan, M. (1998). Ultrafiltration and Microfiltration Handbook. CRC Press, Boca Raton, FL.
5. Blatt, W.F., Dravid, A., Michaels, A.S., & Nelsen, L. (1970). Solute Polarization and Cake Formation in Membrane Ultrafiltration: Causes, Consequences, and Control Techniques. In: Flinn, J.E. (ed.) Membrane Science and Technology: Industrial, Biological, and Waste Treatment Processes. Plenum Press, New York, pp. 47–97.

6. Porter, M.C. (1972). Concentration polarization with membrane ultrafiltration. Industrial & Engineering Chemistry Product Research and Development, 11, 234–248.
7. Ren, J., & Wang, R. (2011). Preparation of Polymeric Membranes. In: Wang, L.K., Chen, J.P., Hung, Y.T., Shammas, N.K. (eds) Membrane and Desalination Technologies. Handbook of Environmental Engineering, vol 13. Humana Press, Totowa, NJ.
8. Fick, A. (1855). On liquid diffusion. The London, Edinburgh, and Dublin Philosophical Magazine and Journal of Science, 10(63), 30–39.
9. Loeb, S., & Sourirajan, S. (1960). Sea water demineralization by means of a semipermeable membrane. UCLA Department of Engineering Report.
10. Reid, C.E., & Breton, E.J. (1959). Water and ion flow across cellulosic membranes. Journal of Applied Polymer Science, 1(1), 133–143.
11. Loeb, S., & Sourirajan, S. (1962). Sea water demineralization by means of an osmotic membrane. Advances in Chemistry, 38, 117–132.
12. Cadotte, J.E., Petersen, R.J., Larson, R.E., & Erickson, E.E. (1980). A new thin-film composite seawater reverse osmosis membrane. Desalination, 32, 25–31.
13. Cadotte, J.E., & Petersen, R.J. (1981). Thin-film composite reverse-osmosis membranes: Origin, development, and recent advances, American Chemical Society, DOI: 10.1021/bk-1981-0153.ch021.
14. Peterson, R., Cadotte, J., & Buttner, J. (1982). Final report of development of FT-30 membrane in spiral wound modules. Contract No 14-34-0001-8547 for US Department of the Interior, October.
15. Blatt, W.F., Dravid, A., Michaels, A.S., & Nelsen, L. (1970). Solute Polarization and Cake Formation in Membrane Ultrafiltration: Causes, Consequences, and Control Techniques. In: Flinn, J.E. (ed.) Membrane Science and Technology. Springer, Boston, MA.
16. Lalia, B.S., Kochkodan, V., Hashaikeh, R., & Hilal, N. (2013). A review on membrane fabrication: Structure, properties and performance relationship. Desalination, 326, 77–95.
17. Duong, H.C., Tran, T.L., Ansari, A.J. et al. (2019). Advances in membrane materials and processes for desalination of brackish water. Current Pollution Reports, 5, 319–336.
18. Ismail, A.F., Rahman, M.A., Othman, M.H.D., Matsuura, T. (eds) (2018). Membrane Separation Principles and Applications: From Material Selection to Mechanisms and Industrial Uses. Elsevier Science, Dordrecht.
19. Lee, K.P., Arnot, T.C., & Mattia, D. (2011). A review of reverse osmosis membrane materials for desalination—Development to date and future potential. Journal of Membrane Science, 370, 1–22.
20. Ravichandran, S.R., Venkatachalam, C.D., Sengottian, M., Sekar, S., Subramaniam Ramasamy, B.S., Narayanan, M., Gopalakrishnan, A.V., Kandasamy, S., & Raja, R. (2022). A review on fabrication, characterization of membrane and the influence of various parameters on contaminant separation process. Chemosphere, 306, 135629.
21. Urper, G.M., Sengur-Tasdemir, R., Turken, T., Ates Genceli, E., Tarabara, V.V., & Koyuncu, I. (2017). Hollow fiber nanofiltration membranes: A comparative review of interfacial polymerization and phase inversion fabrication methods. Separation Science and Technology, 52(13), 2120–2136.
22. Spiegler, K.S. (1977). Salt-Water Purification (2nd ed.). Plenum Press, New York.

23. Trettin, D.R., & Doshi, M.R. (1980). Limiting flux in ultrafiltration of macromolecular solution. Chemical Engineering Communications, 4, 507–522.
24. Michaels, A.S. (1968). New separation technique for the CPI. Chemical Engineering Progress, 64, 31.
25. Belfort, G., Davis, R.H., & Zydney, A.L. (1994). The behavior of suspensions and macromolecular solutions in crossflow microfiltration. Journal of Membrane Science, 96, 1–58.
26. Wijmans, J.G., Nakao, S., Van Den Berg, J.W.A., Troelstra, F.R., & Smolders, C.A. (1985). Hydrodynamic resistance of concentration polarization boundary layers in ultrafiltration. Journal of Membrane Science, 22, 117–135.
27. Mohammad, S., & Gill, W.N. (1981). Review of reverse osmosis membranes and transport models. Chemical Engineering Communications, 12, 279–363.
28. Liu, C., Rainwater, K., & Song, L. (2011). Energy analysis and efficiency assessment of reverse osmosis desalination process. Desalination, 276, 352–358.
29. Elimelech, M., & Phillip, W.A. (2011). The future of seawater desalination: Energy, technology, and the environment. Science, 333, 712–717.
30. Deshmukh, A., Boo, C., Karanikola, V., Lin, S., Straub, A.P., Tong, T., Warsinger, D.M., & Elimelech, M. (2018). Membrane distillation at the water-energy nexus: Limits, opportunities, and challenges. Energy and Environmental Science, 11(5), 1177–1196.
31. Lotfy, H.R., Staš, J., & Roubík, H. (2022). Renewable energy powered membrane desalination—Review of recent development. Environmental Science and Pollution Research, 29, 46552–46568.
32. Charcosset, C. (2009). A review of membrane processes and renewable energies for desalination. Desalination, 245(1–3), 214–231.

Part I

Fundamentals of Retained Particles

This part of the book deals with the roles of the species retained by the membrane. Permeate fluxes of membrane separation processes can be significantly reduced by the accumulation of retained species. Retained species accumulate in membrane processes in two major states or forms, dependent on their thermodynamic properties. Retained species in different states affect the performance of the membrane processes in different ways and to different extents.

One state of retained species is in the concentration polarization layer over the membrane surface. Concentration polarization occurs in membrane processes when retained species are small in size or when the permeate flux is low. In that case, the hydraulic drag force on the species is not sufficiently high to hold the species on the membrane surface. The retained species would remain in the liquid phase and move with it out of the membrane processes. Another state of retained species is to form a cake layer on the membrane surface. Cake formation occurs under the conditions that the hydraulic drag force is high enough to hold the species on the membrane surface due to the large size of the species or the high permeate flux or both. At this time, the retained species are no longer a part of the liquid and do not move with the liquid anymore. Retained species in the cake layer attach to the membrane and accumulate inside the membrane processes.

The distinct features of retained species in the concentration polarization layer and cake layer are discussed in detail with emphasis on the fundamental natures of an idealized model binary solution. Mathematical models for the impacts of concentration polarization and cake formation on membrane processes are presented with detailed derivations. Thermodynamical

DOI: 10.1201/9781003743569-2

aspects of membrane separation processes, such as the limited role of concentration polarization on performance, critical pressure for cake formation, and nature of cake layer, are essential to understanding the roles of retained species in membrane processes.

The following topics will be covered in this part of the book:

Chapter 2. Roles of Retained Particles

Chapter 3. Concentration Polarization

Chapter 4. Cake Formation

2

Roles of Retained Particles

A consequence of all pressure-driven membrane separation processes is the retention and accumulation of species rejected by the membrane [1–4]. Depending on the types of membranes, these retained species can be suspended particulates, colloids, solutes, and ions. For convenience, all these species rejected by membranes are generally termed "retained particles." In turn, the retained particles can affect membrane processes profoundly [5–10]. In many cases, the behavior and performance of a membrane process are virtually determined by the amount and state of retained species in the process.

The basic filtration equation (Eq. 1.10) introduced in the previous chapter can only be applied to the idealized cases where the influence of retained species is negligible, for example, a membrane process that is operated in a narrow range of very low applied pressure with a feed water of very low particle concentration. Otherwise, the linear relationship between permeate flux and driving pressure as described by the basic filtration equation only exists at the beginning of the filtration operation. The behaviors of most practical membrane processes are observed to deviate significantly from the basic filtration equation because of the influence of retained species. Two commonly observed phenomena and behaviors that deviate from the basic filtration equation are mentioned below as examples.

2.1 Experimental Observations

2.1.1 Permeate Flux Declining with Time

A phenomenon of great significance in membrane processes is that the flux declines with time in membrane processes while the conditions in the membrane process are maintained unchanged. This phenomenon can be observed in all types of membrane processes. It is particularly obvious in microfiltration and ultrafiltration processes, where the fluxes can drop to less than 10% of their initial value in only a few minutes.

As shown in Figure 2.1, fluxes in membrane processes are often observed to decline with time while operating conditions remain constant. Steady state exists in some cases where flux declines from an initial value v_0 to a steady state value v_{ss} in a relatively short period [11 14]. Permeate flux remains constant thereafter if the operating conditions are maintained unchanged. This behavior

DOI: 10.1201/9781003743569-3

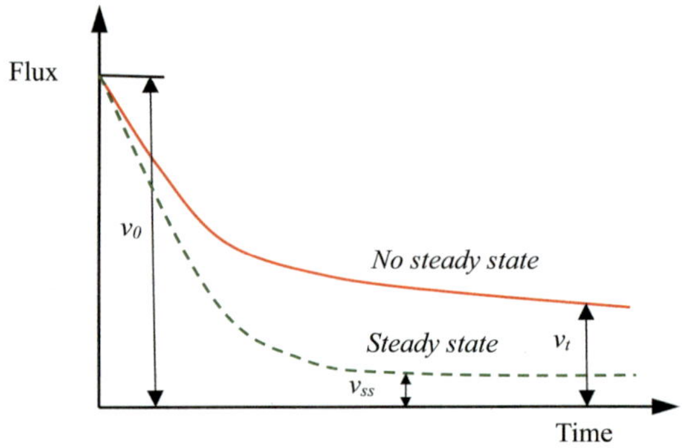

FIGURE 2.1
Permeate fluxes are often observed to decline with time while operating conditions remain constant. Some have steady states to approach, and others do not.

is portrayed by the broken line in the figure. On the other hand, permeate flux in most cases declines all the time [15–18]. It starts with a rapid decline from the initial value v_0 to a certain value. Then the flux v_t decreases slowly with time, for which no steady state can be observed in the practical operation times. This behavior is described with the solid line in the figure. The latter can occur in all types of membrane processes, both dead-end and crossflow filtrations, while the former only happens in crossflow microfiltration and ultrafiltration processes.

2.1.2 Nonlinear Dependence of Fluxes on Pressure

Another interesting observation is the dependence of permeate flux on applied pressure in crossflow membrane processes. Strictly speaking, the linearity as predicted by the basic filtration equation Eq. (1.10) is only true for pure water filtration, as shown with the broken line in Figure 2.2.

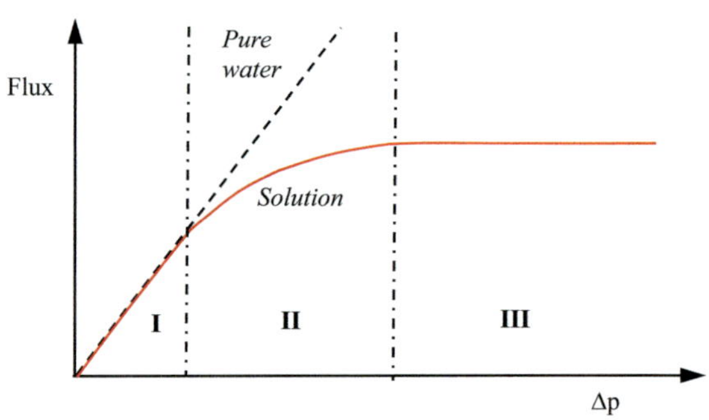

FIGURE 2.2
Permeate flux in some membrane processes can be divided into three pressure regions with different behaviors.

When there are solutes or particles in solutions that are rejected by the membrane, permeate flux is observed to have a rather complex dependence on the applied pressure [19–23]. As shown in Figure 2.2, there are three pressure regions in which the permeate flux behaves differently with pressure. In Region I of very low pressure, flux increases almost linearly with pressure, which is like that of pure water. In Region II of intermediate pressure, flux still increases with pressure but at a reduced rate compared to that in Region I. In Region III of high pressure, flux stops increasing with pressure but remains constant. Different types of membrane processes are primarily operated in different pressure regions. While reverse osmosis processes are always operated in Region I, microfiltration and ultrafiltration processes are commonly working in Regions II and III, especially for the separation of biological products, such as proteins.

In membrane processes with otherwise well-defined conditions, such as feed water property, membrane characteristics, and applied pressure, the amount and state of retained particles are the only parameters that change with time. It is perceivable that the retained particles can add additional resistance to permeate flow in the membrane processes. Careful inspection of retained particles shall provide clues to understanding the above-mentioned behaviors of permeate flux in membrane processes.

The influence of retained particles on permeate flux is one of the most important phenomena occurring in all membrane separation processes. Phenomena associated with the retained particles are by far among the most fascinating and influential incidents in membrane processes. It is impossible to design and operate an efficient membrane process without a thorough understanding of the roles played by the retained particles in the process. In the following section in this chapter, fundamentals of retained particles will be first detailed and delineated. Their impacts on performance (permeate flux) will be discussed in the remaining chapters in this part.

2.2 Interactions on Retained Particles

The states of retained particles in a membrane process are first determined by the amount of retained particles accumulated. It is also controlled by the interactions exerted on the retained particles [24–28]. There are many possible interactions on the retained particles, such as electrical, magnetic, and gravitational interactions. However, these interactions do not play significant roles in most membrane separation processes. The two most important interactions in all membrane processes are the diffusive force due to thermal motion and the drag force due to the flowing liquid on the retained particles. For the sake of convenience and rigor of derivation, particles are assumed to be spherical in shape.

2.2.1 Diffusive Force

A small particle in liquid is in constant thermal motion, and the net effect is the movement from regions of high concentration to regions of low concentration. This phenomenon is called diffusion, which is driven by the gradient of Gibbs free energy or chemical potential of the particle in a liquid.

At constant pressure and temperature, the Gibbs free energy of a particle is

$$G = G^0 + kT \ln\left(\frac{c}{c^0}\right)$$

where G is the Gibbs free energy or chemical potential, and G^0 is the Gibbs free energy at the standard condition, k is the Boltzmann constant, T is the absolute temperature, c is the concentration of retained particles, and c^0 is the particle concentration at the standard condition. The diffusive force is then determined by

$$F_d = -\nabla G = -kT\frac{1}{c}\nabla c \tag{2.1}$$

where F_d is the diffusive force on a particle, and ∇ is the gradient operator.

2.2.2 Drag Force

There is always a permeate flow in a membrane process. The permeate flow in a membrane process is typically very small in velocity and can be safely treated as a laminar flow. In an unbounded liquid, as shown in Figure 2.3, the hydraulic drag force of permeate flow on an isolated spherical-shaped particle can be calculated with Stokes' law

$$F_f = -6\pi_c \mu a_p v \tag{2.2}$$

where F_f is the frictional drag force on a spherical particle, π_c (≈ 3.14) is the circumference ratio, μ is the viscosity of the liquid, a_p is the radius of the

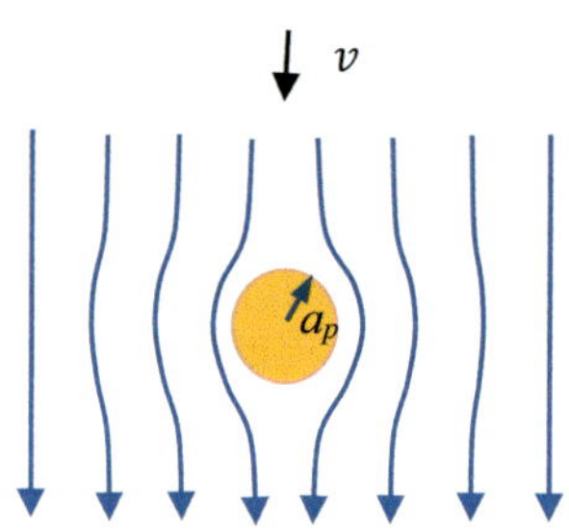

FIGURE 2.3
A laminar flow passing an isolated single spherical particle.

particle, and v is the velocity of permeate flow. The symbol π_c is used in this book to represent the circumference ratio because the standalone symbol π is reserved exclusively for osmotic pressure. There is a negative sign in Eq. (2.2) because the permeate flow is traditionally defined as positive.

2.2.3 Force Balance on a Retained Particle

A most important assumption of a retained particle in a membrane process is that it is stationary with respect to the membrane surface. In other words, it does not move toward or away from the membrane surface. Therefore, the sum of all forces on the retained particle in the direction vertical to the membrane surface must be equal to zero, i.e.,

$$-kT\frac{1}{C}\frac{dC}{dy}-6\pi_c\mu a_p v=0$$

where C is the concentration of retained particles, and y is the coordinate perpendicular to the membrane surface with the positive direction that is opposite to the permeate flow. The capital letter "C" is used here for the concentration of retained particles, while the lowercase "c" represents the concentration of total particles. Because when c_0 is the particle concentration in feed water, the total particle concentration is $c = c_0 + C$.

This is a one-dimensional force balance problem. Concentration gradient of retained particles is the derivative of concentration over coordinate y (i.e., dC/dy). It should be pointed out that the positive direction of permeate flow is defined as toward the membrane surface. A parameter, which is actually the diffusion coefficient, is introduced as

$$D=\frac{kT}{6\pi_c\mu a_p} \tag{2.3}$$

where D is the diffusion coefficient of the particle. Eq. (2.3) is the famous Stokes–Einstein equation. The force balance equation can now be rewritten as

$$D\frac{dC}{dy}+vC=0 \tag{2.4}$$

Equation (2.4) is the governing equation for particle distribution along the y direction over the membrane surface.

Example 2.1

Particles diffuse through a porous membrane. Membrane thickness is 50 μm, and particle radius is 2 nm. Particle concentrations on two sides of the membrane are 1000 and 200 mg/L, respectively. Calculate the diffusive forces and average diffusive velocities of the particles at two membrane surfaces at steady state.

SOLUTION

Concentration gradient across the membrane is constant at steady state because of mass conservation. Therefore,

$$\frac{dc}{dx} = \frac{\Delta c}{l} = \frac{200\,mg/L - 1000\,mg/L}{50 \times 10^{-6}\,m} = -1.60 \times 10^{7}\,(mg/L)/m$$

At the side of the membrane with a concentration of 1000 mg/L, the diffusive force can be calculated with Eq. (2.1) as

$$\begin{aligned} F_d &= -kT\frac{1}{c}\frac{dc}{dx} \\ &= -1.38 \times 10^{-23}\,m^3 \cdot Pa/K \times 293\,K \times \frac{-1.60 \times 10^{7}\,(mg/L)/m}{1000\,mg/L} \\ &= 6.47 \times 10^{-17}\,Newton \end{aligned}$$

By equating the friction drag force to the diffusive force in value but opposite in direction, the average diffusive velocity can be derived from Eq. (2.2) as

$$v = \frac{F_d}{6\pi_c \mu a_p} = \frac{6.47 \times 10^{-17}\,N}{6 \times 3.14 \times 10^{-3}\,Pa \cdot s \times 2 \times 10^{-9}\,m} = 1.72 \times 10^{-6}\,m/s$$

At the side with concentration 100 mg/L,

$$F_d = -1.38 \times 10^{-23}\,m^3 \cdot Pa/K \times 293\,K \times \frac{-1.60 \times 10^{7}\,(mg/L)/m}{200 mg/L}$$

$$= 3.23 \times 10^{-16}\,Newton$$

$$v = \frac{F_d}{6\pi_c \mu a_p} = \frac{3.23 \times 10^{-16}\,N}{6 \times 3.14 \times 10^{-3}\,Pa \cdot s \times 2 \times 10^{-9}\,m} = 8.58 \times 10^{-6}\,m/s$$

2.3 Suspended State

When a certain amount of retained particles are accumulated, the distribution of the retained particles is determined by the force balance equation Eq. (2.4). With careful analysis of force balance, insight into the existing state of retained particles in membrane processes can be obtained, which is one of the keys to understanding the roles of retained particles on permeate flux of membrane processes.

2.3.1 A Thought Experiment on Concentration Polarization

Let us start with a thought experiment on the distribution of retained particles under the influence of permeate flow. Considering a membrane

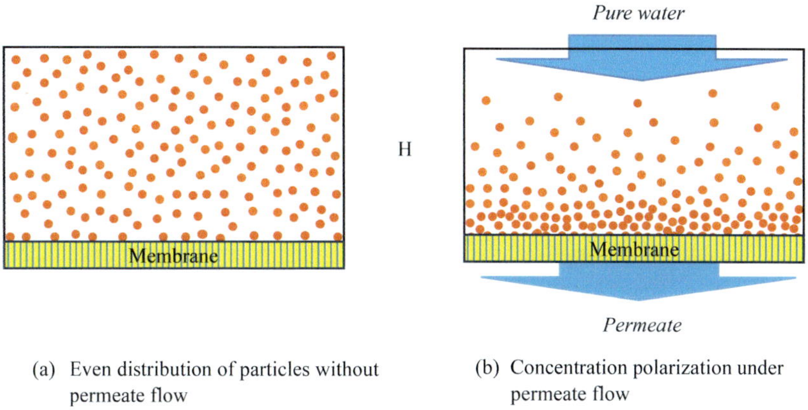

FIGURE 2.4
(a) Vertical concentration distributions of retained particles over the membrane surface **(b)** with and without permeate flow.

chamber of height H as shown in Figure 2.4, in which a certain amount of retained particles is evenly distributed along the height initially with no permeate flow (Figure 2.4(a)). When permeate velocity is induced by pumping pure water to the top of the chamber, the retained particles will be brought toward the membrane surface (Figure 2.4(b)). Higher particle concentration on the membrane surface causes diffusive movement of retained particles away from the membrane surface. At equilibrium, the retained particles become stationary when the diffusive force and the hydraulic drag force by permeate flow on them become equal, i.e., it is governed by Eq. (2.4).

With or without the permeate flow in the membrane chamber, the total amount of particles over a unit area of the membrane is unchanged regardless of the distribution of retained particles along the vertical direction. Therefore, mass conservation along the vertical direction in the membrane chamber can be written as

$$\int_0^H C\,dy = M \tag{2.5}$$

where M is the total amount of retained particles per unit area of membrane. The quantity M must have a unit that is consistent with the unit of C. For instance, the unit of M should be the total number per membrane area for C in number concentration or the total volume per membrane area for c in solid volume fraction.

The general solution of Eq. (2.4) gives the concentration distribution along the vertical direction to the membrane surface as

$$C = C_w e^{-\frac{v}{D}y} \tag{2.6}$$

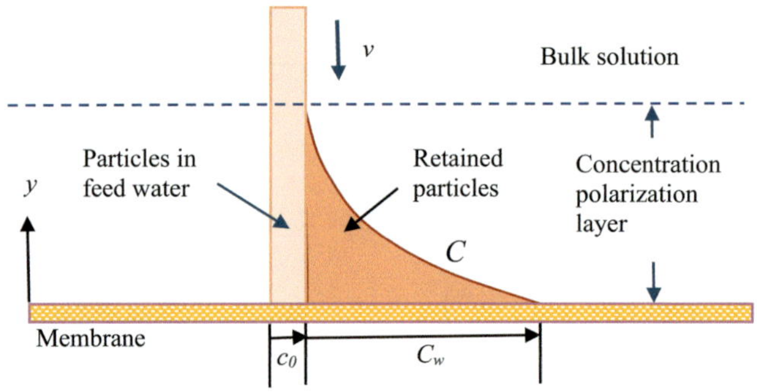

FIGURE 2.5
Schematic of concentration polarization over a membrane surface. Concentration polarization is an existing state of retained particles.

where C_w is the wall concentration of retained particles, i.e., the concentration on the membrane surface. Eq. (2.6) indicates that retained particle concentration attains its maximum value at the membrane surface and decreases exponentially into the bulk solution. The occurrence of uneven vertical concentration distribution over the membrane surface is called "concentration polarization." As will be discussed later, concentration polarization is one of the most profound phenomena in membrane processes that can affect their performance significantly.

A schematic of a concentration polarization layer over a membrane surface is shown in Figure 2.5. The feed water flows perpendicularly toward the membrane at a velocity v, and the original particle concentration in feed water is c_0. The thickness of the concentration polarization layer is indicated by a broken horizontal line within which all retained particles are located. The unevenly distributed retained particles in the concentration polarization layer are indicated by C with the wall concentration C_w on the membrane surface. The concentration of all particles (original and retained) in the concentration polarization layer is $c = c_0 + C$. In this thought experiment, $c = C$ because $c_0 = 0$.

One of the most important observations from Figure 2.5 is that concentration polarization is a result of the accumulation of retained particles. There would be no concentration polarization if there were no accumulation of retained particles over the membrane surface. The amount of retained particles plays an extremely important role in the development of concentration polarization. Under the same permeate velocity and diffusion coefficient, a stronger concentration polarization develops when a larger amount of retained particles accumulates.

2.3.2 Wall Concentration

Wall concentration of retained particles on the membrane surface is a convenient measurement or indicator for the strength of concentration polarization.

It can be determined by carrying out the integration of Eq. (2.5) by substituting Eq. (2.6) for C in the equation, i.e.,

$$\int_0^\infty C_w e^{-\frac{v}{D}y}\,dy = M$$

The upper integration limit H is replaced by infinity in the above integration. The change of integration limit won't deviate from the original problem at all if the thickness of the concentration polarization layer is much smaller than the physical height of the membrane chamber or channel. However, by doing so, the mathematical treatment of the problem becomes much more streamlined and elegant.

Carrying out the integration and rearranging results in

$$C_w = \frac{v}{D}M \tag{2.7}$$

Eq. (2.7) is the most important output of this thought experiment that reveals the essence of concentration polarization. It shows that wall concentration on the membrane surface (the strength of concentration polarization) is linearly dependent on the permeate velocity v and inversely on the diffusion coefficient D of the particles. More importantly, wall concentration is proportional to the amount of retained particles M. The quantification of this rather important parameter will remain a focal point in the rest of this chapter and other chapters in this part.

Example 2.2

A membrane chamber is 1 mm in height with a membrane on the bottom. The particles of radius 5 nm are retained by the membrane up to the amount of 0.1 mg/m^2. A permeate velocity of $1 \times 10^{-5}\,m/s$ is induced by pumping pure water from the top of the membrane chamber. The specific gravity of the particles is assumed to be unity. (1) Calculate particle concentration at the bottom (membrane surface), at the middle height, and at the top of the membrane chamber. (2) Determine the distance from the membrane surface at which the number concentration of particles is smaller than 1.

SOLUTION

Preparations

Using the viscosity $\mu = 10^{-3}\,Pa \cdot s$ at $T = 20°C$, the diffusion coefficient of the particles is

$$D = \frac{kT}{6\pi_c \mu a_p} = \frac{1.38 \times 10^{-23}\,Pa \cdot m^3/K \times 293K}{6 \times 3.14 \times 1.00\,Pa \cdot s \times 5 \times 10^{-9}\,m} = 4.29 \times 10^{-11}\,m^2/s$$

The volume of a particle

$$V_p = \frac{4}{3}\pi_c a_p{}^3 = \frac{4}{3}\times 3.14\times\left(5\times 10^{-9}\ m\right)^3 = 5.24\times 10^{-25}\ m^3$$

The weight (mass) of a particle

$$w_p = V_p\rho = 5.24\times 10^{-25}\ m^3\times 10^9\ mg/m^3 = 5.24\times 10^{-16}\ mg$$

The amount of retained particles in #/m² is

$$M = 1000\ mg/m^2\times\frac{1}{5.24\times 10^{-16}\ mg} = 1.91\times 10^{18}/m^2$$

1. The retained particle concentrations at:

   At the membrane surface

$$C_w = \frac{v}{D}M = \frac{5\times 10^{-6}\ m/s}{4.29\times 10^{-11}\ m^2/s}\times 1.91\times 10^{18}/m^2 = 4.45\times 10^{23}/m^3$$

   At the middle point of the membrane chamber

$$C = C_w e^{-\frac{v}{D}y} = 4.45\times 10^{23}/m^3\times e^{-\frac{5\times 10^{-6}\ m/s}{4.29\times 10^{-11}\ m^2/s}\times 0.5\times 10^{-3}\ m} = 1.08\times 10^{-27}/m^3$$

   At the top of the membrane chamber

$$C = C_w e^{-\frac{v}{D}y} = 4.45\times 10^{23}/m^3\times e^{-\frac{5\times 10^{-6}\ m/s}{4.29\times 10^{-11}\ m^2/s}\times 10^{-3}\ m} = 2.62\times 10^{-78}/m^3$$

2. The distance at which particle concentrations are smaller than 1.

$$y = -\frac{D}{v}\ln\frac{C}{C_w} = -\frac{4.29\times 10^{-11}\ m^2/s}{5\times 10^{-6}\ m/s}\ln\frac{1/m^3}{4.45\times 10^{23}/m^3} = 2.34\times 10^{-4}\ m$$
$$= 0.234\ mm$$

As mentioned earlier, wall concentration is an indicator of the strength of concentration polarization. Because the diffusion coefficient is constant for a given particle size, concentration polarization is jointly controlled by permeate velocity and the amount of retained particles. The impact of permeate velocity and the amount of retained particles on wall concentration can be effectively described by the iso-wall concentration diagram as shown in Figure 2.6. The horizontal coordinate (abscissa) is the permeate velocity, and the vertical coordinate (ordinate) is the amount of retained particles in units of mL/m². Wall concentrations of retained particles on the membrane surface in solid volume fraction are indicated on the curves. The smaller wall concentration appears at a lower position in the figure.

It can be seen from Figure 2.6 that the wall concentration can be increased by either increased permeate velocity or the amount of retained particles. For example, when the amount of retained particles is 15 mL/m², the wall concentration

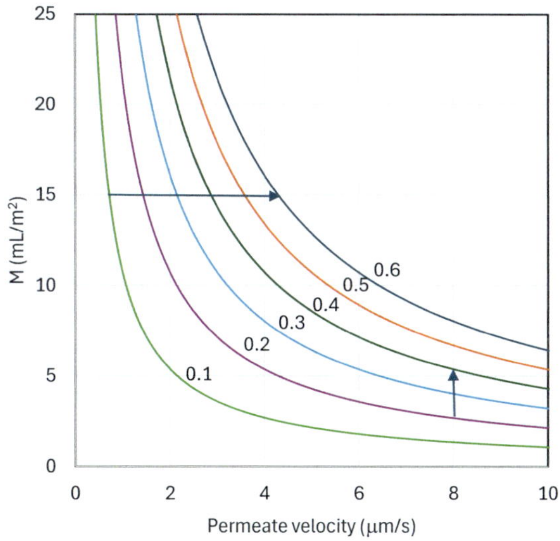

FIGURE 2.6
Curves of isoconcentration on the membrane surface developed based on Eq. (2.7). Particle radius of 2 nm and temperature of 20°C are used in the calculation. Wall concentrations are indicated in the figure.

can be increased from 0.1 to 0.6 when the permeate velocity is increased from roughly 0.7×10^{-6} to 4.6×10^{-6} m/s, as indicated by a horizontal arrow on the figure. Again, at a permeate velocity of 8.0×10^{-6} m/s, the wall concentration changes from 0.2 to 0.4 when the amount of retained particles increases from about 2.7 to 5.1 mL/m^2, as indicated by a vertical arrow in the figure. Both the amount of retained particles and permeate velocity play the same important roles in the development of concentration polarization. In the extreme cases, when any one of two parameters vanishes to zero, the wall concentration of retained particles will be zero, meaning no concentration polarization.

2.3.3 Pressure Difference across a Concentration Polarization Layer

The retained particles are stationary with respect to the membrane surface. Permeate flow has to pass over or squeeze through these stationary particles to reach the membrane. One way to study the impact of the retained particles on the permeate flux is to subtract the pressure difference across the concentration polarization layer from the total applied pressure. As shown in Figure 2.7, the pressure difference across the concentration polarization layer is caused by the drag force asserted by all retained particles on the water flow in the layer.

Because particle concentration in concentration polarization layer can be much higher than in the bulk solution, the other particles in the neighborhood affect water flow of the particle under consideration as shown in Figure 2.8. The drag force on the particle would be bigger than that on the

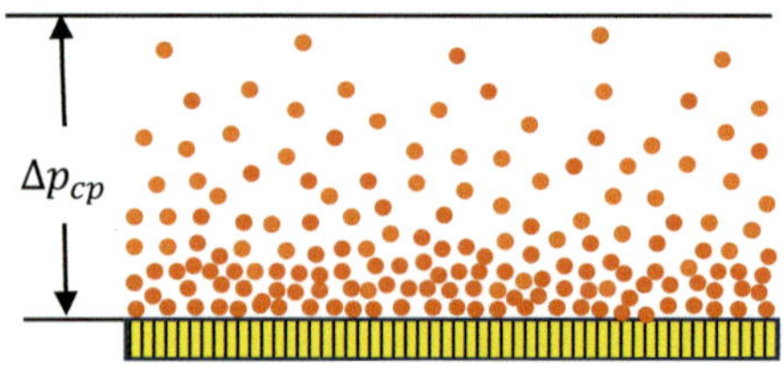

FIGURE 2.7
A pressure difference generated across the concentration polarization layer.

isolated one because higher velocity is needed for water to squeeze through the voids between the particles.

A correction function must be used when Stokes' law is still used for the drag force on a particle, i.e.,

$$F_{drag} = 6\pi_c \mu a_p v A_s$$

where A_s is the correction factor for particles with spherical shape. A correction function based on Happel's cell model [29] below is commonly used

$$A_s = \frac{1 + \frac{2}{3}\theta^5}{1 - \frac{3}{2}\theta + \frac{3}{2}\theta^5 - \theta^6} \tag{2.8}$$

where θ is a dimensionless variable that is defined as one-third of the solid volume fraction of particles in water. In the case of concentration polarization, variable θ is related to the concentration of retained particles by,

$$\theta = C^{1/3}$$

where C is the concentration in solid volume fraction. With the help of Eq. (1.3), the solid volume fraction can be converted to the number concentration

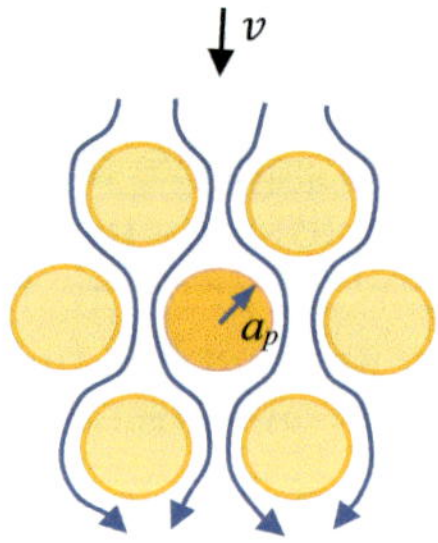

FIGURE 2.8
The drag force on a particle by water flow is increased because of the existence of neighboring particles.

$$C_n = \frac{C}{\frac{4}{3}\pi_c a_p^3}$$

Now, the pressure difference across the concentration polarization layer can be calculated with

$$\Delta p_{cp} = \int_0^{\infty} \left(6\pi_c \mu a_p v\right) A_s \frac{C}{\frac{4}{3}\pi_c a_p^3} dy \tag{2.9}$$

where Δp_{cp} is the pressure difference across concentration polarization layer, the term in the bracket is the Stokes' law for drag force on a single particle, A_s is the correction function for Stokes' law. Because the drag force here is for a single particle, the number concentration should be used in Eq. (2.9). The reason for the use of infinity as the upper integration limit is discussed before.

From Eq. (2.4), the concentration of retained particles can be expressed as

$$C = -\frac{D}{v}\frac{dC}{dy}$$

Substituting this expression into Eq. (2.9) and rearranging results in

$$\Delta p_{cp} = \frac{6\pi_c \mu a_p D}{\frac{4}{3}\pi_c a_p^3} \int_0^{C_w} A_s \, dC \tag{2.10}$$

Because the integration variable changes from y to C, the integration limits have to change correspondingly. The negative sign is removed by shifting the lower and upper integration limits.

According to Stokes–Einstein law Eq. (2.3),

$$6\pi_c \mu a_p D = kT$$

Two dimensionless parameters are introduced in order to further streamline Eq. (2.10). The first one is the filtration number, defined as

$$N_F = \frac{4\pi_c a_p^3 \Delta p_{cp}}{3kT} \tag{2.11}$$

where N_F is the filtration number. The other one is the mean correction factor, defined as

$$A_s^* = \frac{1}{C_w} \int_0^{C_w} A_s \, dC \tag{2.12}$$

where A_s^* is the mean correction factor. It corrects the deviation of hydraulic force from the Stokes equation in the entire concentration polarization layer with wall concentration of C_w.

Now Eq. (2.10) becomes

$$\Delta p_{cp} = kT\left(\frac{C_w}{\frac{4}{3}\pi_c a_p^3}\right)A_s^* = kTC_n A_s^* \tag{2.13}$$

where C_n is the number concentration of retained particles. The combination of Eqs. (2.10) and (2.11) results in

$$N_F = A_s^* C_w \tag{2.14}$$

2.3.4 The Mean Correction Factor

The mean correction factor as defined by Eq. (2.12) can be expressed explicitly as a function of solid volume fraction C as in the equation below:

$$A_s^* = \frac{1}{C_w}\int_0^{C_w} \frac{1+\frac{2}{3}C^{5/3}}{1-\frac{3}{2}C^{1/3}+\frac{3}{2}C^{5/3}-C^{6/3}}\,dC \tag{2.15}$$

For a given wall concentration, the mean correction factor can be determined with numerical integration. Generally speaking, the definite integration of a one-dimensional function is to determine the total area under the curve defined by the function between the two integration limits. When a function cannot be analytically integrated, numerical integration is a simple and straightforward way to deal with it. The concept of numerical integration is to divide the whole integration domain into many small pieces. Then the area of each small piece is calculated by approximating the actual shape with a simple regular shape of known formula for its area. Finally, the total area (i.e., the value of the definite integration) is determined by the summation of the areas of all the small pieces. High accuracy can be obtained with numerical integration when the step number is sufficiently large.

As an example, the equation for numerical integration with the midpoint rule is,

$$\int_a^b f(x)\,dx = \sum_{i=1}^{n} f\left(\frac{x_{i-1}+x_i}{2}\right)\Delta x \tag{2.16}$$

In this method, the integration domain is divided into n pieces of equal width Δx. The area of a small piece is determined by replacing the real shape

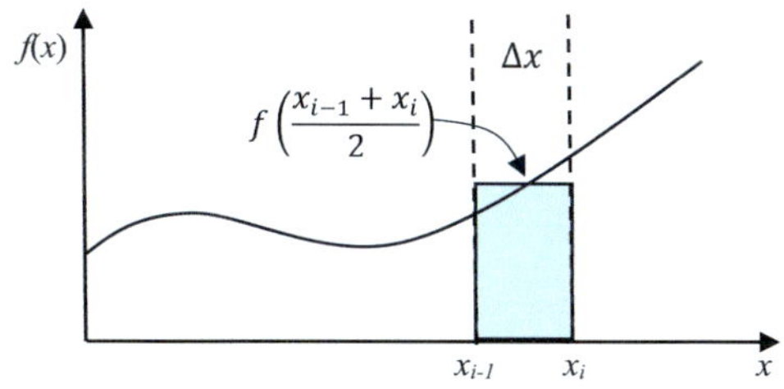

FIGURE 2.9
A small piece of area under the function is approximated with a rectangle in midpoint numerical integration.

as the pieces and a rectangle with the same width Δx with a height equal to the function value at the midpoint of the piece (Figure 2.9).

The mean correction factor A_s^* as expressed in Eq. (2.15) was calculated using midpoint numerical integration with 500 equal divisions of the concentration domain. The mean correction factor A_s^* calculated with Eq. (2.15), together with the correction factor A_s calculated with Eq. (2.8) and filtration number N_F calculated by Eq. (2.11), at the corresponding wall concentration, are presented in Table 2.1.

The maximum random packing density for identical rigid spherical particles is 0.64 [30]. Table 2.1 stops at this concentration with values of 24.19, 123.22, and 15.5 for the mean correction factor, correction factor, and filtration number at this maximum wall concentration, respectively. Table 2.1 can only contain values for a limited number of wall concentrations. For the values that fall in between two wall concentrations, the mean correction factors can be estimated with linear interpolation.

2.3.5 Features of Concentration Polarization Layer

First, retained particles in the concentration polarization layer are suspended in the liquid. There is a concentration gradient in the concentration polarization layer that generates a diffusive force to balance the drag force on the particles to maintain their position [31–33].

There is no physical contact between the particles in the concentration polarization layer. Therefore, they are not restricted from moving freely. When there is tangential water flow along the membrane surface, the particles can be carried away downstream, though their distances remain constant with respect to the membrane surface.

Another very important feature of the concentration polarization layer is that its impact on permeate flux is limited. Because of the existence of maximum wall concentration, the maximum reduction of driving pressure is the pressure difference across the concentration polarization layer. This feature is described by the example below.

TABLE 2.1

Correction Factor and Mean Correction Factor

C_w	A_s^*	A_s	N_F
0.000	1	1	0.000
0.001	1.13	1.18	0.001
0.002	1.17	1.23	0.002
0.004	1.22	1.31	0.005
0.008	1.30	1.43	0.010
0.010	1.33	1.48	0.013
0.02	1.46	1.68	0.03
0.04	1.66	2.04	0.07
0.08	2.02	2.73	0.17
0.10	2.20	3.11	0.23
0.20	3.24	5.64	0.65
0.25	3.90	7.54	0.98
0.30	4.72	10.13	1.42
0.35	5.74	13.75	2.01
0.40	7.05	18.92	2.82
0.45	8.77	26.48	3.95
0.50	11.09	37.91	5.54
0.55	14.30	55.81	7.86
0.60	18.91	85.12	11.3
0.64	24.19	123.22	15.5

Example 2.3

A membrane is used to remove particles with a radius of 10 nm from a suspension. (1) Calculate the maximum possible pressure difference across the concentration polarization layer, and (2) what is the maximum possible deduction on the permeate flux of this membrane process due to the development of concentration polarization if the applied pressure is 10 bar?

SOLUTION

The maximum wall concentration is 0.64 (v/v). That means 0.64 m^3 solids in 1 m^3 suspension.

From Table 2.1, it is found that $A_s^* = 24.19$ for this wall concentration. The volume of one particle is

$$\frac{4}{3}\pi_c a_p^{\,3} = \frac{4}{3} \times 3.14 \times \left(10 \times 10^{-9}\, m\right)^3 = 4.19 \times 10^{-24}\, m^3$$

The number concentration is then

$$C_w = \frac{0.64\ m^3/m^3}{4.19\times 10^{-24}\ m^3} = 1.53\times 10^{23}/m^3$$

The pressure difference across the concentration polarization layer is (assume T = 20°C)

$$\Delta p_{cp} = kTC_w A_s^* = 1.38\times 10^{-23}\ J/K \times 293\ K \times 1.53\times 10^{23}/m^3 \times 24.19$$
$$= 1.49\times 10^4\ Pa = 0.15\ bar\ \left(\text{note } 1\ J = 1\ Pa\cdot m^2\right)$$

The permeate flux is proportional to the net driving pressure. Therefore, the maximum deduction of flux by concentration polarization in this membrane process is about

$$\frac{\Delta J}{J_0} = \frac{\Delta p_{cp}}{\Delta p} = \frac{0.15\ bar}{10\ bar} = 0.015 = 1.5\%.$$

The impact of this magnitude is insignificant to a membrane process.

The net (effective) driving pressure for permeate flux through a membrane is the applied pressure minus the pressure difference across the concentration polarization layer. With the mean correction factor given in Table 2.1, the pressure difference on the concentration polarization layer can be directly calculated with Eq. (2.13) as long as the wall concentration is known. Particularly, when the maximum wall concentration is known a priori, such as the solid volume fraction of 0.64 for the rigid identical spheres, the maximum pressure difference across the concentration polarization can be determined.

2.4 Packed State

The wall concentration of a concentration polarization layer is determined by the balance between the diffusive force and the drag force on retained particles. For single-component solutions or suspensions, the diffusion coefficient is a single-valued constant. It can be seen more clearly from Eq. (2.7) that wall concentration increases with the product of permeate velocity and the amount of retained particles accumulated.

Considering a dead-end membrane chamber, in which a feed water with a particle concentration of 0.01% is filtered through the membrane at a constant velocity of 10^{-6} m/s. The particle radius in the feed water is 1 nm. The amount of retained particles will increase linearly with time as

$$M = c_0 vt$$

Substituting the expression into Eq. (2.7), the wall concentration of retained particles on the membrane surface also increases linearly with time according to

$$C_w = \frac{v^2}{D} c_0 t$$

The diffusion coefficient of particles can be determined with the Stokes–Einstein equation to be $2.15 \times 10^{-11} m^2/s$. The diffusive force on the retained particles increases with wall concentration. Wall concentration will stop to increase if the diffusive force is big enough to balance the drag force of the permeate flow on the retained particles. Otherwise, the wall concentration will continue to increase. With the above equation, it takes about just 1.37 seconds for the wall concentration to reach 0.64, the maximum random packing density if the permeate flux is assumed constant.

At the maximum wall concentration, if the drag force still cannot be balanced by the diffusive force backward, the excessive drag force will push the particles onto the membrane surface and pack them at the maximum packing density. This new phenomenon is commonly called cake formation. A cake is a layer of retained particles packed at a constant density on the membrane surface. The impact of a cake layer on permeate flux in a membrane process can be much more profound than that of a concentration polarization layer because the cake layer is packed with particles at the highest density throughout the whole layer.

2.4.1 Feature of Cake Layer

It is an equilibrium operation when retained particles stay in the concentration polarization layer. The hydraulic force on the particles can be balanced by the diffusive force generated by the concentration gradient. Cake formation is a significant event in membrane processes because it signals the breakup of equilibrium operation. The transition of particles in the concentration polarization layer to the cake layer is analogous to the phase change of water molecules from the liquid state to the ice state. They are freely movable in the former but fixed in space in the latter. Consequently, retained particles in the concentration polarization layer and cake layer affect permeate flux in different ways and to different extents.

The retained particles in the cake layer are packed at the highest random packing density throughout the cake layer, leaving the smallest voids between them for water to pass. The retained particles in the cake layer are no longer part of the water flow and move with the tangential flow. The retained particles attach to the membrane and become a part of it. They no longer move with the water flow.

There is no concentration gradient in the cake layer and, therefore, there is no longer a diffusive force to drive the particles back to the bulk. The physical support (contact) between the particles and eventually with the membrane balances the drag force on the particles by permeate flow.

Furthermore, the cake layer affects permeate flux in different ways than the concentration polarization layer does. As discussed previously, the impact of the concentration polarization layer on permeate flux can be conveniently handled by deduction of the pressure drop caused by it from the applied driving pressure. Because the cake layer is attached to the membrane, it may be more reasonable to treat it as an increment of the total resistance to the permeate flow by adding the cake resistance to the membrane resistance.

For the same amount of retained particles, the cake layer reduces the permeate flux much more than the concentration polarization layer because of the much greater resistance of the cake layer. The cake layer also distinguishes itself from the concentration polarization layer in that there is no limit on its impact on the permeate flux, while there is a maximum pressure difference across the concentration polarization layer. Therefore, the impact of the cake layer on permeate flux can be more profound than that of the concentration polarization layer.

The term membrane fouling is sometimes used as a synonym for cake formation in literature [34–36]. However, this term is avoided in the current part of this book because it is not a rigorously defined term. Historically, membrane fouling was used to indicate performance deterioration (mainly flux decline) of a membrane process caused by any, single or multiple, specifiable or unspecifiable, reasons. With such a vague concept of membrane fouling, it is impossible to discuss the behaviors of membrane processes beyond simple superficial descriptions. On the contrary, theories and mathematical models can be rigorously constructed to describe the behaviors associated with the well-defined phenomenon of a cake layer. The term of membrane fouling will be used in the next part of this book because it appears appropriately in the context there.

2.4.2 A Necessary Condition for Cake Formation

From the above discussions, concentration polarization appears first when there is an accumulation of retained particles. Initially, a balance can be maintained between the hydraulic drag force and the diffusive force on the retained particles in the concentration polarization layer. This signals an equilibrium filtration operation where there is only a concentration polarization layer over the membrane surface. A cake layer, which is a sign and result of nonequilibrium operation, may form later in some membrane processes. It may also never form in some other membrane processes. What are the conditions for cake formation? This question, which is of fundamental importance and practical interest in a membrane process, will be deliberated here.

A necessary condition for cake formation can be derived from the distribution of applied pressure in a membrane system. The net driving pressure for permeate flow in a membrane system is the applied pressure minus the osmotic pressure of the feed water. As shown in Figure 2.10, the net driving

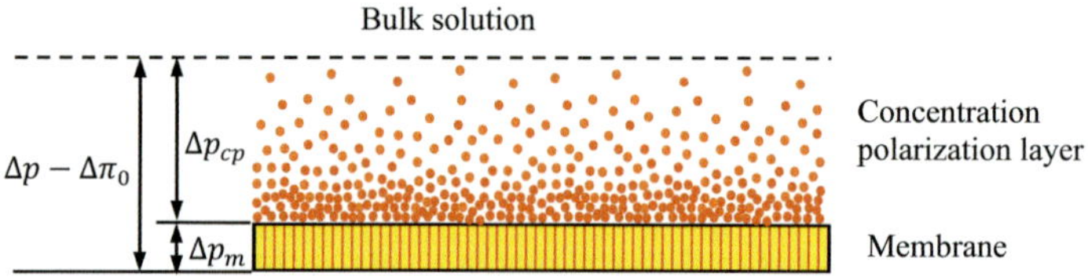

FIGURE 2.10
The net driving pressure is split between the membrane and the concentration polarization layer.

pressure in a membrane process can be expressed as the sum of the pressure on the membrane and the pressure on the concentration polarization layer,

$$\Delta p - \Delta \pi_0 = \Delta p_m + \Delta p_{cp} \tag{2.17}$$

where Δp is the applied pressure, $\Delta \pi_0$ is the osmotic pressure of feed water, Δp_m is the pressure to drive water through the membrane, and Δp_{cp} is the pressure consumed on the concentration polarization layer.

At the beginning of the separation operation in a dead-end process, the net driving pressure is all applied on the membrane because there is no concentration polarization layer yet. As time passes, the pressure across the concentration polarization increases and the pressure on the membrane decreases accordingly as the concentration polarization layer grows with time. If only the two pressure components exist in the membrane system, the pressure component on the membrane will eventually diminish, and the pressure component on the concentration polarization layer takes over completely the net driving pressure, i.e.,

$$\lim_{t \to \infty} \Delta p_{cp} = \Delta p - \Delta \pi_0$$

Similar analysis can be applied to a crossflow process by replacing the time t with the distance x from the entrance of the membrane channel.

Therefore, the necessary condition for cake formation is that the net driving pressure is greater than the maximum pressure on the concentration polarization layer, i.e.,

$$\Delta p > \Delta p_{cp}^{Max} + \Delta \pi_0 = \Delta p_c \tag{2.18}$$

where Δp_c is the critical pressure for cake formation, at which the wall concentration reaches maximum random packing density. Please be reminded that the critical pressure for cake formation is equal to the sum of the maximum pressure difference across the concentration polarization layer and the feed osmotic pressure. Therefore, it is the real particle concentration c, not the retained particle concentration C, that reaches the maximum packing density for cake formation.

A cake layer will not form if $\Delta p < \Delta p_c$ in a membrane process. The condition $\Delta p > \Delta p_c$ means that the applied pressure is too high to be absorbed even by the fully developed concentration polarization layer. A cake layer must form beneath the concentration polarization layer to adsorb the additional pressure. Critical pressure for cake formation is a thermodynamic property of suspension, which is independent of the membrane processes. On the other hand, the more frequently used concept "critical flux" for cake formation is process-specific. It not only depends on the particle size, but also varies with operation conditions (e.g., crossflow velocity) and device geometry (channel length) of the membrane processes. Therefore, critical pressure is superior to critical flux to characterize the condition for cake formation.

Although critical pressure is a necessary condition, it is not a sufficient condition for cake formation. To reach the maximum packing density for cake to form, according to Eq. (2.7), the amount of retained particles has to reach a certain amount. This is the reason that cake only forms after a period of time in a dead-end filtration device or beyond a distance from the entrance of a crossflow membrane channel. In other words, a cake layer might not form in some membrane processes even though they are operated above the critical pressure for cake formation.

2.4.3 Critical Pressure for Cake Formation

According to Eq. (2.14), filtration number N_F is a function of a sole independent variable, wall concentration C_w, because the other parameter A_s^* in the equation is a function of C_w, too. Therefore, when the maximum packing density of retained particles is known, the filtration number N_F for cake formation will be known. For example, for the closest packing density of 0.64, from Table 2.1, it is known that the cake layer starts to form when the filtration number N_F is 15.5.

Sometimes, it is usually of great practical interest to know the pressure at which cake starts to form for particles of given sizes. The critical pressure can be calculated with Eq. (2.13) by replacing the wall concentration with the cake density

$$\Delta p_{cl} = kT \frac{c_g}{\frac{4}{3}\pi_c a_p^3} A_s^* \tag{2.19}$$

where Δp_{cl} is the critical pressure for cake formation, and c_g is the solid volume fraction of particles at maximum packing density. Eq. (2.19) shows that, for a given solid volume fraction, the critical pressure for cake formation is a strong function of particle size. This is a significant difference from the filtration number, which is invariant of particle size.

The critical pressure at 20°C is calculated with Eq. (2.19) for various particle sizes at the closest random packing solid volume fraction $c_g = 0.64$. The

TABLE 2.2

Critical Pressure for Cake Formation (Maximum Pressure Difference Across Concentration Polarization Layer)[a]

a_p (μm)	Δp_{cl} (bar)
1.00×10^{-3}	149
2.00×10^{-3}	18.6
4.00×10^{-3}	2.33
8.00×10^{-3}	0.291
1.00×10^{-2}	0.149
2.00×10^{-2}	1.86×10^{-2}
4.00×10^{-2}	2.33×10^{-3}
6.00×10^{-2}	6.90×10^{-4}
8.00×10^{-2}	2.91×10^{-4}
0.100	1.49×10^{-4}
0.200	1.86×10^{-5}
0.400	2.33×10^{-6}
0.600	6.90×10^{-7}
0.800	2.91×10^{-7}
1.000	1.49×10^{-7}

[a] $c_g = 0.64$ and $T = 20°C$ are used in the calculation.

corresponding mean correction factor $A_s^* = 24.19$ used in the calculation is found from Table 2.1. The results are presented in Table 2.2.

It can be seen from Table 2.2 that the pressure across the concentration polarization layer is small ($\Delta p_c < 0.15\ bar$) for particle radius $a_p > 0.01\,\mu m$. Because the critical pressure for cake formation is actually the maximum possible pressure difference across the concentration polarization layer, the impact of concentration polarization on the permeate flux in most microfiltration/ultrafiltration is negligibly small because the applied pressures are at least several bars in these membrane processes.

The sufficient conditions for cake formation are presented in Figure 2.11 for various particle sizes. When the working points are located above the curves of particle size as indicated on the figure, cake layers are destined to form in the membrane processes. Points A, B, and C in the figure represent three working points of membrane processes. Point A on the cake layer of 2 nm indicates that any particles of larger sizes form the cake layer. Only particles of 4 nm or greater form a cake layer at working point B, though it has a larger amount of retained particles. Particles of sizes greater than 8 nm can form a cake layer at point C, though it has the highest permeate velocity. It should be pointed out that it is impossible for a membrane process to work in the region for cake formation if

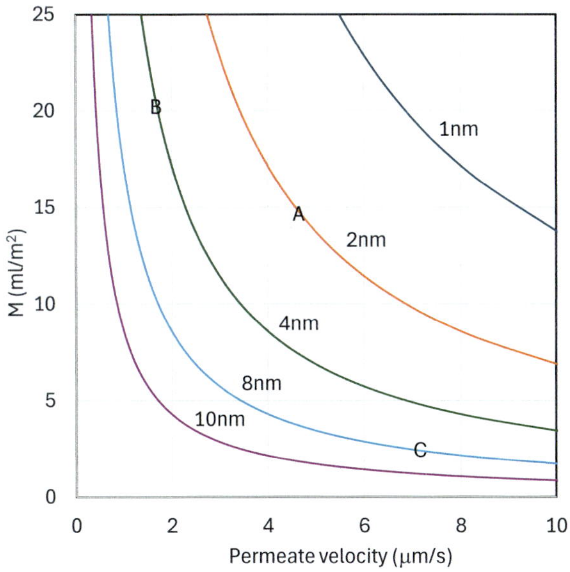

FIGURE 2.11
Cake formation regions for different particle sizes. Cake layer only forms when the working point is above the curve for that particle size.

the necessary condition is not satisfied. The reason is that the permeate velocity would reduce to zero before the cake concentration is reached in that case.

Example 2.4

A suspension contains biological particles with a hydrodynamic radius of 10 nm. The particles are expected to form a gel layer of solid volume fraction 0.4. What is the critical pressure and filtration number for gel formation in ultrafiltration of this suspension?

SOLUTION

From Table 2.1, the mean correction factor $A_s^* = 7.05$ for $c_g = 0.4$ v/v. Then the critical pressure for gel formation can be calculated with Eq. (2.19) (T = 20°C is assumed),

$$\Delta p_{cl} = kT \frac{c_g}{\frac{4}{3}\pi_c u_p^3} A_s^*$$

$$= 1.38 \times 10^{-23}\, J/K \times 293\, K \times \frac{0.4}{\frac{4}{3} \times 3.14 \times \left(10^{-8}\, m\right)^3} \times 7.05$$

$$= 2722\ Pa = 0.027\ bar$$

The filtration number for cake formation is calculated with Eq. (2.14) by replacing the wall concentration with the cake density,

$$N_F = A_s^* c_g = 7.05 \times 0.4 = 2.82$$

2.4.4 Critical Particle Size for Cake Formation

An inverse question about the critical pressure for a given particle size can be asked sometimes: What is the particle size that will form a cake layer at a given applied pressure? Especially, when the retained particles are not uniform in size, it is better to know which fraction of the particles will form a cake layer and which fraction will not. The critical particle size is defined as the size of particles for a given applied pressure, above which particles will form a cake on the membrane surface. The particles smaller than the critical size will remain in the liquid.

The critical particle size for cake formation can be found by rearranging Eq. (2.19) as

$$a_p = \sqrt[3]{\frac{3}{4}\frac{kTc_g A_s^*}{\pi_c \Delta p}} \tag{2.20}$$

The critical particle sizes at various applied pressures are presented in Table 2.3. Again, the solid volume fraction $c_g = 0.64$ and mean correction factor $A_s^* = 24.19$ are used in the calculations.

It can be seen in Table 2.3 that the critical particle size at 0.1 bar is 11.4 nm. Because all microfiltration and most ultrafiltration processes are operated at much higher pressure with the target species equal to or greater than this size, it is safe to say that cake formation is inevitable in these membrane processes. The situations in reverse osmosis processes are just the opposite. The major target species in the reverse osmosis process is sodium chloride. The diffusion coefficient of sodium chloride is 1.49×10^{-9} m^2/s. Using Stokes and Einstein equation, the nominal or hydrodynamic radius of sodium chloride is 0.165 nm. It can be seen in Table 2.3 that sodium chloride does not form a cake layer even at the pressure as high as 256 bars, which is way higher than the applied pressure in any practical reverse osmosis processes.

Example 2.5

The test for silt density index (SDI) of feed water to the reverse osmosis process is done at a pressure of 2 bars. The silt in water is assumed to be rigid spherical particles in a wide size range. The packing density of the particles is further assumed to be 0.60. What is the critical particle size to form a cake on the membrane?

SOLUTION

From Table 2.1, the mean correction factor $A_s^* = 18.91$ for $c_g = 0.6$ v/v. Then the critical particle to form a cake layer is calculated with Eq. (2.20) (T = 20°C is assumed),

TABLE 2.3

Critical Particle Size for Cake Formation at Various Pressures

	Particle Radius	
Pressure (Bar)	**m**	**nm**
0.05	1.44×10^{-8}	14.4
0.10	1.14×10^{-8}	11.4
0.20	9.08×10^{-9}	9.08
0.40	7.20×10^{-9}	7.20
0.80	5.72×10^{-9}	5.72
1.00	5.31×10^{-9}	5.31
2.00	4.21×10^{-9}	4.21
4.00	3.34×10^{-9}	3.34
8.00	2.65×10^{-9}	2.65
16.00	2.11×10^{-9}	2.11
32.00	1.67×10^{-9}	1.67
64.00	1.33×10^{-9}	1.33
128.00	1.05×10^{-9}	1.05
256.00	8.36×10^{-10}	0.836

$$a_p = \sqrt[3]{\frac{3}{4}\frac{kTc_g A_s^*}{\pi \Delta p}} = \sqrt[3]{\frac{3}{4} \times \frac{1.38 \times 10^{-23}\, J/K \times 293\, K \times 0.6 \times 18.91}{3.14 \times \left(2 \times 10^5\, Pa\right)}}$$

$$= 3.80 \times 10^{-9}\, m = 3.80\, nm$$

The pressure unit of Pa should be used in the equation to keep the unit consistent with other variables. The unit for particle radius in the equation is meters.

2.4.5 Specific Resistance of Cake Layer

The impact of the cake layer on permeate flux can be accounted for by considering the additional resistance of the cake layer on the membrane. Because the concentration (packing density) is constant for the entire thickness of a cake layer, the resistance of the cake layer can be calculated by

$$R_{cl} = r_{cl}\delta \tag{2.21}$$

where R_{cl} and r_{cl} are the resistance and specific resistance of the cake layer, respectively, and δ is the thickness of the cake layer. The specific resistance is defined as the resistance per unit thickness of the cake layer.

Specific resistance of a cake layer is a function of particle size and packing density. One of the popular models for specific resistance of cake layers consisting of identical spherical particles is the Carman–Kozeny equation [37], which is

$$r_{cl} = \frac{45\mu c_g^{\ 2}}{a_p^2\left(1-c_g\right)^3} \tag{2.22}$$

Alternatively, the Happel model can also be used to calculate the specific resistance of a cake layer. The model was used previously to calculate the drag force on particles in a concentration polarization layer, which is closely linked to the resistance of the cake layer. Because packing density is constant in the cake layer, the pressure difference across a cake layer can be determined by

$$\Delta p_{cl} = \left(6\pi_c \mu a_p\right) v c_n \delta A_s = \left(\frac{kT}{D}\right) v c_n \delta A_s \tag{2.23}$$

where Δp_{cl} is the pressure difference across the cake layer, and c_n is the number concentration. Because of Stokes–Einstein law, the two expressions for Δp_{cl} in Eq. (2.23) are equivalent. One expression can be more convenient than the other in certain situations. The correction factor A_s is directly calculated from the Happel model Eq. (2.8). The number concentration is related to the more convenient solid volume fraction c_g by Eq. (1.3) as

$$c_n = \frac{c_g}{\frac{4}{3}\pi_c a_p^{\ 3}}$$

Substituting this relationship in Eq. (2.23), we obtain

$$\Delta p_{cl} = \left(6\pi_c \mu a_p\right) v \left(\frac{c_g}{\frac{4}{3}\pi_c a_p^{\ 3}}\right) \delta A_s = \frac{9}{2}\frac{\mu v c_g}{a_p^{\ 2}} \delta A_s$$

According to the definition of the resistance in the membrane filtration, the resistance of the cake layer is the ratio of the pressure difference to the permeate velocity, i.e.,

$$R_{cl} = \frac{\Delta p_{cl}}{v} = \frac{9}{2}\frac{\mu c_g}{a_p^{\ 2}} \delta A_s \tag{2.24}$$

Therefore, taking the derivative of R_{cl} in Eq. (2.24) with respect to the thickness of the cake layer, the specific resistance of the cake layer is obtained as

$$r_{cl} = \frac{dR_{cl}}{d\delta} = \frac{9}{2}\frac{\mu c_g}{a_p^{\ 2}} A_s \tag{2.25}$$

TABLE 2.4

Specific Resistance of Cake Layer Based on the Carman–Kozeny and Happel Models[a]

c_g	Happel	Carman–Kozeny	Error (%)
0.64	3.55×10^9	3.95×10^9	10.2
0.60	2.30×10^9	2.53×10^9	9.2
0.55	1.38×10^9	1.49×10^9	7.5
0.50	8.53×10^8	9.00×10^8	5.2
0.45	5.36×10^8	5.48×10^8	2.1
0.40	3.40×10^8	3.33×10^8	−2.1

[a] Unit of specific resistance is Pa.s/m^2. Particle radius a_p = 10 μm and viscosity $\mu = 10^{-3}$ Pa·s at 20°C were used in the calculation.

Comparing Eqs. (2.22) and (2.25), both equations have a dependence on the reciprocal of the radius squared. Happel model appears to have a simpler explicit dependence on solid volume fraction in Eq. (2.25). However, because of the correction factor A_s, its dependence on solid volume fraction is more complex than the Carman–Kozeny model.

A sample calculation for the specific resistance of cake layer determined with the Happel model–based Eq. (2.25) and Carman–Kozeny model–based Eq. (2.22) is presented in Table 2.4. The particle radius of 10 μm is used in the calculations. It can be seen that the Carman–Kozeny model generally predicts a higher specific resistance than the Happel model, but the relative error is 11.6% at the maximum, which is not high at all in this case. Trial calculations are done for other particle sizes, and the results showed that the relative errors are not affected by particle size. Happel model–based Eq. (2.25) is recommended for the specific resistance of the cake layer because it will provide a smooth transition of the specific resistance from the cake layer to the concentration polarization layer. The counterpart of the Carman–Kozeny model is unavailable yet for the concentration polarization layer.

Example 2.6

A cake layer is formed by identical spherical particles of radius 10 nm. The cake packing density is 0.60, and the temperature is 20°C. Determine the thickness of the cake layer for a resistance that is equal to the membrane resistance of 10^{10} Pa·s/m.

SOLUTION

For $c_g = 0.6$

$$\theta = c_g^{\,1/3} = 0.6^{1/3} = 0.843$$

Calculate A_s with Eq. (2.3)

$$A_s = \frac{1+\frac{2}{3}\theta^5}{1-\frac{3}{2}\theta+\frac{3}{2}\theta^5-\theta^6} = \frac{1+\frac{2}{3}\times 0.843^5}{1-\frac{3}{2}\times 0.843+\frac{3}{2}\times 0.843^5-0.843^6} = 85.1$$

The specific resistance of the cake layer can be calculated with Eq. (2.25),

$$r_{cl} = \frac{9}{2}\frac{\mu\, c_g}{a_p^{\,2}} A_s = 4.5\times\frac{10^{-3}\, Pa\cdot s\times 0.6}{\left(10^{-8}\, m\right)^2}\times 85.1 = 2.30\times 10^{-15}\, Pa\cdot s/m^2$$

The thickness of the cake layer can be calculated with Eq. (2.21),

$$\delta = \frac{R_{cl}}{r_{cl}} = \frac{10^{10}\, Pa\cdot s/s}{2.30\times 10^{-15}\, Pa\cdot s/m^2} = 4.35\times 10^{-6}\, m = 4.35\,\mu m$$

2.5 Problems

1. A consequence of membrane separation is the accumulation of retained particles on the feed side of the membrane. How can the permeate flux of the membrane be affected by the retained particles?
2. Elaborate possible cause(s) for permeate flux deviating from linear dependence of the net driving force as described by the basic membrane filtration equation.
3. Is it correct to say that a steady state of operation cannot be achieved in a dead-end membrane process? Why?
4. A dead-end microfiltration device is used to filter 1 L of feed water with a particle concentration of 150 mg/L. Permeate of 0.9 L and retentate of 0.1 L are obtained. The particle concentration in the retentate is 250 mg/L, while the permeate is particle-free. Membrane area of the device is 2.5×10^{-4} m^2. Calculate the amount of retained particles per unit area of membrane.
5. Why can a steady state be reached in a crossflow membrane process but not in a dead-end membrane process?
6. Why are the roles of retained particles the key to understanding the behaviors of membrane processes?
7. The permeate flux in a pressure-driven membrane process is usually a function of pressure. For example, permeate flux is always proportional to the net pressure, as described by the basic filtration equation. However, the limiting flux in a membrane process, by definition, is not affected by the applied pressure. What are the possible reasons for the pressure-independent limiting flux?
8. Can a limiting flux be achieved in a dead-end membrane process? Why?

9. Permeate flux in a crossflow membrane device usually decreases downstream while the driving pressure is virtually constant. What are the reasons for this phenomenon? What parameters can affect the declining rate of flux along the membrane channel?
10. From your knowledge or speculation, what are the possible states for the retained particles to exist in membrane processes?
11. A spherical particle of radius 0.1 μm is held still in a creeping flow of velocity 10^{-5} m/s. Calculate the drag force on the particle by the flow.
12. What concentration gradient is needed to balance the hydraulic drag force in Question 1 when the particle concentration is 1000 mg/L at the point of consideration?
13. Calculate the pressure drop of a water flow passing a layer of still particles as shown below. The particle layer is 1 mm thick with a constant concentration of 0.01 mole/L. The particles are assumed to be uniform spheres with a radius of 1 nm. The flow velocity is 10^{-5} m/s.
14. The diffusion coefficient of sodium chloride (NaCl) is 1.29×10^{-9} m^2/s at 20°C. Calculated the dynamic equivalent radius of sodium chloride with Stokes–Einstein equation.
15. A dead-end membrane filtration chamber with a height of 1 cm is filled with an initial sodium chloride (NaCl) concentration of 1500 mg/L. When a permeate flow of 2×10^{-6} m/s is forced through the membrane by pumping pure water from the top of the chamber, derive the expression for salt concentration distribution over the membrane surface at steady state. What is the wall salt concentration on the membrane surface? What is the pressure needed to maintain the flux when the membrane resistance is 10^{11} Pa·s/m?
16. Calculate the Happel correction function for drag force for a suspension with a solid volume fraction of 0.15. How many times the drag force on a particle in this case than that on a single particle in the liquid?
17. Using Stokes–Einstein equation, calculated the diffusion coefficient of a spherical particle of radius 0.1 μm.
18. A water molecule has a diameter of 0.275 nm. Calculate its diffusion coefficient at 20°C.
19. The amount of accumulated particles at a point in a membrane process is 3.07×10^{13}/m^2. Calculated the wall concentration in number concentration and solid volume fraction. Other parameters are that the particle radius is 0.1 μm, the permeate velocity is 10^{-5} m/s, and the temperature is 20°C.
20. The wall concentration of retained sodium chloride (NaCl) in a dead-end reverse osmosis process is 1500 mg/L. The permeate velocity is 3×10^{-6} m/s. Calculated the amount of retained salt.

21. A finite integration of a complicated function can be well approached numerically

$$\int_a^b f(x)dx = \lim_{n\to\infty}\sum_{i=1}^{n} f(x_i)\Delta x$$

where $\Delta x = (b-a)/n$, $x_i = i\Delta x$, $i = 1, 2, \ldots, n$. Use MS Excel to estimate the integration in the equation for A_s^* the above formula at $C_w = 0.5$ for $n = 25$ and 50, respectively. Compare the two values.

22. Calculate the pressure across the concentration polarization layer for particles of radius 15 nm when the wall concentration is 0.6, and the particle concentration in the feed is negligible compared to the wall concentration.
23. What is the filtration number for the case as in Question 12?
24. Compare the features of retained particles in the concentration polarization layer and the cake layer.
25. What is the necessary condition for cake formation in a membrane filtration process?
26. Determine the critical particle size for cake formation at a pressure of 3 bars when the maximum packing density is 0.6.
27. A cake layer is formed with particles of 0.01 μm packing at 0.6 solid volume fraction. Calculate the specific resistance of the cake layer with both the Carman–Kozeny and Happel models.
28. From the flux decline from the clean membrane to the membrane with a cake layer on it, the resistance of the cake layer is determined as 10^9 Pa·s/m. Assume the density of the cake layer is 0.64 and the particle radius is 5 nm. Calculate the thickness of the cake layer.
29. The flux of a microfiltration membrane reduces from the initial value of 1.5×10^4 LMH (liter per square meter per hour) to 2.5×10^3 LMH at the end of the test. The pressure remains constant at 2 bar. Calculate the resistance of the cake layer and the thickness of the particles in the feed water if they are identical spheres of radius 1 μm.
30. Estimate the maximum pressure drop across the fully developed concentration polarization layer in Question 19.

References

1. Nagata, N., Herouvis, K., Dziewulski, D.M., & Belfort, G. (1989). Crossflow membrane microfiltration of a bacterial fermentation broth. Biotechnology and Bioengineering, 44, 161–181.

2. Leonard, E.F., & Vassilieff, C.S. (1984). The deposition of rejected matter in membrane separation processes. Chemical Engineering Communications, 30, 209–217.
3. Belfort, G. (1989). Fluid mechanics in membrane filtration: Recent developments. Journal of Membrane Science, 40, 123–147.
4. Murkes, J., & Carlsson, C.-G. (1988). Crossflow Filtration, Theory and Practice. Wiley, New York.
5. Green, G., & Belfort, G. (1980). Fouling of ultrafiltration membranes: Lateral migration and the particle trajectory model. Desalination, 35, 129–147.
6. Ofsthun, N.J., & Colton, C.K. (1987). Visual evidence of concentration polarization in crossflow membrane plasmapheresis. American Society for Artificial Internal Organs Journal, 10, 510–517.
7. Marshall, A.D., Munro, P.A., & Tragardh, G. (1993). The effect of protein fouling in microfiltration and ultrafiltration on permeate flux, protein retention and selectivity: A literature review. Desalination, 9(1), 65.
8. Arora, N., & Davis, R.H. (1994). Yeast cake layers as secondary membranes in deadend microfiltration of bovine serum albumin. Journal of Membrane Science, 92(C), 247–256.
9. Palecek, S.P., & Zydney, A.L. (1994). Hydraulic permeability of protein deposits formed during microfiltration: Effect of solution pH and ionic strength. Journal of Membrane Science, 95, 71–81.
10. Cogan, N. G., & Chellam, S. (2009). Incorporating pore blocking, cake filtration, and EPS production in a model for constant pressure bacterial fouling during dead-end microfiltration. Journal of Membrane Science, 345, 81–89.
11. Bowen, W.-R., & Gan, Q. Properties of microfiltration membranes. Flux loss during constant pressure permeation of bovine serum albumin. Biotechnology and Bioengineering, 38, 688.
12. Jonsson, G. (1984). Boundary layer phenomena during ultrafiltration of dextran and whey protein solutions. Desalination, 51(I), 61–77.
13. Peppin, S.S.L., & Elliott, J.A.W. (2001). Non-equilibrium thermodynamics of concentration polarization. Advances in Colloid and Interface Science, 92, 1–72.
14. Vilker, V.L., Colton, C.K., Smith, K.A., & Green, D.L. (1984). The osmotic pressure of concentrated protein and lipoprotein solutions and its significance to ultrafiltration. Journal of Membrane Science, 20, 11.
15. Tummons, E.N., Tarabara, V.V., Chew, J.W., & Fane, A.G. (2016). Behavior of oil droplets at the membrane surface during crossflow microfiltration of oil-water emulsions. Journal of Membrane Science, 500, 211–224.
16. Bowen, W.R., & Jenner, F. (1995). Theoretical descriptions of membrane filtration of colloids and fine particles – An assessment and review. Advances in Colloid and Interface Science, 56, 141–200.
17. Bhattacharjee, S., Kim, A.S., & Elimelech, M. (1999). Concentration polarization of interacting solute particles in crossflow membrane filtration. Journal of Colloid and Interface Science, 212, 81–99.
18. Hong, S., Faibish, R.S., & Elimelech, M. (1997). Kinetics of permeate flux decline in crossflow membrane filtration of colloidal suspensions. Journal of Colloid and Interface Science, 196(2), 267–277.
19. McDonogh, R.M., Bauser, H., Stroh, N., & Grauschop, U. (1995) Experimental in situ measurement of concentration polarisation during ultra- and microfiltration of bovine serum albumin and Dextran Blue solutions. Journal of Membrane Science, 104, 51–63.

20. Bacchin, P., Si-Hassen, D., Starov, V., Clifton, M.J., & Aimar, P. (2002). A unifying model for concentration polarization, gel-layer formation and particle deposition in crossflow membrane filtration of colloidal suspensions. Chemical Engineering Science, 57, 77–91.
21. Sim, L., Wang, Z., Gu, J., Coster, H., & Fane, A. (2013). Detection of reverse osmosis membrane fouling with silica, bovine serum albumin and their mixture using in-situ electrical impedance spectroscopy. Journal of Membrane Science, 443, 45–53.
22. Bacchin, P., Aimar, P., & Field, R.W. (2006). Critical and sustainable fluxes: Theory, experiments and applications. Journal of Membrane Science, 281, 42–69.
23. Zydney, A.L., & Colton, C.K. (1986). A concentration polarization model for the filtrate flux in crossflow microfiltration of particulate suspensions. Chemical Engineering Communications, 47, 1–21.
24. Drew, D.A., Schonberg, J.A., & Belfort, G. (1991). Lateral inertial migration of a small sphere in fast laminar flow through a membrane duct. Chemical Engineering Science, 46, 3219–3224.
25. Davis, R.H., & Leighton, D.T. (1987). Shear-induced transport of a particle layer along a porous wall. Chemical Engineering Science, 42, 275–281.
26. Goren, S.L. (1979). The hydrodynamic force resisting the approach of a sphere to a plane permeable wall. Journal of Colloid and Interface Science, 69, 78–85.
27. Sherwood, J.D. (1988). The force on a sphere pulled away from a permeable half-space. Physicochemical Hydrodynamics, 10, 3–12.
28. Thomas, D.G. (1973). Forced convection mass transfer in hyperfiltration at high fluxes. Industrial & Engineering Chemistry Fundamentals, 12, 396–405.
29. Happel, J. (1958). Viscous flow in multiparticle systems: Slow motion of fluids relative to beds of spherical particles. AIChE Journal, 4, 197–201.
30. Jaeger, H. M., & Nagel, S. R. (1992). Physics of the granular state. Science, 255, 1523–1531.
31. Qiu, T.Y., & Davies, P.A. (2015). Concentration polarization model of spiral-wound membrane modules with application to batch-mode RO desalination of brackish water, Desalination, 368, 36–47.
32. Gupta, B.B. (1996). Use of hydrodynamic methods for pressure driven membrane operations. Indian Journal of Chemical Technology, 3, 156–163.
33. Moussy, Y. (2000). A model for high-pressure ultrafiltration of blood. Biotechnology and Bioengineering, 69, 21–30.
34. Karabelas, A.J., & Sioutopoulos, D.C. (2015). New insights into organic gel fouling of reverse osmosis desalination membranes. Desalination, 368, 114–126.
35. Ruan, H.M., Yang, Z.R., Lin, J. et al. (2015). Biogas slurry concentration hybrid membrane process: Pilot-testing and RO membrane cleaning, Desalination, 368, 171–180.
36. Hao, L., Liss, S.N., & Liao, B.Q. (2017). Effect of solids retention time on membrane fouling in membrane bioreactors at a constant mixed liquor suspended solids concentration. Membrane Water Treatment, 8, 337–353.
37. Carman, P.C. (1997). Fluid flow through granular beds. Chemical Engineering Research & Design, 75, 150–166.

3

Concentration Polarization

With knowledge on the fundamentals of retained particles developed in the previous chapter, their enormous impact on the performance of membrane processes can be delineated and described with authority and awesomeness. Mathematical models are developed in this and the next chapters for permeate fluxes in membrane processes that are affected or controlled by retained particles [1–5]. The membrane processes with only concentration polarization are considered in this current chapter, while the impact of the cake layer will be added later in the next chapter. In order to treat this subject rigorously and to develop mechanistic models from fundamental principles, a model membrane separation system is first introduced. The model membrane system will enable us to study membrane processes and associated phenomena with scientific methodologies and strictly defined concepts. It will provide a solid foundation for the development of theories and models for the phenomena occurring in the membrane processes.

3.1 A Model Membrane System

The real-world phenomena are complex. Idealization and simplification are necessary to scientifically delineate the fundamental principles of these phenomena. For example, the concept of "point mass" in classical mechanics is a very useful term to describe the essence of complex interactions between objects or to express the fundamental mechanics principles. The Earth can be treated as a point mass when its movement around the Sun is being considered. The second Newton's law can only have its awesomely simple form because the object is treated as a point mass. The response of a real-world object with a certain volume and shape to a force can be too hard a challenge to describe accurately. It would depend on the force direction with respect to the mass center of the object. The trajectory of the object under a force can undoubtedly become too complex with a little complication, such as on a supporting plane that is not frictionless. The "point charge," similar to the point mass in classical mechanics, is one of the essential "bricks" that are used to build up the classical electrostatics.

Membrane processes can be very complex, in which permeate flux is affected by many factors. The particles can coagulate to grow in size, which

DOI: 10.1201/9781003743569-4

would reduce the resistance in the cake layer. Particles can intrude into the pores of the membrane and block them so as to increase the resistance of the membrane. The membrane resistance can also be increased due to the adsorption of organic matter or the scaling of minerals onto the membrane. Obviously, many of these phenomena only occur in some cases, not in others, which means that they are not essential to all membrane separation processes and therefore only need to be considered case by case. Even worse, the "including everything" approach may distract attention away from the key fundamental principles of membrane separation processes.

Because the feed waters in practical membrane processes may contain species with various physical and chemical properties. Some of them exist only in the concentration polarization layer, while some others exist in both the concentration polarization and cake layers. It makes accurate modeling of membrane systems extremely difficult, if not completely impossible. In order to focus on the essence of membrane separation, a concept of a model membrane system is introduced here. The model membrane system only has the features that are intrinsic to all membrane separation processes. The properties of the model membrane system are highly idealized and simplified so that analysis and modeling of membrane systems from first principles become possible.

3.1.1 Binary Solution

The model membrane system is first characterized with a binary solution (or suspension). There are only two components in the binary solution. One component is a continuous medium, and the other component is particles dispersed in the former. The particles can be in a dissolved state (true solution) or a suspended state (suspension). The continuous component is usually water, but can be other liquids, though the term "water" is generally used. The dispersed component can be solutes, ions, particulates, or other species scattered in the liquid.

The particles in the binary solution are further assumed to be identical spheres characterized by size and concentration. The particles are only subjected to the hydraulic drag force asserted by the water flow and the diffusive force induced by the concentration gradient in the liquid. Both forces can be quantified by the classic theories of fluid dynamics and thermodynamics. The density of particles is only needed to convert between solid volume fraction and number concentration. Otherwise, it is a trivial property because the particles are so small that the effects of inertia and gravity on the movement of these particles are negligible. The particles retain their individuality during the filtration process, and there are no other interactions among them.

Furthermore, the osmotic pressure of the binary solution can be determined by the van't Hoff equation at any concentration without correction for the impact of neighboring particles. In other words, the correction factor is unity ($A_s = 1$) in the binary solution regardless of the concentration of retained particles.

3.1.2 Other Ideal Properties of the Model Membrane System

First, the membrane has a perfect rejection of the particles in the binary solution. This property makes it very convenient to quantify the retained particles, which is the total amount of particles originally carried by the permeate. There are also no interactions (e.g., electrical attraction or repulsion, chemical affinity) between particles and the membrane. Pore openings of the membrane should be smaller than the particles in the binary solution so that no particles can intrude into the membrane. Therefore, the membrane has a constant permeability (or resistance) to liquid during the separation process.

As for the membrane chamber in the dead-end device or membrane channel in the crossflow device, the height of the chamber or channel is ideally unlimited (mathematically infinite) in the upper half-space over the membrane surface. Therefore, the hydraulic conditions (e.g., tangential flow velocity or shear rate) are maintained unchanged throughout the operation and unaffected by the accumulation of retained particles. A practical membrane device approximates the ideal case if its chamber or channel is sufficiently high so that the concentration polarization layer and cake layer can be fully developed without restriction. In other words, the thickness of the concentration polarization layer, or cake layer, or a combination of both layers is much smaller than the height of the membrane chamber or channel.

Some of the above properties of the model membrane system may not be mandatory for the development of mathematical models. However, with these ideal properties, models can be developed more elegantly with less effort and without loss of general applicability. In the rest of this chapter and the next chapter, the model membrane system will be referred to in the study of concentration polarization and cake formation on system performance.

3.2 The Fundamental Cause for Concentration Polarization

The most commonly mentioned factors that affect concentration polarization in membrane processes are convective movement of solution and diffusion of particles in it [6–13]. However, an even more important factor for concentration polarization is the accumulation of retained particles. No retained particles, no concentration polarization! Actually, concentration polarization is a phenomenon or existing state of retained particles controlled by hydraulic conditions over the membrane surface. Under the same hydraulic conditions, concentration polarization with different strengths can be developed according to the amount of retained particles [14–16].

Because of the important roles of retained particles for concentration polarization, any attempt to analyze concentration polarization from convection and diffusion only is just like a figure (person) in two-dimensional

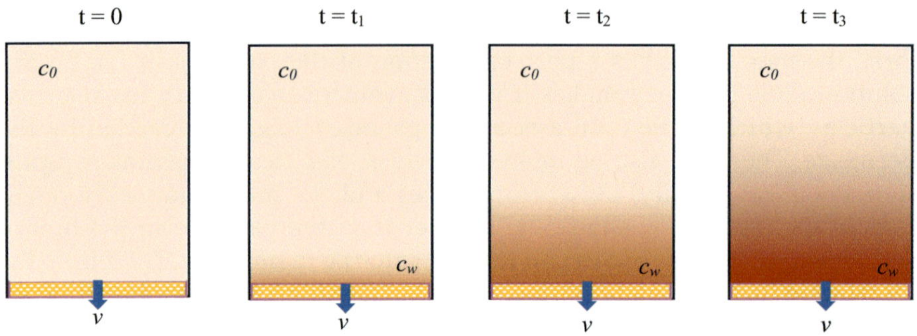

FIGURE 3.1
Concentration polarization develops in a dead-end filtration. Concentration polarization intensifies with time as the amount of retained particles increases.

worlds to explain a phenomenon occurring in three-dimensional worlds. It hardly makes any sense because the most important dimension (the amount of retained particles) is missing. The figures from the world with only two dimensions of convection and diffusion cannot truly comprehend concentration polarization because they have no idea about the third dimension of this phenomenon, which is the amount of retained particles.

The role of retained particles on concentration polarization in dead-end filtration is described in Figure 3.1. The figure presents a series of snapshots of particle concentration distribution in a dead-end membrane device at different times, $0 = t_0 < t_1 < t_2 < t_3$. The membrane process is operated under constant permeate flux. There is obviously no concentration polarization at $t = 0$ because there is no accumulation of retained particles yet. At time t_1, a thin layer of higher concentration than the feed concentration c_0 forms over the membrane surface when there is a small amount of retained particles. The thickness of the concentration polarization layer, as well as the wall concentration, increases with time as the separation operation is going on. Because diffusion coefficient and permeate velocity do not change with time, the series of snapshots demonstrates that concentration polarization intensifies only due to the increased amount of retained particles with time.

The development of concentration polarization as a consequence of the accumulation of retained particles is also obvious in crossflow filtration. As schematically shown in Figure 3.2, there is no concentration polarization at the entrance of a membrane channel $x = 0$ because there is no retained particle at that point. Concentration polarization starts to develop after the entrance as retained particles accumulate near the membrane surface. Concentration polarization intensifies downstream as the amount of retained particles increases. In the crossflow filtration case, the height of the membrane channel can be made sufficiently large so that the hydraulic conditions can be regarded as constant along the channel length. The only reason

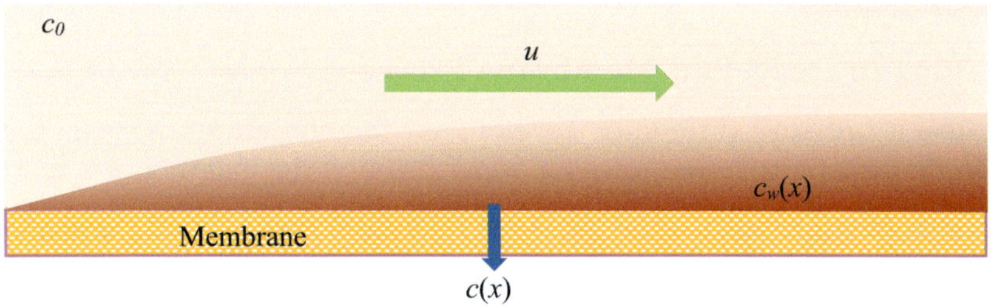

FIGURE 3.2
Concentration polarization develops downstream along a crossflow filtration channel as the amount of retained particles increases.

for concentration polarization to intensify is the increase in the amount of retained particles along the membrane channel.

Some key observations of concentration polarization, as presented in Figures 3.1 and 3.2, are:

1. The initial particle concentration in feed water is not directly related to concentration polarization. It is the background value on which concentration polarization is developed by the addition of retained particles. There is no concentration polarization if there are no retained particles.
2. The extent or strength of concentration polarization is determined by the amount of retained particles under the same hydraulic conditions. Therefore, a key step to quantify concentration polarization is to track the amount of retained particles accurately.
3. There is always a net particle flux into the concentration polarization layer because of permeate flow. It is evident that the amount of retained particles increases with time in the dead-end filtration. Even in the crossflow filtration at a steady state, the amount of retained particles increases downstream as more permeate is produced along the membrane channel.

When the applied pressure is smaller than the critical pressure for cake formation, all retained particles stay in the concentration polarization layer. Even in some cases where the applied pressure is greater than the critical pressure for cake formation, the retained particles can only exist in concentration polarization layer if the amount of retained particles does not reach the value for cake formation due to the limited operation time or channel length.

It is of great interest to know how the retained particles in the concentration polarization layer affect permeate flux in the membrane process. This rather important issue will be discussed in the rest of this chapter for dead-end filtration and crossflow filtration, respectively.

3.3 Flux in Dead-End Membrane Processes

Because a pressure difference develops on the concentration polarization layer, the effective driving force for water through the membrane is reduced [17–20]. When there is a concentration polarization layer but no cake layer in a membrane process, the basic filtration equation Eq. (1.11) can be modified by subtraction of the pressure difference across the concentration polarization layer from the applied pressure,

$$v = \frac{\Delta p - \Delta \pi_0 - \Delta p_{cp}}{R_m} \tag{3.1}$$

When retained particles are assumed to be stationary with respect to the membrane surface, it was shown in the previous chapter that wall concentration is linearly related to the amount of retained particles with Eq. (2.7). By tracking the amount of retained particles in the membrane processes quantitatively, the equations for permeate fluxes affected by concentration polarization (only) are derived below for dead-end membrane processes.

The pressure difference on concentration polarization can be obtained by combining Eqs. (2.7) and (2.13),

$$\Delta p_{cp} = kT \frac{v}{D} M \tag{3.2}$$

The correction factor A_s^* does not appear in Eq. (3.2) because it is unity in the model membrane system. Because of the involvement of the Boltzmann constant k in the equation, the amount of retained particles should be expressed in units of number of particles per unit area (#/m^2).

Substituting Eq. (3.2) into Eq. (3.1) yields

$$v = \frac{\Delta p - \Delta \pi_0 - kT \dfrac{vM}{D}}{R_m}$$

The permeate velocity appears on both sides of this equation. Solving the equation for v results in

$$v = \frac{\Delta p - \Delta \pi_0}{R_m + kT \dfrac{M}{D}} \tag{3.3}$$

There are two unknowns of v and M in Eq. (3.3). Another relationship between these two unknowns is required to finally define this problem.

In a dead-end membrane process, the total amount of retained particles is those that are carried originally in the permeate, i.e.,

$$M = c_0 \int_0^t v\, dt \tag{3.4}$$

where c_0 is the particle concentration in the feed water, and t is the operation time of the filtration. Again, the perfect rejection rate property of the model membrane process is used to reach Eq. (3.4).

Substituting Eq. (3.3) in Eq. (3.4) to eliminate v, we have

$$M = c_0 \int_0^t \frac{\Delta p - \Delta\pi_0}{R_m + kT\dfrac{M}{D}} dt \tag{3.5}$$

The only unknown in this integration equation is the amount of retained particle M. Now the problem is defined, and a solution can be pursued.

Taking the derivative of both sides of Eq. (3.5) with respect to time t results in

$$\frac{dM}{dt} = c_0 \frac{\Delta p - \Delta\pi_0}{R_m + kT\dfrac{M}{D}}$$

Integrating the differential equation with the variable separation method

$$\int_0^M \left(R_m + kT\frac{M}{D} \right) dM = c_0 \left(\Delta p - \Delta\pi_0 \right) \int_0^t dt$$

Carrying out the integrations, the amount of retained particle is obtained as

$$M = \frac{D}{kT} \left\{ \left[R_m^2 + \frac{2\Delta\pi_0}{D} \left(\Delta p - \Delta\pi_0 \right) t \right]^{1/2} - R_m \right\} \tag{3.6}$$

The van't Hoff equation $(\Delta\pi_0 = kTc_0)$ is used to reach the above equation. Substituting this expression for M in Eq. (3.3) results in

$$v = \frac{\Delta p - \Delta\pi_0}{\left[R_m^2 + \dfrac{2\Delta\pi_0}{D} \left(\Delta p - \Delta\pi_0 \right) t \right]^{1/2}} \tag{3.7}$$

The second term in square brackets in the denominator represents the impact of concentration polarization on permeate flux of dead-end membrane filtration. When the concentration polarization term outweighs the membrane resistance term, permeate flux declines roughly according to the square root of operation time ($\sim t^{-1/2}$). A sample calculation for permeate flux as a function of time with Eq. (3.7) is presented in Figure 3.3. Parameters used in the calculation are shown in the figure. It can be seen that the flux declines very rapidly at the beginning of filtration, followed by a long, slowly declining tail.

Example 3.1

In a dead-end membrane device, a suspension of 0.02% (v/v) is filtered under a pressure of 2.5 bars at a temperature of 20°C. The membrane resistance is 1.3×10^{10} Pa·s/m, and the particle radius is 2 nm. Calculate the wall concentration of retained particles as a function of time and present the results graphically.

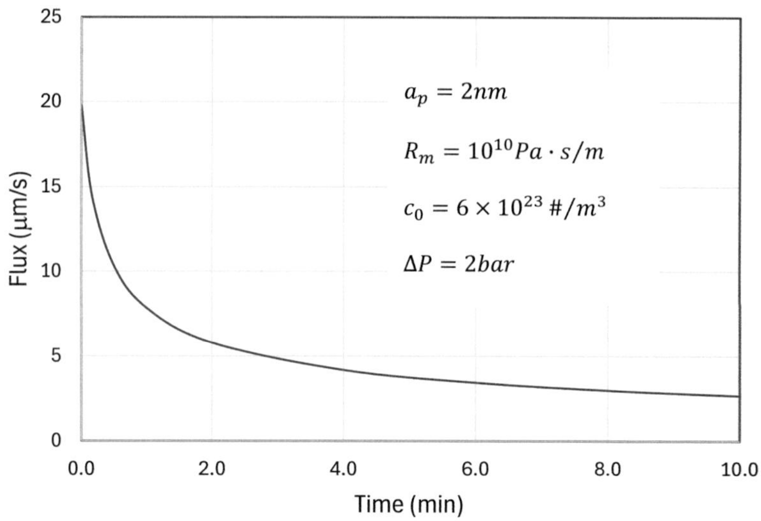

FIGURE 3.3
Flux declines with time in a dead-end membrane device with parameters given in the figure.

SOLUTION

The diffusion coefficient of the particles is

$$D = \frac{kT}{6\pi_c \mu a_p} = \frac{1.38 \times 10^{-23}\, Pa \cdot m^3 \times 293\, K}{6 \times 3.14 \times 10^{-3}\, Pa \cdot s \times 2 \times 10^{-9}\, m} = 1.07 \times 10^{-10}\, m^2/s$$

The number concentration can be converted from the solid volume fraction with

$$c_n = \frac{c_g}{\frac{4}{3}\pi_c a_p^{\ 3}} = \frac{0.0002}{\frac{4}{3} \times 3.14 \times \left(2 \times 10^{-9}\, m\right)^3} = 5.97 \times 10^{21}/m^3$$

The initial osmotic pressure is

$$\Delta\pi_0 = kTc_n = 1.38 \times 10^{-23}\, Pa \cdot m^3 \times 293\, K \times 5.97 \times 10^{22}/m^3 = 24.1\, Pa$$

To calculate wall concentration, permeate velocity and amount of retained particles have to be determined first with Eqs. (3.6) and (3.7), respectively

$$M = \frac{D}{kT}\left\{\left[R_m^{\ 2} + \frac{2\Delta\pi_0}{D}\left(\Delta p - \Delta\pi_0\right)t\right]^{1/2} - R_m\right\}$$

$$v = \frac{\Delta p - \Delta\pi_0}{\left[R_m^{\ 2} + \frac{2\Delta\pi_0}{D}\left(\Delta p - \Delta\pi_0\right)t\right]^{1/2}}$$

Then the wall concentration is calculated with Eq. (2.7)

$$C_w = \frac{v}{D} M$$

The calculation results are presented in the table below.

t (min)	M (#/m²)	v (m/s)	C_w (#/m³)	C_w (v/v)
0	0	1.92×10^{-5}	0	0.00
0.25	1.72×10^{18}	1.91×10^{-5}	3.06×10^{23}	0.01
0.50	3.43×10^{18}	1.90×10^{-5}	6.08×10^{23}	0.02
1.00	6.82×10^{18}	1.89×10^{-5}	1.20×10^{24}	0.04
2.00	1.35×10^{19}	1.85×10^{-5}	2.33×10^{24}	0.08
2.50	1.68×10^{19}	1.83E $\times 10^{-5}$	2.87×10^{24}	0.10
5	3.29×10^{19}	1.76×10^{-5}	5.38×10^{24}	0.18
10	6.31×10^{19}	1.63×10^{-5}	9.56×10^{24}	0.32
15	9.13×10^{19}	1.52×10^{-5}	1.29×10^{25}	0.43
20	1.18×10^{20}	1.43×10^{-5}	1.57×10^{25}	0.53
25	1.43×10^{20}	1.36×10^{-5}	1.81×10^{25}	0.61

The wall concentrations calculated with Eq. (2.7) were in number concentrations, as listed in the second column from the right side in the table. They were converted to solid volume fraction as listed in the last column on the right side of the table.

The wall concentration in solid volume fraction is plotted versus time in the figure below.

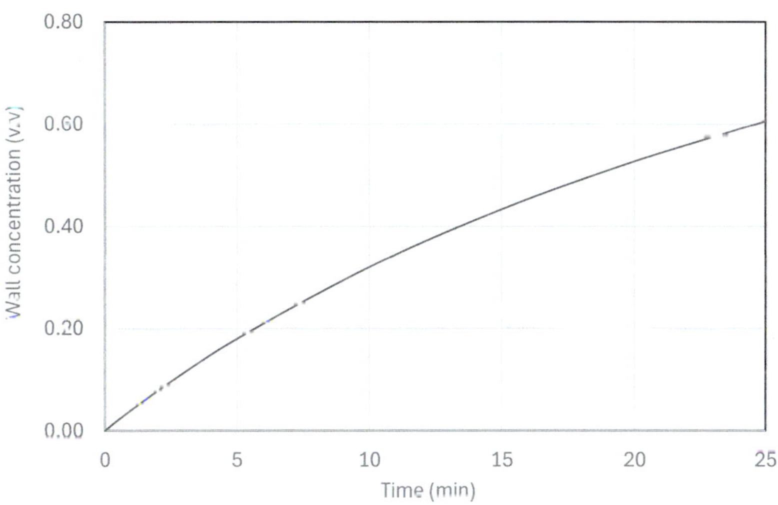

3.4 Flux in Crossflow Membrane Processes

Concentration polarization in crossflow membrane processes is a more complex phenomenon. In a crossflow membrane channel, feed water flows tangentially along the membrane surface from the entrance to the exit [21–28]. This tangential flow brings retained particles downstream and eventually carries them out of the membrane channel. As a result, a steady state can be achieved in a crossflow membrane channel. At steady state, the thickness of the concentration polarization layer grows from the entrance toward the exit of the membrane channel as the amount of retained particles increases, which is depicted in Figure 3.4.

In this figure, in addition to the y coordinate vertical to the membrane surface, another coordinate x parallel to the membrane surface is used to indicate the distance from the entrance along the flow direction. A profile of permeate velocity is shown underneath the membrane. The permeate velocity is anticipated to decrease downstream due to the development of concentration polarization.

The concentration polarization layer in crossflow filtration is much thinner than the velocity boundary layer. Therefore, a shear tangential flow can be assumed in the concentration polarization layer

$$u = \gamma y \tag{3.8}$$

where u is the flow velocity of feed water tangential to the membrane surface, and γ is the shear rate. As shown in Figure 3.4.

3.4.1 Principal Mass Balance Equation

There are two quantities of essential importance in a crossflow membrane channel. The first one is the collection rate of retained particles, defined as

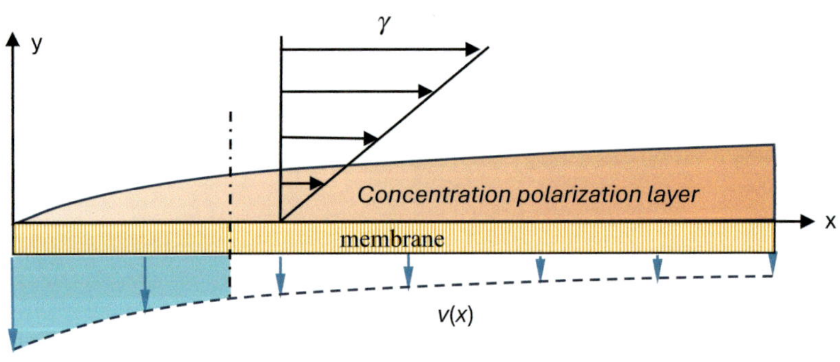

FIGURE 3.4
Concentration polarization grows along a crossflow membrane channel. Permeate flux declines correspondingly.

$$N_1 = c_0 \int_0^x v\, dx' \tag{3.9}$$

where N_1 is the collection rate of retained particles, c_0 is the particle concentration in feed water, and x' is the dummy integration variable. It is the amount of retained particles collected from the entrance to point x in the membrane channel per unit of time.

The second quantity is the carrying capacity of tangential flow defined as

$$N_2 = \int_0^\infty uC\, dy \tag{3.10}$$

where N_2 is the carrying capacity of tangential flow, u is the tangential flow velocity, and C $(= c - c_0)$ is the concentration of retained particles in the concentration polarization layer. Both tangential flow velocity and concentration of retained particles are functions of the y coordinate. It is the amount of retained particles that can be carried by the tangential flow per unit of time passing point x in the membrane channel.

One of the most important relationships for retained particles in crossflow membrane processes at steady state is the principal mass balance equation. Because the concentration of retained particles in a crossflow membrane channel does not change with time, the following mass balance relationship must be satisfied [14,16,24],

$$c_0 \int_0^x v\, dx' = \int_0^\infty uC\, dy \tag{3.11}$$

Eq. (3.11) states that the collection rate and the carrying capacity are equal to each other everywhere in crossflow filtration at steady state. Eq. (3.11) is of essential importance to concentration polarization in crossflow filtration. It will be shown later that this principal mass balance equation also plays a central role in the study of the cake layer in crossflow filtration.

3.4.2 Governing Equation for Permeate Flux

Now let us construct the governing equation for permeate flux in a crossflow membrane channel affected by concentration polarization. The starting point is the carrying capacity (Eq. 3.10).

Substituting Eqs. (3.8) and (2.6) for u and C in the right-hand side of Eq. (3.10), the carrying capacity of retained particle by tangential flow at position x is,

$$N_2 = \int_0^\infty uC\, dy = \int_0^\infty \gamma y c_w e^{-\frac{v}{D}y}\, dy = \gamma c_w \frac{D^2}{v^2}$$

Substituting the carrying capacity into Eq. (3.11), the wall concentration of retained particles at position x can be expressed as,

$$c_w = \frac{c_0}{D^2\gamma} v^2 \int_0^x v\, dx' \tag{3.12}$$

With the wall concentration of retained particles, the pressure difference across the concentration polarization layer can be calculated with,

$$\Delta p_{cp} = kTc_w = \frac{\Delta\pi_0}{D^2\gamma} v^2 \int_0^x v\, dx' \tag{3.13}$$

The correction factor $A_s^* = 1$ in the model membrane system is used here. Please note that $\Delta\pi_0$ replaces kTc_0 the gain in Eq. (3.13) using the van't Hoff equation. Now, the pressure difference on the concentration polarization layer is a variable of x.

Rearranging Eq. (3.1), we have

$$\Delta p = \Delta\pi_0 + vR_m + \Delta p_{cp}$$

Replacing ΔP_{cp} in the above equation with Eq. (3.13) results in

$$\Delta p = \Delta\pi_0 + vR_m + \frac{\Delta\pi_0}{D^2\gamma} v^2 \int_0^x v\, dx' \tag{3.14}$$

Eq. (3.14) is the governing equation for permeate flux in crossflow filtration. The three terms on the right-hand side of Eq. (3.14) represent allocations of the applied pressure on feed water (to overcome the osmotic pressure of feed water), the membrane, and the concentration polarization layer. It can be seen that there is only one unknown, the permeate flux v, in Eq. (3.14) and, therefore, the permeate flux in the crossflow membrane channel is well defined by this governing equation.

3.4.3 Solutions of the Governing Equation

Although it is well defined, the permeate flux is implicitly involved in Eq. (3.14). There is a need to find out the explicit expression of permeate flux. This can be done separately for three cases where concentration polarization is either negligible, predominant, or comparable to membrane resistance, respectively.

1. *Negligible concentration polarization.* This case occurs for feed water containing a very low concentration of particles or in the initial stage of operation with a small amount of retained particles. In such cases, the pressure difference across the concentration polarization layer is

negligible compared to other pressure components. The last term on the right-hand side of Eq. (3.14) can be taken out of the equation. The permeate flux is then determined as

$$v = \frac{\Delta p - \Delta\pi_0}{R_m}$$

This is the basic filtration equation. The permeate velocity is a constant along the entire membrane channel length. Pure water filtration also follows this equation, with the initial osmotic pressure of feed water being zero.

2. *Predominant concentration polarization.* This case occurs with a significant accumulation of retained particles so that the pressure on the membrane is negligibly small compared with the other two pressure components. In this case, the second term on the right-hand side of Eq. (3.14) can be removed from the equation. The governing equation becomes

$$\Delta p = \Delta\pi_0 + \frac{\Delta\pi_0}{D^2\gamma} v^2 \int_0^x v\,dx' \tag{3.15}$$

The integration equation Eq. (3.15) can be completed by the method of variable replacement. According to this method, a new variable is introduced as

$$w = \int_0^x v\,dx' \tag{3.16}$$

It is obvious that the permeate velocity v can be obtained by taking the derivative of w with respect to x as,

$$v = \frac{dw}{dx}$$

Replacing v and its integration in Eq. (3.15) with dw/dx and w yields

$$\Delta p = \Delta\pi_0 + \frac{\Delta\pi_0}{D^2\gamma}\left[\frac{dw}{dx}\right]^2 w$$

Now the equation becomes a nonlinear differential equation of the new variable $w(x)$. Rearranging the equation results in,

$$\frac{dw}{dx} w^{1/2} = \left[\frac{D^2\gamma(\Delta p - \Delta\pi_0)}{\Delta\pi_0}\right]^{1/2}$$

The differential equation is readily integrable with the method of variable separation, and the solution is

$$w = \left(\frac{3}{2}\right)^{2/3} \left[\frac{D^2\gamma\left(\Delta p - \Delta\pi_0\right)}{\Delta\pi_0}\right]^{1/3} x^{2/3}$$

Taking the derivative of $w(x)$, we have

$$v = \left(\frac{2}{3}\right)^{1/3} \left(\frac{\Delta p - \Delta\pi_0}{\Delta\pi_0}\right)^{1/3} \left(\frac{D^2\gamma}{x}\right)^{1/3} \tag{3.17}$$

All parameters in Eq. (3.17) are constant except for x. Eq. (3.17) shows that permeate flux declines downstream with one-third order of the distance from the entrance of the membrane channel ($\sim x^{1/3}$).

The average permeate flux in a membrane channel of length L is

$$\bar{v} = \frac{1}{L}\int_0^L v\,dx = \left(\frac{3}{2}\right)^{2/3} \left(\frac{\Delta p - \Delta\pi_0}{\Delta\pi_0}\right)^{1/3} \left(\frac{D^2\gamma}{L}\right)^{1/3} \tag{3.18}$$

where $\bar{v}$ is the average permeate flux of the crossflow membrane channel. The last factor $\left(D^2\gamma/L\right)^{1/3}$ in Eq. (3.18) is the so-called Leveque factor. This linear dependence of the average permeate flux on the Leveque factor has been experimentally verified. It is interesting to see that the Leveque factor appears naturally in Eq. (3.18), which is rigorously derived from the first principles.

Example 3.2

The applied pressure and osmotic pressure of feed water in a crossflow membrane channel are 10 and 4 bars, respectively. Because of the small membrane resistance, the applied pressure is primarily used to overcome the feed water osmotic pressure and the pressure difference across the concentration polarization layer. The membrane channel is 6 m long and has a tangential flow with a shear rate of 100/s over the membrane surface. The diffusion coefficient of retained salt is 1.0×10^{-10} m^2/s. (1) Determine the Leveque factor and (2) find the average flux of this membrane channel.

SOLUTION

The Leveque factor is

$$\left(\frac{D^2\gamma}{L}\right)^{1/3} = \left(\frac{\left(10^{10}\,m^2/s\right)^2 \times 100/s}{6m}\right)^{1/3} = 5.5\times 10^{-7}\,m/s$$

The average permeate velocity

$$\bar{v} = \left(\frac{3}{2}\right)^{2/3} \left(\frac{\Delta p - \Delta\pi_0}{\Delta\pi_0}\right)^{1/3} \left(\frac{D^2\gamma}{L}\right)^{1/3}$$
$$= \left(\frac{3}{2}\right)^{2/3} \left(\frac{10\ bar - 4\ bar}{4\ bar}\right)^{1/3} \times 5.5\times 10^{-7}\ m/s$$
$$= 8.25\times 10^{-7}\ m/s$$

3. *Comparable concentration polarization and membrane resistance.* When both concentration polarization and membrane resistance are not negligible, the solution of Eq. (3.14) has to be sought directly without any simplification.

   It will take three steps to find the explicit expression for permeate flux from Eq. (3.14).

   i. The first step is to convert the integration Eq. (3.14) to a differential equation. For this purpose, taking the derivative of Eq. (3.14) with respect to x yields

$$0 = \frac{dv}{dx} R_m + \frac{2}{v}\frac{dv}{dx}\left[\frac{\Delta\pi_0}{D^2\gamma} v^2 \int_0^x v\,dx'\right] + \frac{\Delta\pi_0}{D^2\gamma} v^3 \tag{3.19}$$

There is still an integration term in Eq. (3.19) in the square brackets. This integration term can be obtained by rearranging Eq. (3.14),

$$\left[\frac{\Delta\pi_0}{D^2\gamma} v^2 \int_0^x v\,dx'\right] = \Delta p - \Delta\pi_0 - vR_m$$

Substituting the above equation into Eq. (3.19) to eliminate the integration term, we have

$$0 = \frac{dv}{dx} R_m + \frac{2}{v}\frac{dv}{dx}\left(\Delta p - \Delta\pi_0 - vR_m\right) + \frac{\Delta\pi_0}{D^2\gamma} v^3$$

A differential equation can be obtained by rearranging,

$$\frac{dv}{dx} = -\frac{1}{2\left(\Delta p - \Delta\pi_0\right) - R_m v} \frac{\Delta\pi_0}{D^2\gamma} v^4 \tag{3.20}$$

   ii. The second step is to seek a solution of the differential equation. Using the variable separation method, we have

$$\int_{v_0}^{v}\left[\frac{2\left(\Delta p - \Delta\pi_0\right)}{v'^4} - \frac{R_m}{v'^3}\right] dv' = -\frac{\Delta\pi_0}{D^2\gamma}\int_0^x dx'$$

where v' is the dummy integration variable and v_0 is the permeate velocity at the entrance of the membrane channel. Because there is no accumulation of retained particles at the entrance, the permeate flux there is simply calculated with the basic filtration equation,

$$v_0 = \frac{\Delta p - \Delta \pi_0}{R_m}$$

By completing the integration, a cubic algebraic equation is obtained as follows,

$$\left[\frac{1}{v_0^{\ 3}} + \frac{6}{\Delta p - \Delta \pi_0}\frac{\Delta \pi_0}{D^2 \gamma} x\right] v^3 + \frac{3R_m}{\Delta p - \Delta \pi_0} v - 4 = 0 \tag{3.21}$$

iii. The third step is to solve the cubic algebraic equation. Existence and forms of solution for the cubic equation are determined by the coefficients of the unknown v in different orders. The coefficients in Eq. (3.21) satisfy the condition to have a real root and a real root only [29].

Using Cardano's formula [29], the permeate velocity along the membrane channel can be obtained from Eq. (3.21) as [25]

$$v = \frac{v_0}{(1+\alpha x/L)^{1/3}} \left\{ \left[\left(\frac{1}{1+\alpha x/L} + 4 \right)^{1/2} + 2 \right]^{1/3} - \left[\left(\frac{1}{1+\alpha x/L} + 4 \right)^{1/2} - 2 \right]^{1/3} \right\} \tag{3.22}$$

where α is a dimensionless parameter that characterizes the membrane system defined by

$$\alpha = 6\left(\frac{\Delta p - \Delta \pi_0}{R_m}\right)^2 \frac{\Delta \pi_0}{R_m}\left(\frac{L}{D^2 \gamma}\right) = 6 v_0^{\ 2} \frac{\Delta \pi_0}{R_m}\left(\frac{L}{D^2 \gamma}\right) \tag{3.23}$$

For this reason, this parameter can be named *the crossflow filtration number* of the membrane channel without a cake layer. Eq. (3.22) is the solution to the governing equation Eq. (3.14) for the permeate flux in crossflow filtration.

As always, average permeate flux is sometimes the preferred indicator for the performance of a membrane process. Its determination can be started by replacing the distance x by channel length L in Eq. (3.14). Rearranging to find the integration

$$\int_0^L v\,dx = \frac{D^2 \gamma}{\Delta \pi_0 v(L)^2}\left(\Delta p - \Delta \pi_0 - v(L) R_m\right)$$

By the definition of average permeate velocity, it is calculated with

$$\bar{v} = \frac{1}{L}\int_0^L v\,dx = \frac{1}{\Delta\pi_0 v(L)^2}\left(\frac{D^2\gamma}{L}\right)\left(\Delta p - \Delta\pi_0 - v(L)R_m\right) \tag{3.24}$$

The value of $v(L)$ in Eq. (3.24) is calculated with Eq. (3.22).

Example 3.3

A crossflow reverse osmosis channel for seawater desalination is 8 m long. A shear flow with a shear rate of 100/s runs over the membrane surface. The applied pressure and initial osmotic pressure are 60 and 25 bars, respectively. The membrane resistance is 3×10^{11} Pa·s/m, and the diffusion coefficient of retained salt is 1.29×10^{-9} m²/s. Develop a graph for permeate flux in the channel. Also, calculate the average flux of this membrane channel.

SOLUTION

The permeate flux at the entrance of the membrane channel x = 0

$$v_0 = \frac{\Delta p - \Delta\pi_0}{R_m} = \frac{(60\,bar - 25\,bar)\times\dfrac{10^6\,Pa}{1\,bar}}{3\times 10^{11}\,Pa\cdot s/m} = 1.17\times 10^{-5}\,m/s$$

The dimensionless parameter

$$\alpha = \frac{6{v_0}^2\Delta\pi_0}{R_m}\left(\frac{L}{D^2\gamma}\right)$$

$$= \frac{6\times\left(1.17\times 10^{-5}\,m/s\right)^2\times 25\times 10^5\,Pa}{3\times 10^{11}\,Pa\cdot s/m}\left(\frac{8\,m}{\left(1.29\times 10^{-9}\,m^2/s\right)^2\times 100/s}\right) = 327.17$$

Sample calculation for permeate flux at x =0.1 m

$$v = \frac{v_0}{(1+\alpha x/L)^{1/3}}\left\{\left[\left(\frac{1}{1+\alpha x/L}+4\right)^{1/2}+2\right]^{1/3}\left[\left(\frac{1}{1+\alpha x/L}+4\right)^{1/2}-2\right]^{1/3}\right\}$$

$$= \frac{1.17\times 10^{-5}\,m/s}{(1+327.17\times 0.1\,m/8\,m)^{1/3}}\left\{\left[\left(\frac{1}{1+327.17\times 0.1\,m/8\,m}+4\right)^{1/2}+2\right]^{1/3}\right.$$

$$\left.-\left[\left(\frac{1}{1+327.17\times 0.1\,m/8\,m}+4\right)^{1/2} 2\right]^{1/3}\right\}$$

$$= 8.34\times 10^{-6}\,m/s - 8.34\,\mu m/s$$

The permeate fluxes at other points along the channel are listed in the table below.

x (m)	0	0.10	0.25	0.50	1.00	2.00	4.00	6.00	8.00
v (μm/s)	11.7	8.34	6.82	5.72	4.73	3.86	3.14	2.77	2.53

It can be seen in the table that the permeate flux at the exit end of the membrane channel $v(8\ m) = 2.53 \times 10^{-6}\ m/s$. Then the average flux is

$$\bar{v} = \frac{\left(\frac{D^2\gamma}{L}\right)}{\Delta\pi_0 v(L)^2}\left(\Delta p - \Delta\pi_0 - v(L)R_m\right)$$

$$= \frac{\left(\frac{\left(1.29\times10^{-9}\ m^2/s\right)^2 \times 100/s}{8\ m}\right)}{25\times10^5\ Pa \times \left(2.53\times10^{-6}\ m/s\right)^2}$$

$$\times\left(60\times10^5\ Pa - 25\times10^5\ Pa - 2.53\times10^{-6}\ m/s \times 3\times10^{11}\ Pa\cdot s/m\right)$$

$$= 3.56\times10^{-6}\ m/s = 3.56\ \mu m/s$$

The results are graphically presented below.

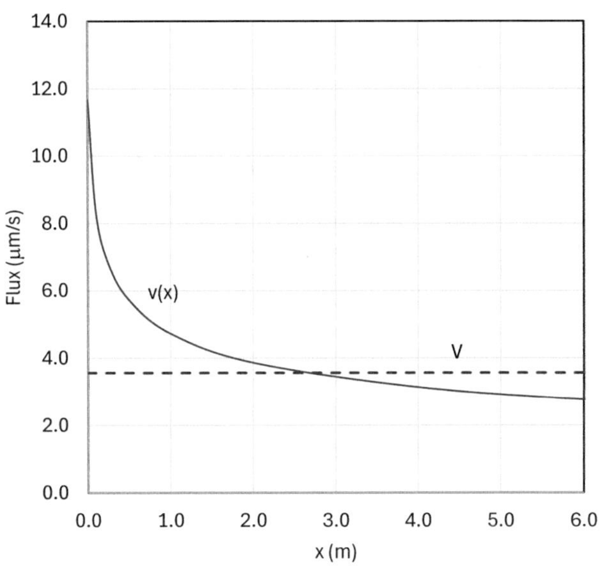

One of the consistent efforts in the development of new RO membranes is to reduce membrane resistance with new materials or technologies. It would be wondered what the limiting flux that can be reached in RO processes by reduction of membrane resistance.

Example 3.4

Calculated permeate flux in a RO process with the same parameters as in Example 3.3 except for membrane resistance, which is approaching zero, i.e., $R_m \to 0$ in this case.

SOLUTION

When the membrane resistance is negligible compared to the concentration polarization layer, the flux along a membrane channel is calculated with Eq. (3.17)

$$v = \left(\frac{2}{3}\right)^{1/3}\left(\frac{\Delta p - \Delta\pi_0}{\Delta\pi_0}\right)^{1/3}\left(\frac{D^2\gamma}{x}\right)^{1/3}$$

Sample calculation for permeate flux at x = 0.1 m

$$v = \left(\frac{2}{3}\right)^{1/3}\left(\frac{60\,bar - 25\,bar}{25\,bar}\right)^{1/3}\left(\frac{\left(1.29\times10^{-9}\,m^2/s\right)^2 \times 100/s}{0.1\,m}\right)^{1/3}$$

$$= 11.58\times10^{-6}\,m/s = 11.58\,\mu m/s$$

The permeate fluxes at other points along the channel are listed in the table below.

x (m)	0.10	0.25	0.50	1.00	2.00	4.00	6.00	8.00
v (μm/s)	11.6	8.53	6.77	5.38	4.27	3.39	2.96	2.69

The average flux is calculated with Eq. (3.18)

$$\bar{v} = \left(\frac{3}{2}\right)^{2/3}\left(\frac{\Delta p - \Delta\pi_0}{\Delta\pi_0}\right)^{1/3}\left(\frac{D^2\gamma}{L}\right)^{1/3}$$

$$= \left(\frac{3}{2}\right)^{2/3}\left(\frac{60\,bar - 25\,bar}{25\,bar}\right)^{1/3}\left(\frac{\left(1.29\times10^{-9}\,m^2/s\right)^2 \times 100/s}{8\,m}\right)^{1/3}$$

$$= 4.03\times10^{-6}\,m/s = 4.03\,\mu m/s$$

The results are presented in the figure below.

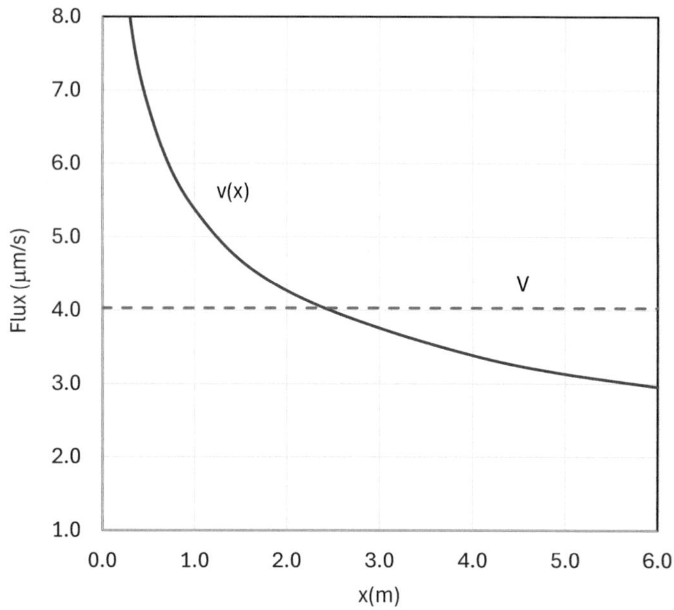

Comparing the average fluxes in the two examples, it can be seen that the maximum flux increase by resistance deduction of RO membranes in seawater desalination is marginal, from 3.56 $\mu m/s$ to 4.03 $\mu m/s$, an increase of about 13%.

3.5 Problems

1. What is a model membrane system? Why can the roles and states of retained particles in membrane processes be more rigorously delineated with the model membrane system than in a common membrane system?
2. Discuss the advantages of a binary solution in the fundamental studies of membrane separation processes over a solution of multiple components.
3. What is the original cause for concentration polarization in a membrane process?
4. The parameters for a dead-end model membrane process are diffusion coefficient = 10^{-11} m^2/s, membrane resistance = 10^{10} Pa·s/m, pressure = 2 bar, and the osmotic pressure of the feed water = 0.02 bar.

(1) Calculate the permeate velocities at the beginning and at 5 minutes after the operation. (2) Determine the amount of retained particles at 5 minutes after the operation.

5. A dead-end membrane device is used to concentrate a solution. The parameters are membrane resistance $R_m = 1 \times 10^{10}$ pa·s/m, initial product concentration $c_0 = 50$ mg/L, applied pressure $\Delta p = 2$ bar, and the height of the membrane chamber is 5 mm. What volume of permeate is required for the average product concentration in the membrane chamber to be concentrated up to 100 times? Determine the time needed for the concentration operation. The osmotic coefficient and diffusion coefficient of the biomedical product are 20 Pa/(mg/L) and 2×10^{-10} m^2/s, respectively.
6. In a dead-end membrane device with a nanofiltration membrane, a solution of $MgSO_4$ of 1000 mg/L is filtered at a pressure of 5 bar for 10 minutes at a temperature of 20°C. The initial permeate flux at the start is 1.5×10^{-5} m/s. What would be the final permeate flux by the end of the filtration (the diffusion coefficient of $MgSO_4$ is 8.5×10^{-10} m^2/s)?
7. Derive the expression of permeate flux in a dead-end membrane process as a function of time, in which the osmotic pressure is negligible compared to the applied pressure.
8. In a dead-end nanofiltration of a suspension of particles of radius 2 nm, the driving pressure and initial osmotic pressure of the suspension are 4 and 0.1 bar, respectively. What time will it take for the permeate flux to decline to 1% of its initial value when the membrane resistance is 10^{10} Pa·s/m?
9. The permeate flux of a membrane process has a permeate velocity of 2.5×10^{-5} m/s under a pressure of 6 bar. It is known that the membrane resistance is 10^{10} Pa·s/m, and the initial osmotic pressure of the feed water is 0.1 bar. What is the pressure difference (drop) across the concentration polarization layer?
10. The pressure difference across a concentration polarization layer of nanoparticles is 5 bar. Determine the wall concentration of particles on the membrane surface if the particle radius is 2.5 nm.
11. The membrane resistance is constant along a crossflow membrane channel. The pressure in the membrane channel can be generally treated as a constant, too. However, the permeate flux decreases downstream along the membrane channel. What is the reason for the decline of permeate flux?
12. If the permeate velocity along a membrane is described by $v(x)$ for $0 < x < L$. The feed concentration is c_0, and the tangential velocity is a constant value u. Write out or derive the expressions for (1) the accumulation rate of retained particles at point X, and (2) the accumulation rate for the channel section from the beginning up to the point X.

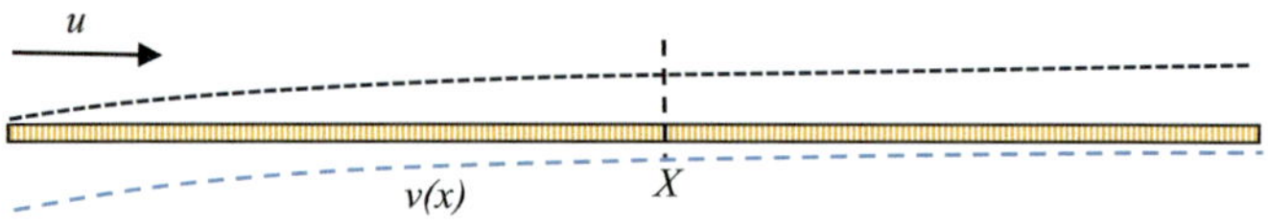

13. In a crossflow membrane process, permeate flux is described by

$$v(x) = \frac{10^{-5}}{\sqrt[3]{2+x}}$$

and feed salt concentration $c_0 = 5000$ mg/L, calculate the accumulation rate of retained salt from the entrance to x = 1, where x is the dimensionless distance from the entrance.
14. At steady state, determine the rate of retained salt moving downstream at x = 2 m in Question 13.
15. In a crossflow membrane process at steady state with the same permeate flux pattern and feed salt concentration as in Question 13, determine the wall salt concentration at L in the membrane channel. The diffusion coefficient and shear rate are 1.29×10^{-9} m²/s and 100/s, respectively.
16. When the flux in a crossflow membrane process with sufficiently large height is controlled predominantly by concentration polarization (the membrane resistance is negligible compared to concentration polarization resistance), calculate the average permeate flux for a membrane channel with the following parameters: pressure = 10 bar, feed salt concentration $c_0 = 2500$ mg/L, osmotic coefficient = 80 Pa/(mg/L), diffusion coefficient = 1.29×10^{-9} m²/s, shear rate = 100/s, and channel length = 6 m.
17. Calculate the local permeate flux along the channel as given in Question 16 and present the results graphically.
18. Derive the expression for wall salt concentration along a crossflow membrane channel at steady state.
19. Calculate the wall salt concentration along the channel as given in Question 16 and present the results graphically.
20. Calculate the filtration parameter a in Eq. (4.20) for a crossflow membrane process with the following parameters:

Parameter	**Value**	**Unit**
Feed osmotic pressure	2	bar
Membrane resistance	1.3×10^{11}	Pa·s/m
Diffusion coefficient	1.29×10^{-9}	m²/s
Shear rate	200	1/s
Pressure	5	bar
Channel length	6	m

21. Calculate the local permeate flux in the membrane channel as described in Question 20 and present the result graphically.
22. Determine the average permeate flux in the membrane channel as described in Question 20.
23. Discuss the impact of diffusion coefficient, shear rate, and channel length on the average flux of a crossflow membrane process when the flux is predominated by concentration polarization.
24. The average permeate flux of a crossflow membrane process is 2.5×10^{-5} m/s. What will be the average permeate flux when the particle size is doubled, with other parameters remaining unchanged?
25. In a crossflow membrane process with membrane resistance 10^{11} P·s/m and feed water osmotic pressure of 1 bar. The permeate flux at a point in the membrane channel is 1.5×10^{-5} m/s under an applied pressure of 5 bar. Determine the pressure difference across the concentration polarization layer at that point.
26. Determine the wall particle concentration at the point of interest in Question 25 when the particle radius is 3 nm.
27. Why does the impact of concentration polarization on permeate flux in a membrane process become insignificant when the particle radius is greater than 10 nm?

References

1. Bouchard, C.R., Carreau, P.J., Matsuura, T., & Sourirajan, S. (1994). Modeling of ultrafiltration: Predictions of concentration polarization effects. Journal of Membrane Science, 97, 215–229.
2. Kim, S., & Hoek, E.M.V. (2005). Modeling concentration polarization in reverse osmosis processes. Desalination, 186, 111–128.
3. Sablani, S.S., Goosen, M.F.A., Al Belushi, R., & Wilf, M. (2001). Concentration polarization in ultrafiltration and reverse osmosis: A critical review. Desalination, 141, 269–289.
4. Rohlfs, W., Thiel, G.P., & Lienhard V, J.H. (2016). Modeling reverse osmosis element design using superposition and an analogy to convective heat transfer. Journal of Membrane Science, 512, 38–49.
5. Herath, G., Yamamoto, K., & Urase, T. (2000). The effect of suction velocity on concentration polarization in microfiltration membranes under turbulent flow conditions. Journal of Membrane Science, 169, 175–183.
6. Bai, W., Samineni, L., Chirontoni, P., Krupa, I., Kasak, P., Popelka, A., Saleh, N.B., & Kumar, M. (2023). Quantifying and reducing concentration polarization in reverse osmosis systems. Desalination, 554, 116480.
7. Jung, O., Saravia, F., Wagner, M. et al. (2019). Quantifying concentration polarization – Raman microspectroscopy for in-situ measurement in a flat sheet cross-flow nanofiltration membrane unit. Scientific Reports, 9, 15885.

8. Brian, P.L.T. (1965). Concentration polar zation in reverse osmosis desalination with variable flux and incomplete salt rejection. Industrial & Engineering Chemistry Fundamentals, 4(4), 439–445.
9. Redkar, S., Kuberkar, V., & Davis, R.H. (1996). Modeling of concentration polarization and depolarization with high-frequency backpulsing. Journal of Membrane Science, 121, 229–242.
10. Parvatiyar, M.G. (1996). Interaction of dispersed phase with concentration polarization. Journal of Membrane Science, 115, 121–127.
11. Bhattacharjee, S., Kim, A.S., & Elimelech, M. (1999). Concentration polarization of interacting solute particles in crossflow membrane filtration. Journal of Colloid and Interface Science, 212, 81–99.
12. Chung, K.Y., Brewster, M.E., & Belfort, G. (1998). Dean vortices with wall flux in a curved channel membrane system: 3. Concentration polarization in a spiral reverse osmosis slit. Journal of Chemical Engineering of Japan, 31, 683–693.
13. Zydney, A.L. (1997). Stagnant film model for concentration polarization in membrane systems. Journal of Membrane Science, 130, 275–281.
14. Song, L., & Elimelech, M. (1995). Theory of concentration polarization in crossflow filtration. Journal of the Chemical Society, Faraday Transactions, 91(19), 3389–3398.
15. Elimelech, M., & Bhattacharjee, S. (1998). A novel approach for modeling concentration polarization in crossflow membrane filtration based on the equivalence of osmotic pressure model and filtration theory. Journal of Membrane Science, 145, 223–241.
16. Song, L., & Yu, S. (1999). Concentration polarization in crossflow reverse osmosis. AIChE Journal, 45(5), 921–928.
17. Mahlab, D., Yosef, N.B., & Belfort, G. (1980). Interferometric measurement of concentration polarization profile for dissolved species in unstirred batch hyperfiltration (reverse osmosis). Chemical Engineering Communications, 6, 225–243.
18. Peppin, S.S.L., & Elliott, J.A.W. (2001). Non-equilibrium thermodynamics of concentration polarization. Advances in Colloid and Interface Science, 92, 1–72.
19. Gill, W.N., Tien, C., &Zeh, D.W. (1965). Concentration polarization effects in a reverse osmosis system. Industrial & Engineering Chemistry Fundamentals, 4(4), 433–439.
20. Kim, S., & Park, H. (2002). Applicability assessment of subcritical flux operation in crossflow microfiltration with a concentration polarization model. Journal of Environmental Engineering, 128, 335–340.
21. Bacchin, P., Si-Hassen, D., Starov, V., Clifton, M.J., & Aimar, P. (2002). A unifying model for concentration polarization, gel-layer formation and particle deposition in crossflow membrane filtration of colloidal suspensions. Chemical Engineering Science, 57, 77–91.
22. Agashichev, S.P. (2007). Modeling of the concentration polarization in crossflow microfiltration of highly viscous media. Theoretical Foundations of Chemical Engineering, 41, 205–212.
23. Wang, X.-M., & Li, X.-Y. (2012). A unified model for quantification of concentration polarization (CP) of particles during crossflow membrane filtration. Colloids and Surfaces A: Physicochemical and Engineering Aspects, 407, 99–107.
24. Zhoug, W., Song, L., & Tay, K.G. (2006). A numerical study on concentration polarization and system performance of spiral wound RO membrane modules. Journal of Membrane Science, 271, 38–46.

25. Song, L., Liu, C., & Liang, S. (2013). Reverse osmosis. In: Tarbara, V.V., Hoek, E.M.V. (eds) Encyclopedia of Membrane Science and Technology. Wiley, Hoboken, NJ, pp. 1361–1393.
26. Denisov, A. (1994). Theory of concentration polarization in cross-flow ultrafiltration: Gel-layer model and osmotic-pressure model. Journal of Membrane Science, 91, 173–187.
27. Airey, D., Yao, S., Wu, J., Chen, V., Fane, A.G., & Pope, J.M. (1998). An investigation of concentration polarization phenomena in membrane filtration of colloidal silica suspensions by NMR micro-imaging. Journal of Membrane Science, 145, 145–158.
28. Chang, S., Jamal, S., Chen, H., Zhou, H., Hong, Y., & Adams, N. (2014). A novel approach to analyzing concentration polarization of polysaccharide solutions. Water Science and Technology, 69, 1344–1348.
29. Spiegel, M.R. (1968). Mathematical Handbook of Formulas and Tables. McGraw Hill, New York.

4

Cake Formation

There is only a concentration polarization layer of retained particles in a membrane when the applied pressure is smaller than the critical pressure for cake formation. When the applied pressure exceeds the critical pressure, there is a potential for a cake layer to form on the membrane surface [1–5]. The cake layer is packed at maximum density and therefore can provide greater resistance to permeate flow than the concentration polarization layer with the same amount of retained particles. Cake layer has significantly different properties from the concentration polarization layer, and therefore its impact on the performance of the membrane process would also be quite different [6–9]. The impact of the cake layer on permeate flux will be investigated in this chapter.

4.1 Governing Equations

When there is a cake layer that forms beneath the concentration polarization layer, the impact of the cake layer on the permeate flux can be treated similarly to that of the concentration polarization layer. That is to deduct the pressure difference on the cake layer from the applied pressure. For this purpose, the applied pressure in the membrane system can be expressed as the sum of the following pressure components,

$$\Delta p = \Delta \pi_0 + \Delta p_m + \Delta p_{cp} + \Delta p_{cl} \tag{4.1}$$

where Δp_{cl} is the pressure difference across the cake layer. On the right-hand side of the equation, in addition to the osmotic pressure of feed water, the pressure difference across the concentration polarization layer is also a constant. This is because there is only one type of particle in both the concentration polarization layer and the cake layer in a model membrane system. The pressure difference across the concentration polarization layer must reach the maximum value when there is a cake layer. The other two components, Δp_m and Δp_{cl}, are variables that are dependent on permeate flux v, which is unknown.

DOI: 10.1201/9781003743569-5

It is sometimes more convenient to handle the problem by modifying the basic filtration equation to,

$$v = \frac{\Delta p - \Delta \pi_0 - \Delta p_{cp}}{R_m + R_{cl}} \tag{4.2}$$

where R_{cl} is the resistance of the cake layer. In this way, the impact of the cake layer is treated as an increment of system resistance, while the impact of the concentration polarization layer is treated as a deduction of the driving pressure. With such treatments, the numerator on the right-hand side of Eq. (4.2) is constant for a given set of applied pressure and feed water. In the denominator, membrane resistance is constant, and the resistance of the cake layer R_{cl} is the only variable. Besides, the resistance of the cake layer is independent of the permeate flux.

Both Eqs. (4.1) and (4.2) are the governing equations to calculate permeate flux when there is a cake layer in the membrane processes. The two equations are generally applicable to both dead-end and crossflow filtrations.

4.2 Dead-End Filtration

The impact of a cake layer on permeate flux is first investigated for dead-end filtration. The general methods to handle the cake layer will be developed with this relatively simple case. More complicated cases with crossflow filtration will be discussed later.

4.2.1 Growth of Cake Layer

Before the discussion of cake formation and growth, the time for the concentration polarization layer to develop to full strength in a dead-end membrane process is estimated. With a constant permeate flux, the accumulation of retained particles in the concentration polarization layer is

$$M = c_0 v_0 t \tag{4.3}$$

By the way, solid volume fraction is the most convenient concentration unit for retained particles when a cake layer is involved. The solid volume fraction will be used hereafter unless otherwise specified. Substituting Eq. (4.3) into Eq. (2.7), the wall concentration of retained particles on the membrane surface is related to filtration time by

$$c_w = \frac{c_0 v_0^2}{D} t \tag{4.4}$$

Under normal operation in microfiltration and ultrafiltration, it takes a very short time for concentration polarization to develop to full strength, which means to reach the maximum wall concentration. This can be shown with an example below.

Example 4.1

An ultrafiltration is operated at a temperature of 20°C with a permeate flux of 10^{-4} m/s. The feed water contains 0.01% (solid volume fraction) of particles of 0.1 μm in radius. Assuming the maximum concentration of the particles is 0.64, how long will it take for the wall concentration to reach the maximum concentration?

SOLUTION

Rearranging Eq. (4.3), the time to reach a given wall concentration is

$$t = \frac{DC_w}{c_0 v_0^2}$$

Einstein–Stokes equation is used to determine the diffusion coefficient of the particles

$$D = \frac{kT}{6\pi_c \mu a_p} = \frac{1.38 \times 10^{-23}\ J/K \times 293\ K}{6 \times 3.14 \times 10^{-3}\ Pa \cdot s \times 10^{-7}\ m} = 2.15 \times 10^{-12}\ m^2/s$$

Substituting the known values for all parameters in the above equation, the time required for the concentration to fully develop is

$$t = \frac{Dc_w}{c_0 v_0^2} = \frac{2.15 \times 10^{-12}\ m^2/s \times 0.64}{10^{-4} \times \left(10^{-4}\ m/s\right)^2} = 1.37\ s$$

It takes 1.37 seconds for the wall concentration to reach the maximum solid fraction of 0.64.

The time for concentration polarization to develop fully (i.e., to reach the maximum wall concentration) is even shorter for larger particles and/or higher concentrations. Practically, concentration polarization has been fully developed right after the start of the filtration process, and its impact on permeate flux cannot be reasonably measured. Therefore, the observed declining fluxes with time in microfiltration and ultrafiltration cannot be attributed to concentration polarization.

After the concentration polarization layer is fully developed, a cake layer forms between the concentration polarization layer and the membrane surface. The thickness of the cake layer grows as the amount of retained particles increases [10–13]. The thickness of the cake layer is determined by

$$\delta = \frac{M}{c_g} = \frac{c_0}{c_g} \int_0^t v dt' \tag{4.5}$$

where δ is the thickness of the cake layer, c_0 and c_g are the particle concentrations in the feed water and in the cake layer, respectively, v is the permeate flux, and t' is the dummy integration variable.

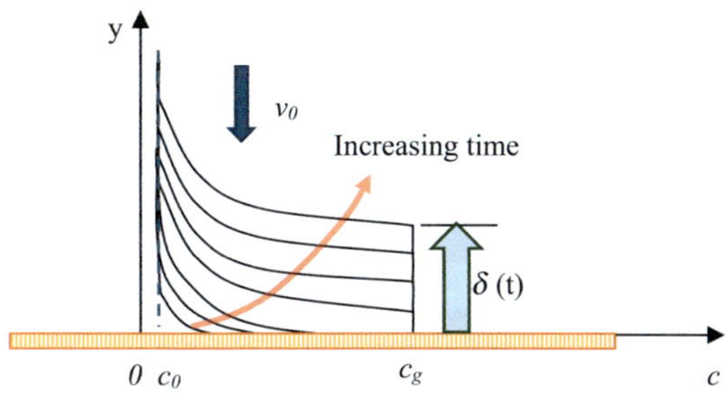

FIGURE 4.1
A cake layer forms as the wall concentration reaches the maximum packing density. Thereafter, the thickness of the cake layer grows with time.

When the permeate flux is constant, the thickness of the cake layer grows linearly with time, i.e.,

$$\delta = \frac{c_0}{c_g} v_0 t$$

Figure 4.1 schematically shows the changes in the concentration profile with time. Wall concentration initially grows with time. When the wall concentration reaches the maximum concentration c_g, a cake layer of constant concentration or density c_g starts to form on the membrane surface. Then thickness of the cake layer grows with time. As a result, the concentration polarization layer is translated upward with an invariant concentration profile in the layer.

4.2.2 Calculation of Permeate Flux

Resistance of cake layer R_{cl} is the product of specific resistance and layer thickness as shown in Eq. (2.21). Substituting Eq. (2.21) into Eq. (4.2), the equation for permeate flux becomes

$$v = \frac{\Delta p - \Delta \pi_0 - \Delta p_{cp}}{R_m + r_{cl}\delta} \tag{4.6}$$

Thickness of cake layer can be determined by substituting Eq. (4.6) into Eq. (4.5). The resultant equation is,

$$\delta = \frac{c_0}{c_g} \int_0^t \frac{\Delta p - \Delta \pi_0 - \Delta p_{cp}}{R_m + r_{cl}\delta} d\tau$$

This expression implicitly defines the thickness of the cake layer because thickness appears on both sides of the equation. An explicit expression for the thickness can be obtained by solving this integration equation.

First, taking the derivative of the above equation with respect to t results in

$$\frac{d\delta}{dt} = \frac{c_0}{c_g} \frac{\Delta p - \Delta\pi_0 - \Delta p_{cp}}{R_m + r_{cl}\delta}$$

Then moving all terms containing thickness to the left side of the equation and all the rest to the right side of the equation as

$$\left(R_m + r_{cl}\delta\right) d\delta = \frac{c_0}{c_g}\left(\Delta p - \Delta\pi_0 - \Delta p_{cp}\right) dt$$

Integrating the equation

$$\int_0^{\delta}\left(R_m + r_{cl}\delta'\right) d\delta' = \frac{c_0}{c_g}\left(\Delta p - \Delta\pi_0 - \Delta p_{cp}\right)\int_0^{t} dt'$$

The symbol δ is the upper integration limit corresponding to the operation time t. In this equation, both δ and t are fixed values in the integration operation with δ' and t as the dummy integration variables. The result is

$$r_{cl}\delta^2 + 2R_m\delta - 2\frac{c_0}{c_g}\left(\Delta p - \Delta\pi_0 - \Delta p_{cp}\right)t = 0$$

This is a quadratic equation of thickness δ, which has a positive root and a negative root. Because the negative thickness does not make any physical sense, the solution is the positive root of this quadratic equation, which is

$$\delta = \frac{1}{r_{cl}}\left[R_m^{\ 2} + 2r_{cl}\frac{c_0}{c_g}\left(\Delta p - \Delta\pi_0 - \Delta p_{cp}\right)t\right]^{1/2} - \frac{1}{r_{cl}}R_m \tag{4.7}$$

Substituting Eq. (4.7) into Eq. (4.6) gives the permeate flux as

$$v = \frac{\Delta p - \Delta\pi_0 - \Delta p_{cp}}{\left[R_m^{\ 2} + 2r_{cl}\frac{c_0}{c_g}\left(\Delta p - \Delta\pi_0 - \Delta p_{cp}\right)t\right]^{1/2}} \tag{4.8}$$

In most microfiltration and ultrafiltration processes, the feed osmotic pressure and the pressure difference across the concentration polarization layer are negligible compared to the driving pressure, i.e., $\Delta\pi_0 \ll \Delta p_{cp} \ll \Delta p$. Then the permeate flux can be calculated by

$$v = \frac{\Delta p}{\left(R_m^{\ 2} + 2r_{cl}\frac{c_0}{c_g}\Delta pt\right)^{1/2}} \tag{4.9}$$

An example below describes the impact of cake formation on the permeate flux in an ultrafiltration process.

Example 4.2

A feed water containing 0.1% (solid volume fraction) of particles with a radius of 0.1 μm is separated with an ultrafiltration membrane at 20°C. The applied pressure is 2 bars, and the membrane resistance is 10^9 Pa·s/m. The closest packing density of the particles is assumed to be 0.64. (1) Calculate the osmotic pressure and the maximum pressure difference across the concentration polarization layer; (2) Calculate the permeate velocities at time $t = 0$ and 25 min.

SOLUTION

1. Feed osmotic pressure and maximum pressure difference across the concentration polarization layer

   The number concentrations of particles in feed water and cake layer can be calculated from the solid volume fraction with Eq. (1.3)

$$\frac{c_0}{\frac{4}{3}\pi_c a_p^{\ 3}} = \frac{10^{-3}}{\frac{4}{3}\times 3.14\times\left(10^{-7}\ m\right)^3} = 2.39\times 10^{17}/m^3$$

$$\frac{c_g}{\frac{4}{3}\pi_c a_p^{\ 3}} = \frac{0.64}{\frac{4}{3}\times 3.14\times\left(10^{-7}\ m\right)^3} = 1.53\times 10^{20}/m^3$$

Using the van't Hoff equation Eq. (1.5), the feed osmotic pressure

$$\Delta\pi_0 = kTc_0 = 1.38\times 10^{-23}\ J/K\times 293\ K\times 2.39\times 10^{16}/m^3$$
$$= 9.66\times 10^{-4}\ Pa$$

From Table 2.1, the mean correction factor at a solid fraction of 0.64 is

$$A_s^{\ *} = 24.19$$

Using Eq. (2.12), the maximum pressure difference across the concentration polarization layer is

$$\Delta p_{cp} = kTc_g A_s^{\ *} = 1.38\times 10^{-23}\ J/K\times 293K\times 1.53\times 10^{20}/m^3\times 24.19$$
$$= 14.95\ Pa$$

It can be seen that $\Delta\pi_0 \ll \Delta p_{cp} \ll \Delta p$. (The use of $A_s^{\ *} = 24.19$ in the calculation of Δp_{cp} is to play safely)

2. Permeate flux declines with time

   The correction factor for solid fraction of 0.64 found from Table 2.1 is $A_s = 123.22$. The specific resistance can be calculated using the Happel model–derived equation Eq. (2.24),

$$r_{cl} = \frac{9}{2}\frac{\mu c_g}{a_p^2}A_s = \frac{9\times 10^{-3}\ Pa\cdot s\times 0.64}{2\times\left(10^{-7}\ m\right)^2}\times 123.22 = 3.55\times 10^{13}\ Pa\cdot s/m^2$$

   The correction factor true A_s should always be used in the calculation of cake specific resistance, even in the model membrane system. The permeate flux at any time is calculated with Eq. (4.9) with the negligible $\Delta\pi_0$ and ΔP_{cp}

$$v = \frac{\Delta p}{\left(R_m^2 + 2r_{cl}\frac{c_0}{c_g}\Delta pt\right)^{1/2}}$$

   At t = 0,

$$v = \frac{\Delta p}{R_m} = \frac{2\times 10^5\ Pa}{10^9\ Pa\cdot s/m} = 2.0\times 10^{-4} m/s$$

   At t = 25 minutes

$$v = \frac{\Delta p}{\left(R_m^2 + 2r_{cl}\frac{c_0}{c_g}\Delta pt\right)^{1/2}}$$

$$= \frac{2\ bar}{\left[\left(10^9\ Pa\cdot s/m\right)^2 + 2\times 3.55\times 10^{13}\ Pa\cdot s/m^2\times\frac{0.001}{0.64}\times 2\times 10^5\ Pa\times 25\ min\times\frac{60s}{1\ min}\right]^{1/2}}$$

$$= 3.42\times 10^{-5}\ m/s$$

   Done.

The example shows that the impact of concentration polarization on permeate flux in this case is negligible because $\Delta p_{cp} \ll \Delta p$. On the contrary, cake formation can reduce the permeate flux substantially. The permeate flux loses about 83% of its initial value in 25 minutes due to cake formation in this example.

The permeate flux in a dead-end ultrafiltration process with the same parameters as in Example 4.2 is calculated for a period of 30 minutes. The results are graphically presented in Figure 4.2. The curve of permeate flux features a rapid decrease initially in a short time and then is followed by a long tail of moderate to slow decrease. Permeate flux drops about 50% of its initial value in the first 2.5 minutes, and the rate of decline decreases significantly

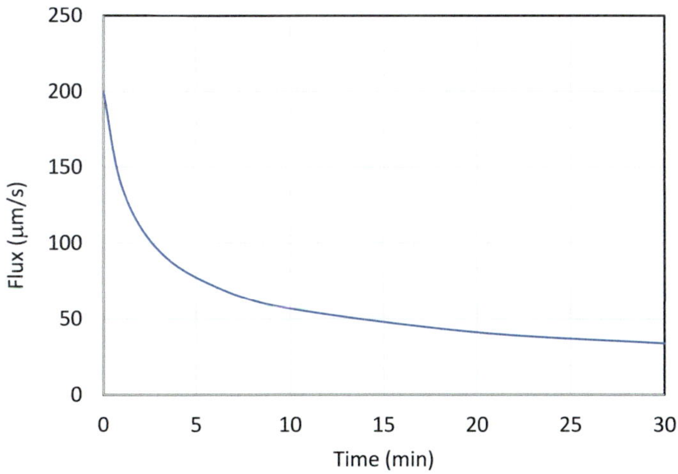

FIGURE 4.2
Permeate velocity declines with time in a dead-end ultrafiltration. The parameters are: $c_0 = 0.1\%$, $c_g = 0.64$, $\Delta P = 2$ bar, $R_m = 10^9$ Pa·s/m, and $T = 25°C$.

with time. The figure also shows that there is no need for other mechanisms to cause the rapid decline of permeate flux in the very initial stage. The rapidly declining flux can be a feature purely caused by cake formation.

4.3 Crossflow Filtration at Steady State

A steady state can be reached in a crossflow filtration with the presence of a cake layer. Same as in a channel without a cake layer, tangential flow in a crossflow membrane channel with a cake layer also has a certain carrying capacity to take the retained particles downstream until out of the membrane channel. A cake layer forms initially when the accumulation rate is greater than the carrying capacity. The thickness of a cake layer grows at a point in the channel where the accumulation rate of retained particles due to permeate flow exceeds the carrying capacity of tangential flow. As the thickness of the cake layer grows, the accumulation rate reduces as a result of declining permeate velocity. At the same time, the carrying capacity increases with the declining permeate velocity. The reason is that the thickness of the concentration polarization layer increases as the permeate velocity decreases. A steady state is reached when the two quantities become equal to each other as a result of the cake layer growing. This is a fundamental difference of crossflow filtration from the dead-end filtration, in which the cake layer can grow infinitely [14–17]. The permeate flux in the crossflow membrane channel with a cake layer at steady state will be studied first in this section. The dynamic process of cake formation in a crossflow membrane channel will be dealt with later in the next section.

The principal mass balance relationship in crossflow filtration at steady state, which is Eq. (3.11), remains the same with or without a cake layer, i.e.,

$$c_0 \int_0^x v\,dx' = \int_0^\infty uC\,dy$$

The permeate flux in a crossflow membrane channel at steady state with the presence of a cake layer will be derived from this principal mass balance relationship. Dependent on the assumptions of cake layer properties, two different scenarios are considered below.

4.3.1 Scenario 1: Undisturbed Tangential Flow Field

This scenario assumes that the tangential flow on the membrane surface is unaffected by the cake layer, as shown in Figure 4.3(a). The cake layer is assumed sufficiently porous so that tangential flow can pass through it easily. As a result, the tangential flow velocity on the cake surface is greater than zero [18]. Tangential velocity is usually assumed to be zero at a solid surface (non-slip boundary condition). It seems at first glance that the non-zero tangential velocity on the cake surface is overestimated. However, some cake layers are not completely immobile but may move slowly downstream on the drag force of tangential flow. A nonzero tangential flow on the cake surface can be regarded as a compensation for the carrying capacity of retained particles by the mobile cake layer because it is not explicitly considered.

As shown in Figure 4.3(a), the y coordinate starts from the membrane surface. The thickness of the cake layer is δ, and the concentration polarization layer develops for $y > \delta$. The shear flow over the membrane surface is assumed to be unaffected by the existence of the cake layer.

A key step in modeling permeate flux is to track the amount of retained particles M along the membrane channel. First, the vertical concentration

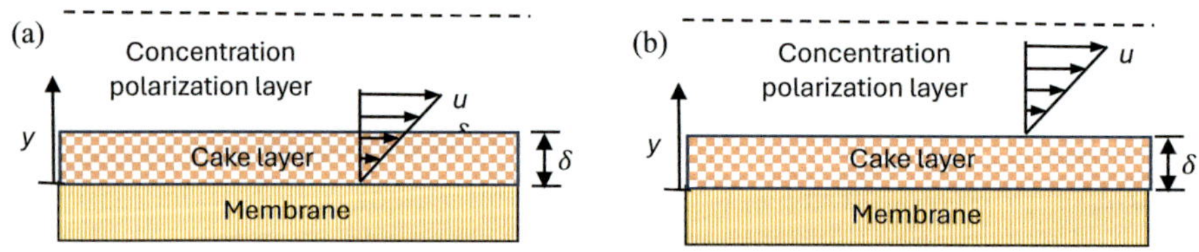

FIGURE 4.3
Two scenarios (assumptions) of cake layer impact on flow field over membrane surface. (a) No impact, and (b) translation of the flow field upward.

distribution of retained particles at an arbitrary point x in the membrane channel with a cake layer should be modified from Eq. (2.6) as

$$C = \frac{Mv}{D}\exp\left[-\frac{v(y-\delta)}{D}\right], \textit{ for } y \geq \delta \tag{4.10}$$

It can be seen from Figure 4.3(a) that $y-\delta = 0$ is the bottom of the concentration polarization layer. It means that the wall concentration of the concentration polarization layer occurs at $y = \delta$, where Eq. (4.10) reduces to Eq. (2.7). For convenience to link with pressure, number concentration will be used in the derivations below.

Substituting Eq. (4.10) into Eq. (3.10), the carrying capacity of the tangential flow is determined,

$$N_2 = \int_\delta^\infty uC\,dy = \int_\delta^\infty \gamma y \frac{Mv}{D}\exp\left[-\frac{v(y-\delta)}{D}\right]dy = \frac{\gamma MD}{v}\left[1+\delta\frac{v}{D}\right]$$

The lower limit of the integration is changed from 0 to δ because the concentration polarization layer starts from $y = \delta$.

With the integration result, now the principal mass balance equation Eq. (3.11) becomes

$$c_0\int_0^x v\,dx' = \frac{\gamma MD}{v}\left[1+\delta\frac{v}{D}\right]$$

The accumulated amount of retained particles appears in the equation. Rearranging the equation results in,

$$M = \frac{c_0}{\gamma D\left[1+\delta\dfrac{v}{D}\right]}v\int_0^x v\,dx' \tag{4.11}$$

From Eq. (4.11), the governing equation for permeate flux can be constructed. When there is a cake layer, the wall concentration of concentration polarization layer is always the cake concentration (density). Then the cake concentration C_g is related to the amount of retained particle M by Eq. (2.7),

$$C_g = \frac{v}{D}M$$

Substituting Eq. (4.11) for M in this equation, we have

$$C_g = \frac{c_0}{\gamma D^2\left[1+\delta\dfrac{v}{D}\right]}v^2\int_0^x v\,dx' \tag{4.12}$$

In Eq. (4.12), because cake density C_g is constant, cake thickness δ is another unknown in addition to permeate velocity v. Another relationship between these two unknowns is required to define permeate flux in the membrane channel.

A quantitative relationship that directly relates thickness δ and permeate velocity v is not available. However, the thickness of the cake layer is related to the pressure drop across it. From Eq. (2.23), we have

$$\delta \frac{v}{D} = \frac{\Delta p_{cl}}{kTA_s C_g}$$

Please note that the number concentration should be used from now on to be consistent with the term kT. As announced at the beginning of the derivation, the number concentration unit was used for particle concentration C_g. There is no need for concentration unit conversion here.

Substituting this expression into Eq. (4.12) results in

$$C_g = \frac{c_0}{\gamma D^2 \left[1 + \dfrac{\Delta p_{cl}}{kTA_s C_g}\right]} v^2 \int_0^x v\,dx'$$

Rearranging,

$$\Delta p_{cl} = kTC_g A_s \left[\frac{c_0}{C_g} \frac{v^2}{\gamma D^2} \int_0^x v\,dx' - 1\right] \tag{4.13}$$

Now the unknown δ in Eq. (4.12) is replaced by unknown Δp_{cl} in Eq. (4.13).

A quantitative relationship exists between the two unknowns Δp_{cl} and v, which is Eq. (4.1). Therefore, the permeate velocity in the crossflow channel is virtually controlled by Eqs. (4.1) and (4.13). Two equations for two unknowns! The permeate flux is finally mathematically defined.

Substituting (4.13) for Δp_{cl} in Eq. (4.1), together with Eqs. (2.13) for Δp_{cp} and vR_m for Δp_m, results in

$$\Delta p = \Delta \pi_0 + kTC_g \left(A_s^* - A_s\right) + vR_m + \frac{\Delta \pi_0 A_s}{\gamma D^2} v^2 \int_0^x v\,dx' \tag{4.14}$$

Comparing Eq. (4.14) with Eq. (3.14) for the membrane channel without cake layer, the only difference is the constant term. It is $\Delta \pi_0$ in Eq. (3.14) but $\left[\Delta \pi_0 + kTC_g \left(A_s^* - A_s\right)\right]$ in Eq. (4.14). Therefore, permeate flux in the crossflow membrane channel with the cake layer can be determined by analogy to Eq. (3.22) for membrane channels without the cake layer

$$v = \frac{v_1}{\left(1+\beta x/L\right)^{1/3}} \left\{ \left[\left(\frac{1}{1+\beta x/L} + 4\right)^{1/2} + 2\right]^{1/3} \left[\left(\frac{1}{1+\beta x/L} + 4\right)^{1/2} - 2\right]^{1/3} \right\} \tag{4.15}$$

where v_1 and β are parameters defined, respectively, by

$$v_1 = \frac{\Delta p - \Delta\pi_0 - kTC_g\left[A_s^* - A_s\right]}{R_m}$$
$$\beta = 6v_0^{\ 2}\frac{\Delta\pi_0}{R_m}A_s\left(\frac{L}{D^2\gamma}\right) \tag{4.16}$$

The parameter v_1 is the initial permeate velocity at the entrance of the membrane channel, where the concentration polarization is fully developed but without the cake layer. The dimensionless parameter β is *the crossflow filtration number* of the membrane process with the presence of a cake layer.

The average permeate flux can be calculated from the following equation

$$\bar{v} = \frac{1}{L}\int_0^L v\,dx = \frac{\gamma D^2}{kTA_s c_0 L v(L)^2}\left\{\Delta p - \Delta\pi_0 - kTC_g\left[A_s^* - A_s\right] - v(L)R_m\right\} \tag{4.17}$$

Example 4.3

Calculate the initial permeate velocity at the entrance and the crossflow filtration number with the cake layer. The parameters of a crossflow membrane channel are listed in the table below.

Parameter	Symbol	Value	Unit
Particle size	a_p	1×10^{-8}	m
Applied pressure	Δp	5×10^5	Pa
Shear rate	γ	100	1/s
Membrane resistance	R_m	1.2×10^{10}	Pa·s/m
Feed concentration	c_0	0.01%	v/v
Cake concentration	c_g	0.60	v/v
Channel length	L	2	m
Temperature	T	20	°C

SOLUTION

From Table 2.1, it can be found that $A_s = 85.12$ and $A_s^* = 18.91$ for $c_g = 0.60$ (v/v),

1. Determine if there will be a cake layer.

   The critical pressure can be determined with Eq. (2.13)

$$\Delta p_{cp} = kTA_s^* \frac{C_g}{\frac{4}{3}\pi_c a_p^{\ 3}}$$

$$= 1.38 \times 10^{-23}\ Pa\cdot m^3/K \times 293\ K \times 18.91 \times \frac{0.6}{\frac{4}{3}\times 3.14 \times \left(10^{-8}\right)^3}$$

$$= 10{,}958\ Pa$$

Because the applied pressure is much greater than the critical pressure for cake formation, a cake layer will form in the membrane channel.

2. Determine the initial permeate velocity at the channel entrance.

$$\Delta\pi_0 = kT\frac{c_0}{\frac{4}{3}\pi_c a_p^{\ 3}} = 1.38\times10^{-23}\,Pa\cdot m^3/K\times293\,K\times\frac{0.0001}{\frac{4}{3}\times3.14\times\left(10^{-8}\right)^3}$$

$$= 0.10\ Pa$$

Convert wall concentration into number concentration unit

$$C_g = \frac{0.6}{\frac{4}{3}\pi_c a_p^{\ 3}} = \frac{0.6}{\frac{4}{3}\times3.14\times\left(10^{-8}\right)^3} = 1.43\times10^{23}/m^3$$

Convert wall concentration into number concentration unit

$$v_1 = \frac{\Delta p - \Delta\pi_0 - kTC_g\left[A_s^* - A_s\right]}{R_m}$$

$$= \frac{5\times10^5\ Pa - 0.10\ Pa}{1.2\times10^{10}\ Pa\cdot s/m}$$

$$-\frac{1.38\times10^{-23}\ Pa\cdot m^3/K\times293\ K\times1.43\times10^{23}/m^3\times[18.91-85.12]}{1.2\times10^{10}\ Pa\cdot s/m}$$

$$= 4.49\times10^{-5}\ m/s$$

3. Determine the crossflow filtration number with the cake layer

$$v_0 = \frac{\Delta p - \Delta\pi_0}{R_m} = \frac{5\times10^5\ Pa - 0.10\ Pa}{1.2\times10^{10}\ Pa\cdot s/m} = 4.17\times10^{-5}\ m/s$$

$$D = \frac{kT}{6\pi_c\mu a_p} = \frac{1.38\times10^{-23}\ Pa\cdot m^3/K\times293\ K}{6\times3.14\times10^{-3}\ Pa\cdot s\times10^{-8}\ m} = 2.15\times10^{-11}\,m^2/s$$

$$\frac{L}{D^2\gamma} = \frac{2\ m}{\left(2.15\times10^{-11}\ m^2/s\right)^2\times100/s} = 4.34\times10^{19}\,s^3/m^3$$

$$\beta = 6v_0^{\ 2}\frac{\Delta\pi_0}{R_m}A_s\left(\frac{L}{D^2\gamma}\right)$$

$$= 6\times\left(4.17\times10^{-5}\ m/s\right)^2\frac{0.10\ Pa}{1.2\times10^{10}\ Pa\cdot s/m}\times85.12\times4.34\times10^{19}\ s^3/m^3$$

$$= 309.85$$

4.3.2 Scenario 2: Non-Slip Boundary on Cake Surface

The other scenario of the tangential flow field on the cake layer is just the opposite of Scenario 1. The cake layer is immobile and impermeable to the tangential flow. A non-slip boundary condition [19,20] is held on the surface of the cake layer, as shown in Figure 4.3(b).

Mathematical treatment of this scenario is much simpler than Scenario 1. In this case, because of the existence of a cake layer, the wall concentration of retained particles in Eq. (3.12) equals the constant concentration C_g of the cake layer,

$$C_g = \frac{c_0}{D^2\gamma} v^2 \int_0^x v\,dx' \tag{4.18}$$

An explicit expression for permeate flux can be derived from Eq. (4.18) with the same method to solve Eq. (3.15) in the previous chapter. Using the function Eq. (3.16) and its derivative

$$w = \int_0^x v\,dx'$$

and

$$\frac{dw}{dx} = v$$

Eq. (4.18) can be transformed into

$$C_g = \frac{c_0}{D^2\gamma}\left(\frac{dw}{dx}\right)^2 w$$

Rearranging yields a differential equation

$$\frac{d}{dx} w^{3/2} = \frac{3}{2}\left(\frac{C_g}{c_0} D^2\gamma\right)^{1/2}$$

The solution of the differential equation is

$$w = \left(\frac{3}{2}\right)^{2/3} \left(D^{2/3}\gamma^{1/3}x^{2/3}\right)\left(\frac{C_g}{c_0}\right)^{1/3}$$

The explicit expression of permeate flux is obtained by taking the derivative of the equation with respect to x, which results in

$$v = \frac{dw}{dx} = \left(\frac{2}{3}\right)^{1/3} \left(D^{2/3}\gamma^{1/3}x^{-1/3}\right)\left(\frac{C_g}{c_0}\right)^{1/3} \tag{4.19}$$

The average permeate flux of the membrane channel is then determined as

$$\bar{v} = \frac{1}{L}\int_0^L v\,dx = \left(\frac{3}{2}\right)^{2/3} \left(\frac{D^2\gamma}{L}\right)^{1/3}\left(\frac{C_g}{c_0}\right)^{1/3} \tag{4.20}$$

Because Eq. (4.20) does not contain the driving pressure and the resistance of the membrane [21], it is actually the limiting flux of the crossflow membrane channel!

Liming flux plays an important role in the design and operation of microfiltration and ultrafiltration. Eq. (4.20) is rigorously derived from the principle mass balance equation on retained particles. It naturally reproduces the Leveque factor on permeate flux. One marked feature of the new expression is its one-third order dependence of permeate flux on the ratio of gel concentration to feed concentration $\left(C_g/c_0\right)^{1/3}$, which is fundamentally different from the commonly reported logarithmic dependence $\ln\left(C_g/c_0\right)$.

The definition of limiting flux is the ultimate flux in a crossflow membrane process that cannot be further increased by increasing applied pressure [22–24]. Limiting flux is not only independent of the applied pressure, but also independent of the membrane resistance. These features of the limiting flux are truly spectacular because the applied pressure and membrane resistance are two principal factors that govern flux according to the basic filtration equation.

Instead, liming flux can be reasonably explained with the behavior of retained particles in the membrane process. Limiting flux is controlled by the hydraulic condition over the membrane surface and the geometry of the membrane channel. In other words, the carrying capacity of tangential flow determines the limiting flux in a crossflow membrane channel. It is well established by experiments that limiting flux is positively related to tangential velocity or shear rate of the tangential flow over the membrane surface. Higher tangential velocity or shear rate is commonly used to increase limiting flux in crossflow membrane processes. Diffusivity of retained particles is one of the important factors for limiting flux. A higher flux is always obtained for the species with a bigger diffusion coefficient under the otherwise same hydraulic conditions. Larger diffusion coefficient increases the carrying capacity of tangential flow because the concentration polarization becomes wider.

In the cases of limiting flux, the thickness of the cake layer can be determined by rearranging Eq. (4.6) as

$$\delta = \frac{\Delta p - \Delta\pi_0 - \Delta p_{cp}}{r_{cl} v} - \frac{R_m}{r_{cl}}$$

The second term on the right-hand side of the equation is insignificant in the limiting flux case. Substituting Eq. (4.19) into the equation and eliminating the membrane resistance–related term result in

$$\delta_{eq}(x) = \frac{\Delta p - \Delta\pi_0 - \Delta p_{cp}}{\left(\frac{2}{3}\right)^{1/3} r_{cl}\left(D^{2/3}\gamma^{1/3}\right)\left(\frac{C_g}{c_0}\right)^{1/3}} x^{1/3} \tag{4.21}$$

where δ_{eq} is the equilibrium thickness of the cake layer.

Example 4.4

Calculate the thickness of the cake layer along a crossflow membrane channel of 1 m long for three feed concentrations of 0.01, 0.1, and 1%. The cake concentration is 0.6, particle radius is 10 nm, pressure is 10 bars, and shear rate is 100/s.

SOLUTION

From Table 2.1, $A_s = 85.12$ for $c_g = 0.60$ (v/v),

The specific resistance of the cake layer is calculated with Eq. (2.25)

$$r_{cl} = \frac{9}{2}\frac{\mu c_g}{a_p^{\ 2}} A_s = \frac{9}{2}\frac{\times 10^{-3}\ Pa \cdot s \times 0.6}{\left(10^{-8}\ m\right)^2} \times 85.12 = 2.3 \times 10^{15}\ Pa \cdot s/m^2$$

From Example 4.3, it can be seen that $D = 2.15 \times 10^{-11}\ m^2/s$. It can be seen that $\Delta\pi_0 \ll \Delta P$ and $\Delta P_{cp} \ll \Delta P$ and therefore can be neglected in the equation to calculate the thickness of the cake layer in Eq. (4.21), which becomes

$$\delta(x) = \frac{\Delta p}{\left(\frac{2}{3}\right)^{1/3} r_{cl}\left(D^{2/3}\gamma^{1/3}\right)\left(\frac{C_g}{c_0}\right)^{1/3}} x^{1/3}$$

A sample calculation of thickness of cake later for c_0 = 0.01% and x = 0.1 m is

$$\delta = \frac{10 \times 10^5\ Pa \times \left(0.001\ m\right)^{1/3}}{\left(\frac{2}{3}\right)^{1/3} 2.3 \times 10^{15}\ Pa \cdot s/m^2 \times \left[\left(2.15 \times 10^{-11}\ m^2/s\right)^{2/3}\left(100/s\right)^{1/3}\right]\left(\frac{0.6}{0.0001}\right)^{1/3}}$$

$$= 7.6 \times 10^{-6}\ m = 7.6\mu m$$

The calculation results for all three feed concentrations along the membrane channel are listed below.

Thickness of Cake Layer (μm)			
x(m)	0.01%	0.10%	1%
0.001	7.6	16.5	35.5
0.002	9.6	20.8	44.7
0.004	12.1	26.2	56.3
0.008	15.3	32.9	71.0
0.016	19.3	41.5	89.4
0.032	24.3	52.3	112.7
0.064	30.6	65.9	142.0
0.128	38.5	83.0	178.9
0.256	48.6	104.6	225.4
0.512	61.2	131.8	283.9
1.024	77.1	166.1	357.7

The thicknesses of cake layers are shown graphically below.

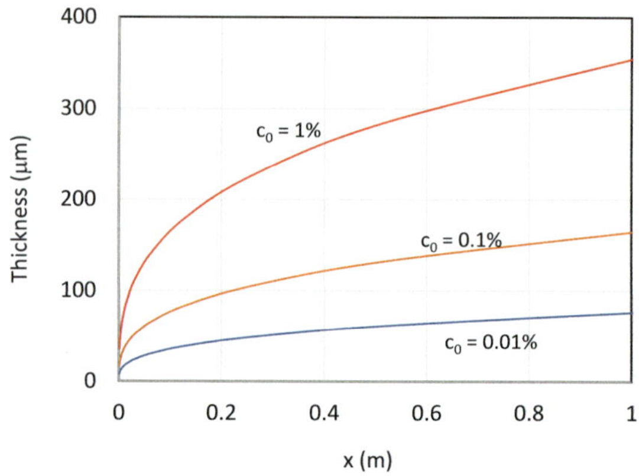

4.3.3 Average Permeate Flux in Crossflow Filtration

Up to now, the permeate flux in crossflow filtration in the whole range of applied pressure can be calculated. A sample calculation of permeate flux is done for a wide range of applied pressures. The particle radius of 2.5 nm is selected in this sample calculation because both concentration polarization and cake formation can be shown for this particle size in a moderate pressure range.

For a given set of parameters including channel length, shear rate, membrane resistance, particle concentrations in feed water and in the cake layer, permeate flux for a pressure Δp can be calculated. First, the filtration number is calculated with Eq. (2.11) with the pressure drop Δp_{cp} replaced by Δp, which is the maximum possible pressure drop across the concentration polarization layer. At small pressure, the filtration number should be smaller than 15.5. There will not be a cake layer in the membrane channel. The average permeate flux is calculated with Eq. (3.24). As applied pressure grows larger, the filtration number can reach and exceed 15.5, and there will be a cake layer in the membrane channel [14]. Depending on scenarios, either Eq. (4.17) or Eq. (4.20) can be used for the average flux of the membrane channel.

The sample calculation is presented in Figure 4.4(a) in linear scale and Figure 4.4(b) in double logarithm scales, respectively. The packing density of 0.64 is used for the cake layer. The transition pressure from pure concentration polarization to cake formation is 9.34 bar, which is indicated in the figure by a broken vertical line with the filtration number "$N_F = 15.1$." It can be seen from the graphs in Figure 4.4 that the flux calculated with Eq. (4.17) increases with applied pressure, as does the pressure range below the critical pressure for cake formation. On the other hand, the flux calculated with Eq. (4.20) is an invariant with applied pressure, the limiting flux. It means the

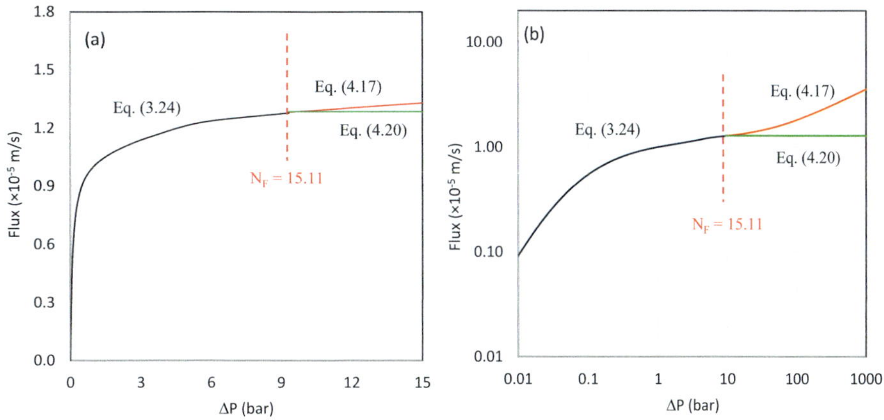

FIGURE 4.4
Average permeate velocity as a function of pressure. (a) In linear scale, and (b) in logarithmic scale. The parameters used in the calculation: a_p = 2.5 nm, c_0 = 0.1% (v/v), γ = 100/s, L = 0.5 m, c_g = 64% (v/v), $R_m = 10^9$ Pa·s/m, T = 293K.

flux cannot be increased by increasing pressure when there is a cake layer. In other words, the cake layer provides a mechanism to adsorb the pressure beyond the transition (critical) pressure. It is a striking surprise to find that the two equations, Eqs. (4.17) and (4.20), produce exactly the same value at the transition pressure, though the two equations are derived from different assumptions and are presented in completely different forms.

At the transition pressure, the permeate flux calculated with Eq. (3.24) prior the cake formation connects smoothly to those calculated by Eqs. (4.17) and (4.20) with the cake layer. This is another amazing finding, considering the equations are derived independently with different conditions.

4.4 Dynamic Process of Cake Formation

In contrast to the quick formation of a concentration polarization layer, cake formation is a relatively slow process. Though a steady state exists for every crossflow membrane channel in principle, it may not be achievable in many practical membrane processes. Steady state may need too long a time to achieve in some cases, so that it cannot be observed in practical operations. In some other cases, the flux already becomes too small before the steady state is reached, so that the operation has to be stopped, and backwash is required to remove the accumulated retained particles. Actually, the most commonly observed gradually declining fluxes in membrane processes are reflections of the dynamic process of cake formation before the steady state is reached.

4.4.1 Two Regions of Cake Layer

Cake formation is a complex process in a crossflow membrane channel. First, to simplify the situation, imagine a membrane channel that extends infinitely from both ends, over which a feed water is flowing tangentially. Permeate flow brings particles to the membrane surface. Assume the applied pressure is greater than the critical pressure for cake formation, and the concentration polarization layer is fully developed instantly. The filtration process can be simply viewed as a growing process of cake layer, just as what happens in a dead-end filter. Theoretically, the thickness of the cake layer grows evenly along the membrane channel without limit. Practically, the cake layer continues to grow until flux dies out (flux becomes zero). There is no steady state in such a crossflow membrane channel of infinite length.

When a crossflow membrane channel has a finite length, growth of the cake layer is strongly affected and somehow restricted by the entrance of the membrane channel. Let us first have a close look at what happens at the entrance $x = 0$. It is clear that there is no concentration polarization and no cake formation at this point because there is no accumulation of retained particles at that point. A concentration polarization layer develops immediately downstream from the entrance as retained particles start to accumulate. A cake layer forms further downstream when the wall concentration of retained particles in the concentration polarization layer reaches the maximum packing density. Cake layer grows in thickness as more retained particles accumulate. An equilibrium thickness will be reached at the starting point of the cake layer, and the region of the cake layer with equilibrium thickness grows downstream with time [25].

The thickness of the cake layer at equilibrium is described by Eq. (4.21), which is reproduced below

$$\delta_{eq}(x) = \frac{\Delta p - \Delta\pi_0 - \Delta p_{cp}}{\left(\frac{2}{3}\right)^{1/3} r_{cl}\left(D^{2/3}\gamma^{1/3}\right)\left(\frac{C_g}{c_0}\right)^{1/3}} x^{1/3}$$

The cake layer in a crossflow membrane channel with finite length can be described in Figure 4.5. There are two regions of the cake layer. The cake layer is schematically depicted in each region with a concentration polarization layer above it. Profiles of permeate flux distributions in both regions are shown in the figure below the membrane.

As depicted in the figure, the equilibrium region is on the left side of the entrance to a point X in the membrane channel. The cake layer in this region has reached equilibrium thickness and will not change with time. Both cake layer thickness and permeate flux in this region are functions of location x, as indicated by $\delta_{eq}(x)$ and $v(x)$ in the figure. Though the cake layer reaches the equilibrium thickness, there is still an accumulation of retained particles in the equilibrium region because the permeate flow is not zero there. However, the retained particles accumulated in this region cannot deposit on the cake

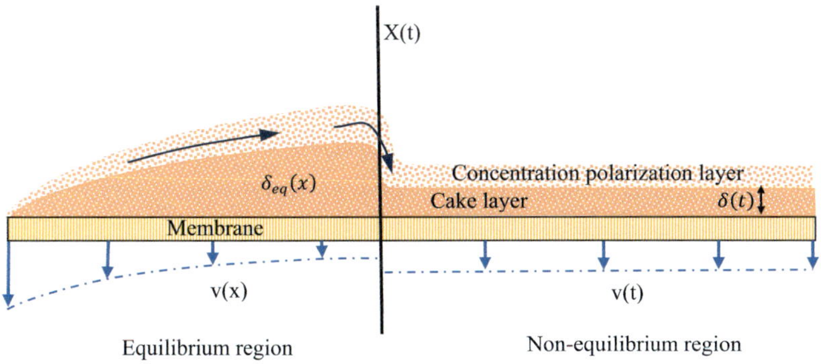

FIGURE 4.5
Schematic for fouling development in a crossflow membrane channel. There are an equilibrium section and a non-equilibrium section in the channel. The equilibrium section grows downstream with time.

layer because it is already at equilibrium. These accumulated retained particles stay in the concentration polarization layer. They are then carried downstream by tangential flow all the way over the whole equilibrium region and finally dumped at the front of the equilibrium region, as indicated in Figure 4.5. These retained particles collected in the equilibrium region cannot move further into the non-equilibrium region because the cake layer has not reached equilibrium thickness there.

The rest of the membrane channel on the left side is the non-equilibrium region. Influence of the entrance of the membrane channel has not arrived in this region. The cake layer here behaves just as in the infinite membrane channel. Both cake layer thickness and permeate flux, which are uniform throughout the whole non-equilibrium region, change with time, as indicated by $\delta(t)$ and $v(t)$. These features of the cake layer are just the opposite of those in the equilibrium region.

The front of the equilibrium region is indicated with a capital letter X in the figure. The front moves downstream as the equilibrium region grows with more retained particles accumulated in the membrane channel. A steady state can be achieved in the membrane channel when the front moves to the end of the membrane channel, which means the equilibrium cake thickness has been reached on the entire membrane channel.

4.4.2 Governing Equation for the Movement of the Division Line

Location of the division line between the two regions, which is also known as the front of the equilibrium region, moves downstream as a result of the deposition of the retained particles collected in the equilibrium region at the front of the region. The movement of the division line can be modeled by considering mass balance at the dumping location. As shown in

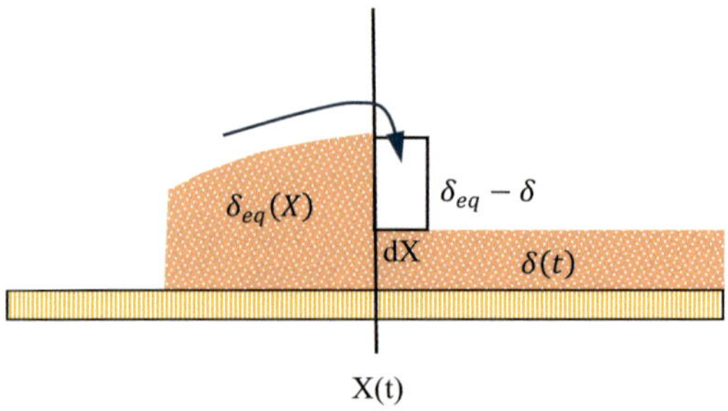

FIGURE 4.6
A small box is drawn at the front of the equilibrium section for mass balance analysis. Width and height of the box are dX and $\delta_{eq} - \delta$, respectively. The particles retained in the equilibrium section are dumped in this box.

Figure 4.6, a small box is drawn in the non-equilibrium region right at the front of the equilibrium region. The height of the small box is the difference between the thicknesses of the two regions $\delta_{eq} - \delta$, while the box width is dX. This box will become a part of the equilibrium region when it is fully filled with retained particles. The mass balance of retained particles in this box is

$$C_g\left[\delta_{eq}(X)-\delta(t)\right]dX=\left[c_0\int_0^X v_{eq}(x)dx\right]dt$$

The term on the left-hand side of the equation is the total amount of retained particles to fully fill up the box. The term in square brackets on the right-hand side of the equation is the accumulation rate of retained particles in the equilibrium region. From the above mass balance equation, the speed at which the division line moves downstream can be derived as,

$$\frac{dX}{dt}=\frac{1}{\left[\delta_{eq}(X)-\delta(t)\right]}\frac{c_0}{C_g}\int_0^X v_{eq}(x)dx$$

The integration in the above equation was done in Eq. (4.20) with X replaced by L. Substituting that integration into the current equation results in,

$$\frac{dX}{dt}=\frac{1}{\left[\delta_{eq}(X)-\delta(t)\right]}\left(\frac{3}{2}\right)^{2/3}\left(D^2\gamma X^2\right)^{1/3}\left(\frac{C_0}{c_g}\right)^{2/3} \tag{4.22}$$

Eq. (4.22) is the governing equation for the movement of the division line between the two regions. However, this governing equation in its current form cannot be solved routinely with the commonly used variable separation methods. The reason is that the variables X and t in the square brackets in Eq. (4.22) cannot be separated into terms exclusive of each other.

4.4.3 Solution of the Governing Equation

Since the two thicknesses are developed in the same membrane channel, they must have some intrinsic relations between them. A bold assumption is made between these two thicknesses as follows:

$$\delta(t) = a\delta_{eq}(X) \tag{4.23}$$

where a is a constant that is independent of t and X. Eq. (4.22) becomes readily solvable with Eq. (4.23). The assumption has to be verified after the solution of Eq. (4.22) is obtained.

Substituting Eq. (4.23) into Eq. (4.22), we have

$$\frac{dX}{dt} = \frac{1}{(1-a)\delta_{eq}(X)}\left(\frac{3}{2}\right)^{2/3}\left(D^2\gamma X^2\right)^{1/3}\left(\frac{C_0}{c_g}\right)^{2/3}$$

The cake thickness in the equilibrium region was derived earlier as Eq. (4.21). Substituting it into the above equation, we have

$$\frac{dX}{dt} = \frac{r_{cl}}{(1-a)\left(\Delta p - \Delta\pi_0 - \Delta p_{cp}\right)}\left(\frac{3}{2}\right)^{1/3} D^{4/3}\gamma^{2/3}\left(\frac{C_0}{c_g}\right)^{1/3} X^{1/3}$$

Integrating the equation results in

$$X = \frac{2}{3}D^2\gamma\left(\frac{C_0}{c_g}\right)^{1/2}\left[\frac{r_{cl}}{(1-a)\left(\Delta p - \Delta\pi_0 - \Delta p_{cp}\right)}t\right]^{3/2} \tag{4.24}$$

Eq. (4.24) is the solution of the governing equation. There are two issues with the solution. First, verify that the parameter a is constant as assumed. Second, the value of parameter a needs to be determined.

To verify the assumption that Eq. (4.23) is correct, substituting Eq. (4.24) into Eq. (4.21), we have

$$\delta_{eq}(X) = \frac{\Delta p - \Delta\pi_0 - \Delta p_{cp}}{\left(\frac{2}{3}\right)^{1/3} r_{cl}\left(D^{2/3}\gamma^{1/3}\right)\left(\frac{C_g}{c_0}\right)^{1/3}}$$

$$\times\left\{\left[\frac{r_{cl}}{(1-a)\left(\Delta p - \Delta\pi_0 - \Delta p_{cp}\right)}t\right]^{3/2}\frac{2}{3}D^2\gamma\left(\frac{C_0}{c_g}\right)^{1/2}\right\}^{1/3}$$

$$-\left(\frac{1}{1-a}\right)^{1/2}\left(\frac{\Delta p - \Delta\pi_0 - \Delta p_{cp}}{r_{cl}}\frac{C_0}{c_g}t\right)^{1/2}$$

When cake layer resistance is predominant, the thickness of the cake layer in the non-equilibrium region is described by Eq. (4.7) with the membrane-resistance-related terms eliminated, i.e.,

$$\delta(t)=2^{1/2}\left(\frac{\Delta p-\Delta\pi_0-\Delta p_{cp}}{r_{cl}}\frac{C_0}{c_g}t\right)^{1/2}$$

Substituting the two expressions for $\delta_{eq}(X)$ and $\delta(t)$ into Eq. (4.23), we have

$$\left[\frac{a}{(1-a)^{1/2}}-2^{1/2}\right]\left(\frac{\Delta p-\Delta\pi_0-\Delta p_{cp}}{r_{cl}}\frac{C_0}{c_g}t\right)^{1/2}=0$$

Because the term in the second brackets is nonzero except $t = 0$, the following algebra equation must hold true

$$\frac{a}{(1-a)^{1/2}}-2^{1/2}=0$$

This equation can be rewritten into a quadratic equation,

$$a^2+2a-2=0$$

The only positive root of this quadratic equation is

$$a=\sqrt{3}-1(\approx 0.732) \tag{4.25}$$

Eq. (4.25) verifies that the parameter a is truly a constant that does not depend on t and X. Therefore, Eq. (4.24) is valid as the solution of the governing equation. At the same time, the value of parameter a is also determined.

Substituting Eq. (4.25) into Eq. (4.24), we have

$$\begin{aligned}X&=\frac{2}{3}D^2\gamma\left(\frac{C_0}{c_g}\right)^{1/2}\left[\frac{r_{cl}}{(2-\sqrt{3})(\Delta p-\Delta\pi_0-\Delta p_{cp})}t\right]^{3/2}\\&=4.81D^2\gamma\left(\frac{C_0}{c_g}\right)^{1/2}\left[\frac{r_{cl}}{(\Delta p-\Delta\pi_0-\Delta p_{cp})}t\right]^{3/2}\end{aligned} \tag{4.26}$$

Eq. (4.26) shows that the division line of the regions progresses in the membrane channel with a speed in the order of $t^{3/2}$.

As a demonstration of Eq. (4.26), the progress of the division line in a membrane channel is calculated for particles of different radius. The results are presented in Figure 4.7 with other parameters given in the caption. It can be

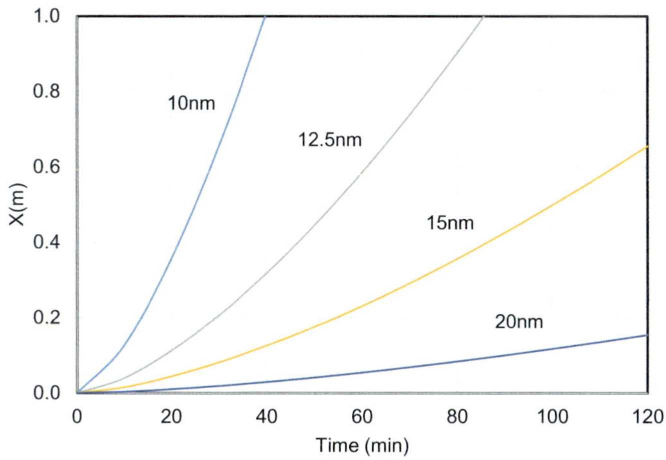

FIGURE 4.7
The division line of regions moves downstream with time. The parameters used in the calculation: $c_0 = 0.05\%$ (v/v), $c_g = 64\%$ (v/v), $\gamma = 100/\text{s}$, $\Delta p = 3$ bar, $\mu = 10^{-3}$ Pa·s, and $T = 293$K. The particle radii are indicated on the figure in units of nanometers.

seen that the speed of the division line is strongly affected by particle size. It takes about 40 minutes for the division line to move 1 m for the particles with a radius of 10 nm. But it moves only about 0.15 m in 120 minutes when the particle size is doubled. Calculations are also done for larger particles with otherwise the same parameters. The progress of the division line is too slow to be shown in the figure. In such crossflow filtrations, a steady state may never be reached in a reasonable operation time, and permeate fluxes would be declining all the time.

4.4.4 Average Permeate Flux

The flux in a membrane channel at steady state is calculated with Eq. (4.19) or Eq. (4.20). Before steady state is reached, the average permeate flux of a membrane channel can be determined by a linear combination of the fluxes in the equilibrium and non-equilibrium regions. Permeate flux in the equilibrium region is the limiting flux and time invariant that can be calculated with the same Eq. (4.19) or Eq. (4.20). The permeate flux in the non-equilibrium region is uniform but time dependent, which can be calculated with Eq. (4.8).

Location of the division line is first calculated with Eq. (4.26). Permeate flux in the equilibrium region is a variable of location. Permeate flow rate is calculated by integration of flux in the region, i.e.,

$$Q_{eq} = \int_0^X v_{eq}(x)\,dx$$

Permeate flux in the non-equilibrium region is constant. Permeate flow rate is therefore the product of the length of the non-equilibrium region and the time variant flux in it, i.e.,

$$Q_{noneq} = (L - X)v(t)$$

Then the average flux of the crossflow membrane channel is calculated by

$$\bar{v} = \frac{1}{L}\left(Q_{eq} + Q_{noneq}\right) = \frac{1}{L}\left[\int_0^X v(x)dx + (L - X)v(t)\right] \tag{4.27}$$

Substituting the expressions for $v(x)$ and $v(t)$ into Eq. (4.27), the average permeate flux in the entire membrane channel at any time can be calculated. Please note that the location of division X is a function of time t. It has to be determined with Eq. (4.26) before the average flux can be calculated with Eq. (4.27).

Time-dependent permeate fluxes in a crossflow membrane channel are calculated for different particle sizes. The results are presented in Figure 4.8 with particle sizes indicated in the figure and other parameters given in the caption. It can be seen that the time for permeate flux to reach steady state is strongly dependent on particle size. Steady state permeate fluxes for particle sizes greater than those shown in Figure 4.8 would take a very long time to

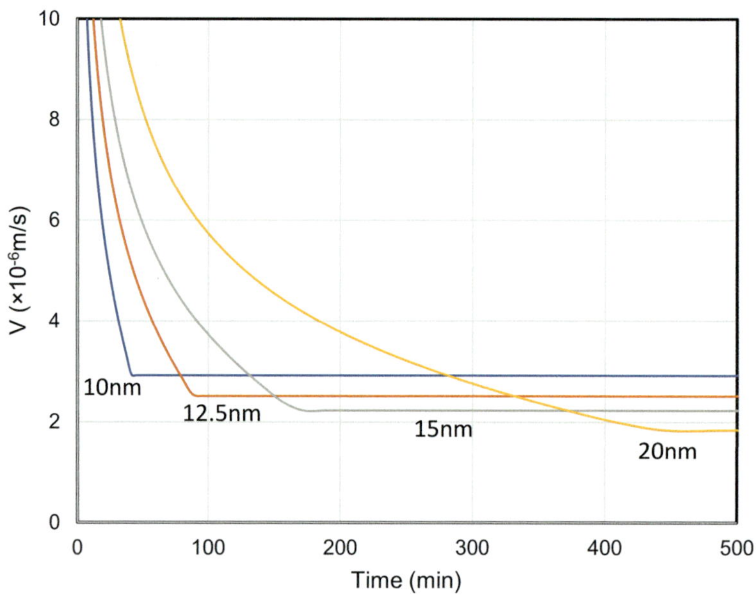

FIGURE 4.8
Average permeate velocities for various particle radii as functions of time. Parameters used in calculation: $c_0 = 0.05\%$ (v/v), $c_g = 64\%$ (v/v), $\gamma = 100$/s, $\Delta P = 3$ bar, L = 1 m, $R_m = 10^{10}$ Pa·s/m, $\mu = 10^{-3}$ Pa·s, and $T = 293$K. Particle radii are indicated on the figure.

reach. Therefore, a steady state may not be seen in some filtration processes because the operations may not last long enough.

4.4.5 Time to Reach Steady State

By replacing variable X in Eq. (4.26) with channel length L, the time to steady state can be derived as

$$t_{ss} = 0.351 \frac{\left(\Delta p - \Delta \pi_0 - \Delta p_{cp}\right)}{r_{cl}} \left(\frac{L}{D^2 \gamma}\right)^{2/3} \left(\frac{C_g}{c_0}\right)^{1/3} \tag{4.28}$$

where t_{ss} is the time to reach steady state.

The dependence of the time to reach steady state on particle size can be analyzed from Eq. (4.28). Both the cake specific resistance r_{cl} and the diffusion coefficient D in the equation are functions of particle size. According to either the Carman–Kozeny equation Eq. (2.22) or the Happel cell model–based equation Eq. (2.25), the specific resistance of the cake layer is inversely proportional to the square of particle radius. On the other hand, the diffusion coefficient is inversely proportional to the particle radius. The overall dependence of the time to reach steady state on particle radius is then

$$t_{ss} \propto \frac{1}{r_{cl}} \left(\frac{1}{D^2}\right)^{2/3} \propto a_p^{\,2} \left(a_p^{\,2}\right)^{2/3} = a_p^{\,10/3}$$

Therefore, the ratio of the times to steady state for two different particle sizes under the otherwise identical conditions is

$$\frac{t_{ss}\left(a_{p2}\right)}{t_{ss}\left(a_{p1}\right)} = \left(\frac{a_{p2}}{a_{p1}}\right)^{10/3} \tag{4.29}$$

The ratios of times to steady state are calculated with Eq. (4.29) for different ratios of particle size and the results are presented in Table 4.1. The time to reach steady state increases to 10 times as the particle size doubles. When the particle size increases by one order of magnitude, the time to the steady state increases by 2154 times!

The time to reach steady state under normal operation conditions for rigid particles with a radius of 10 nm takes about 40 hours to reach steady state (see Figure 4.8). From the time ratio in Table 4.1, the time to reach steady state for particles with a radius of 100 nm (0.1 μm) would take more than 80,000 hours. This time is way too long to be reached in common filtration experiments due to the physical restrictions of the membrane devices. Usually, a steady state cannot be reached technically for particles bigger than 0.1 μm in the practical operation periods.

TABLE 4.1

Ratios of Time to Reach Steady State at Different Particle Ratios

a_{p2}/a_{p1}	$t_{ss}(a_{p2})/t_{ss}(a_{p1})$[a]
1	1
2	10
3	39
4	102
5	214
6	392
7	656
8	1024
9	1516
10	2154

[a] The reported numbers are approximated to the nearest integer.

Example 4.5

Calculate the times for a crossflow membrane channel to reach steady state with suspensions of particle sizes of 10, 30, 50, and 70 nm, respectively. The other parameters are $c_0 = 0.1\%$ (v/v), $c_g = 64\%$ (v/v), $\gamma = 100/\text{s}$, $\Delta p = 2.5$ bar, L = 1 m, $R_m = 10^{10}$ Pa·s/m, and $T = 293$K.

SOLUTION

From Table 2.1, As* = 24.19 and As = 123.22.

The number concentration, osmotic pressure, the maximum pressure difference across the concentration layer, specific cake resistance, and diffusion coefficient are calculated first for a particle size of 10 nm.

Number concentration

$$c_0 = \frac{0.0001}{\frac{4}{3}\times 3.14\times\left(10^{-8}\,m\right)^3} = 2.39\times 10^{20}/m^3$$

Osmotic pressure

$$\Delta\pi_0 = kTc_0 = 1.38\,Pa\cdot m^3/K\times 293\,K\times 2.39\times 10^{20}/m^3 = 9.66\times 10^{-1}\,Pa$$

The pressure difference across the concentration polarization layer

$$\Delta p_c = kT\left(\frac{c_g}{\frac{4}{3}\pi_c a_p^3}\right)A_s^* = 1.38\,Pa\cdot m^3/K\times 293\,K\left(\frac{0.64}{\frac{4}{3}\times 3.14\times\left(10^{-8}\,m\right)^3}\right)\times 24.19$$

$$= 1.5\times 10^4\,Pa$$

Specific resistance of cake layer

$$r_{cl} = \frac{9}{2}\frac{\mu c_g}{a_p^2} A_s = \frac{9}{2} \times \frac{10^{-3}\, Pa \cdot s \times 0.64}{\left(10^{-8}\, m\right)^2} \times 123.22 = 3.55 \times 10^{15}\, Pa \cdot s/m^2$$

Diffusion coefficient

$$D = \frac{kT}{6\pi_c \mu a_p} = \frac{1.38\, Pa \cdot m^3/K \times 293\, K}{6 \times 3.14 \times 10^{-3}\, Pa \cdot s \times 10^{-8}\, m} = 2.15 \times 10^{-11}\, m^2/s$$

With the above parameters, the time to steady state is calculated

$$t_{ss} = 0.351 \frac{\left(\Delta p - \Delta\pi_0 - \Delta p_{cp}\right)}{r_{cl}} \left(\frac{L}{D^2 \gamma}\right)^{2/3} \left(\frac{c_g}{c_0}\right)^{1/3}$$

$$= 0.351 \frac{\left(5 \times 10^5\, Pa - 9.66 \times 10^{-1}\, Pa - 1.5 \times 10^4\, Pa\right)}{3.55 \times 10^{15}\, Pa \cdot s/m^2}$$

$$\left[\frac{1m}{\left(2.15 \times 10^{-11}\, m^2/s\right)^2 \times 100/s}\right]^{2/3} \left(\frac{0.64}{0.001}\right)^{1/3}$$

$$= 982s = 16.4\ min = 0.27\ hour$$

The results for all particle sizes are listed in the table below.

a_p (nm)		10	20	30	40
c_0 ($\times 10^{18}/m^3$)		239	29.9	8.85	3.73
$\Delta\pi_0$ (Pa)		0.97	0.12	0.036	0.015
Δp_{cp} (Pa)		15,000	1870	554	234
r_{cl} ($\times 10^{14}$ Pa·s/m³)		35.5	8.87	3.94	2.22
D ($\times 10^{-12}$ m²/s)		21.5	10.7	7.15	5.37
t_{ss}	**(seconds)**	982	10,451	40,591	106,036
	(minutes)	16.4	174.2	676.5	1767.3
	(hours)	0.27	2.90	11.28	29.45

The results show that, under the conditions simulated, the time to reach steady state increases rapidly with particle sizes.

Flux paradox refers to the phenomenon where the observed fluxes in ultrafiltration of rigid colloids are orders of magnitude higher than the values theoretically predicted at steady state [26–29]. Considering the time required for a membrane channel to reach steady state, the much higher than the predicted steady state flux can be a premature value before steady state. It is not a surprise that fluxes in microfiltration and ultrafiltration can decline by orders of magnitude before steady state is reached.

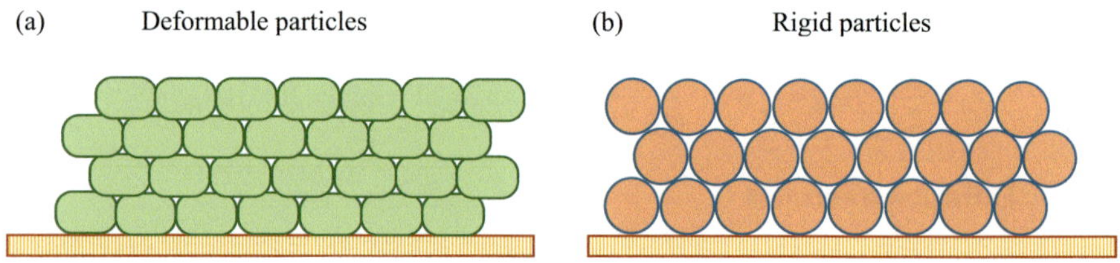

FIGURE 4.9
Schematics for (a) cake layer with deformable organic macromolecules and (b) cake layer with rigid inorganic colloids.

Flux paradox does not occur in the ultrafiltration of proteins or other biological products because these biological particles are not rigid. As shown in Figure 4.9, proteins and other deformable organic particles can provide much higher specific resistance than the rigid nondeformable inorganic colloids. Therefore, steady state can be reached in these processes with much less retained particles than required for the rigid colloidal particles [30–37]. It is more likely that most fluxes at steady state are measured with the deformable particles.

4.4.6 Nature of Cake Formation

There is no cake formation when a membrane process is operated under a pressure that is smaller than the critical pressure for cake formation. Under this condition, the rate of particles brought to the membrane surface can be balanced by back diffusion. However, when pressure is greater than the critical pressure, a cake layer forms on the membrane surface because the convective particle flux to the membrane surface is greater than the maximum diffusive flux away from the membrane surface. Therefore, cake formation is a natural consequence of the unbalanced operation of membrane processes. Under normal operating pressures, cake formation is inevitable in microfiltration and ultrafiltration of macromolecules and colloids with sizes greater than a few nanometers in radius (referring to critical particle sizes in Table 2.3).

In dead-end membrane processes, there is no mechanism to stop the growth of the cake layer when the pressure is greater than the critical pressure. The cake layer will grow with time until the permeate flux diminishes ultimately, as indicated by Eq. (4.8) or Eq. (4.9).

In the crossflow membrane process, tangential flow can take the retained particle out of the membrane channel. Carrying capacity of tangential flow to carry out of retained particles is well described by the principal mass balance relationship of crossflow filtration (Eq. 3.11). The higher the tangential flow velocity and the more extended the concentration polarization layer, the larger the carrying capacity to take retained particles away. The local permeate flux at any location in the membrane channel is virtually determined by this carrying capacity at that location.

When the pressure is smaller than the critical pressure for cake formation, the rate of particles brought to the membrane surface is certainly lower than the maximum carrying capacity of tangential flow. All retained particles can be carried away by the tangential flow. The membrane process in this case is said to work at equilibrium. When the pressure in a membrane process is greater than the critical pressure, the rate of particles brought to the membrane surface can exceed the carrying capacity of tangential flow. This membrane process is not at equilibrium and cannot maintain its operational state. The excessive amount of retained particles above the carrying capacity would deposit on the membrane surface to form a cake layer.

Cake layer creates an additional resistance to permeate flow. Permeate flux decreases as the cake layer grows in thickness. When the permeate flux reduces to the value at which the rate of particles retained by the membrane equals the carrying capacity of tangential flow, an equilibrium is established at that point, and the cake layer stops growing. When equilibrium is established everywhere in the entire membrane channel, the membrane process reaches a steady state. In this sense, cake formation is a natural consequence of non-equilibrium (overburdened) operation of a membrane process, i.e., the initial permeate flux brings too many particles to the membrane surface. Cake formation is a self-corrective behavior of the membrane process to adapt it from unsustainable operation (non-equilibrium) to sustainable operation (equilibrium).

As the original driving force for permeate in a membrane process, the applied pressure plays an interesting role in the permeate flux in a crossflow filtration process when a limiting flux exists. At the start, the initial permeate velocity v_0 is determined by the applied pressure, as shown in Figure 4.10. The flux decreases with time to the liming flux, which is pressure independent. The invariant limiting flux can be experimentally verified with different

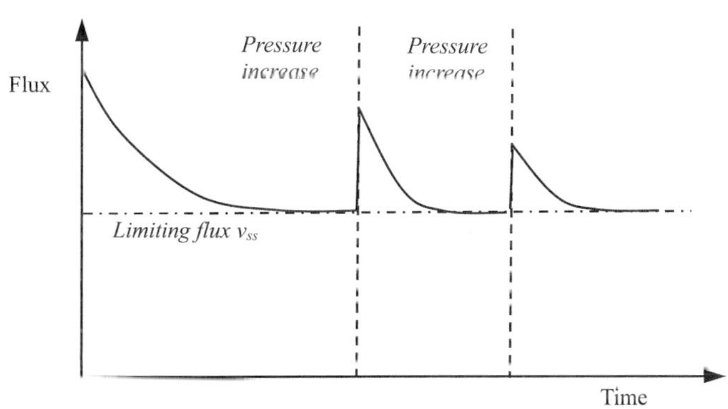

FIGURE 4.10
Flux can only be raised from the limiting flux temporarily by an increase in pressure. Flux tends to go back to the original limiting flux when the steady state is reestablished.

applied pressures. It may take a different time to reach the limiting flux but at the same value v_{ss}.

It does not mean that the membrane process is insensitive to applied pressure when limiting flux is reached. The flux can be raised by an increase in applied pressure. However, the rise in permeate flux is temporary. As shown in Figure 4.10, the same increases of pressure at t_1 and t_2 temporarily boost the flux to v_1 and v_2, respectively. The flux will ultimately reduce to the limiting flux v_{ss} when steady state is restabilized, even though the pressures are kept at the increased value.

The fundamental reason for such interesting behavior is the nature of cake formation. When applied pressure is raised in an initially balanced membrane process, the accumulation rate of retained particles exceeds the carrying capacity of tangential flow. The operation is not at an equilibrium. The thickness of the cake layer grows until the accumulation rate of retained particles is reduced back to equal the carrying capacity. The operation becomes equilibrium again.

By the way, cake formation is sometimes named as "membrane fouling" or is treated as one type of membrane fouling in the literature. That is destructive for the scientific study of this interesting subject. Generally speaking, membrane fouling is a vaguely defined term that generally implies something corrupt and abnormal occurs to the membrane system. It is generally considered to be caused by outside intruders into the system or by abnormal behaviors of the target species to be separated. In many cases, reasons for membrane fouling may not be clearly defined or specified. On the contrary, the cake layer is formed by the species to be separated. Cake formation is a normal behavior of retained particles at the given operation conditions. The distinction of cake formation from the vaguely defined term membrane fouling is essential for the rigorous treatment of this rather important phenomenon that affects membrane processes profoundly.

4.4.7 A Tertiary Solution

A binary solution was used in the discussions in this part up to now. There is only one type of identical particles in the binary solution that are retained by the membrane. The retained particles can exist in the concentration polarization layer or coexist in both the concentration polarization layer and the cake layer, depending on the applied pressure. When the size of particles is greater than a few nanometers, the pressure difference on the concentration polarization layer becomes insignificant compared to the applied pressure. In that case, the performance of the membrane processes is predominantly controlled by the retained particles in the cake layer. The binary solution is an important concept in the construction of a theoretical framework on concentration polarization and cake formation, and to study their influence on permeate flux in membrane processes. If a multi-component solution were involved, with some components staying in the concentration polarization layer only and some existing in both the concentration polarization layer and the cake layer, it would be a mess that would be hardly delineated with authority and rigor.

However, a little relaxation from the strict binary solution may be useful sometimes. A model tertiary solution can be a helpful concept for certain studies of membrane processes. In the model tertiary solution, a liquid contains two types of distinct particles that are retained completely by the membrane. An important feature of this tertiary solution is that the two types of particles can behave completely differently in the membrane system. One type of particle is so small in size that it cannot form the cake layer under the operating conditions. On the contrary, the other type of particles is sufficiently large so that they always form a cake layer under the same operation conditions. In addition, the amount and impact of large particles in the concentration polarization layer are negligible. Therefore, the impact of the large particles on permeate flux can be treated exclusively through the cake layer.

Now, in a crossflow membrane system with the model tertiary solution, concentration polarization layer and cake layer are formed by two distinct particles exclusively. The membrane system seems affected independently by two processes with different time scales. When the smaller particles are dominant in quantity, a strong concentration polarization layer can develop immediately and reduce the permeate flux to a certain extent. Because the retained particles in the concentration polarization layer cannot accumulate with time, the impact of concentration polarization does not change with time. On the contrary, the larger particles are relatively small in quantity. It may initially be insignificant compared to the concentration polarization layer. However, the large particles form a cake layer, and the amount of retained particles increases with time. The impact of the cake layer on the membrane system grows with time as more retained particles are accumulated.

Feed water to reverse osmosis desalination systems can be treated as a tertiary solution. Water passes through the membrane and salts (mainly sodium chloride), and particulate matters are retained. Under the practical range of applied pressure, salts stay in the concentration polarization layer, but particulate matters form a cake layer. The salt usually contributes more than 99% of the impurities in feed water and is the primary target to be removed by the reverse osmosis membrane. The particulate matters are not the intended target of the reverse osmosis membrane. They are intended to be removed from the feed water in the pretreatment stage before the reverse osmosis membrane. In this sense, particulate matters are the intruders to the ideal reverse osmosis desalination system. Cake formation by particulate matters is a deterioration from the normal operation of reverse osmosis desalination processes. Therefore, it is appropriate to use membrane fouling to refer to cake formation in reverse osmosis desalination processes.

4.5 Problems

1. What is the maximum deduction from the applied pressure by the concentration polarization layer formed with particles of radius 10 nm?
2. Is there possibly a cake layer in an ultrafiltration process with a suspension of particles of radius 0.1 μm at a pressure of 3 bars?
3. How many components can the applied pressure be divided into a membrane process with the existence of a cake layer? Which components are invariant with the applied pressure?
4. The permeate flux in an ultrafiltration process is 1.2×10^{-4} m/s, and the particle concentration is 0.05% in volume fraction. If the flux can be treated as a constant in an initial period, calculate the amount of retained particles in 10 minutes.
5. Determine the thickness of the cake layer in Question 4 when the cake concentration is 0.54.
6. A suspension that contains volume fraction 0.01% particles of radius 10 nm flows to the membrane at a velocity of 10^{-5} m/s in a dead-end membrane device. Calculate the wall concentration of particles as a function of time, assuming an invariant permeate velocity. What time will it take for the wall concentration to reach 0.6, the maximum packing density?
7. In the dead-end membrane process described in Question 6, calculate the thickness of the cake layer after 5 minutes of operation.
8. A suspension that contains particles of radius 15 nm forms a cake layer of density 0.55. (1) What is the maximum pressure drop of the applied pressure caused by the concentration polarization layer? (2) What is the specific resistance of the cake layer?
9. When the applied pressure is 2 bars, calculate the permeate flux of the membrane process in Question 8 as a function of time up to 10 minutes (minimum five time points required). The membrane resistance is 1.1×10^{-9} Pa·s/m and feed concentration is 0.1%.
10. Can the feed osmotic pressure and the pressure drop across the concentration polarization layer be neglected in the calculation of permeate flux in Question 9?
11. Is it correct that the feed osmotic pressure and pressure drop across the concentration polarization layer are insignificant in a membrane process for particle radius of 0.1 μm and above?
12. Give the condition for steady state in a crossflow filtration process.

13. The parameters of a crossflow membrane process are given in the following table.

Parameter	Value	Unit
Particle radius	1.5×10^{-8}	m
Feed concentration	1.45×10^{19}	#/m^3
Cake concentration	3.54×10^{22}	#/m^3
Temperature	20	°C
Membrane resistance	1.1×10^{10}	Pa·s/m
Shear rate	200	1/s
Pressure	5	bar
Channel length	2	m

a. What is the Leveque factor of this membrane process?
b. What is the crossflow filtration number of the membrane channel?
c. What is the permeate flux at the entrance of the membrane channel?
d. Calculate the permeate flux along the channel with at least five points. Plot the results.
e. Determine the average flux of this membrane channel.

14. The parameters of a crossflow membrane process that reaches the limiting flux are given in the following table.

Parameter	Value	Unit
Particle radius	1.2×10^{-9}	m
Feed concentration	0.001	v/v fraction
Cake concentration	0.60	v/v fraction
Temperature	20	°C
Shear rate	100	1/s
Channel length	1.5	m

a. What is the Leveque factor of this membrane process?
b. According to the equation for limiting flux, what is the permeate flux at the entrance of the membrane channel?
c. Calculate the permeate flux along the channel with at least five points. Plot the results.
d. Calculate the cake thickness at the same points as in (c) and plot the results.
e. Determine the average flux (limiting flux) of this membrane channel.

15. Describe the two regions of the cake layer in a crossflow membrane channel. Answer the following questions:
    a. What is the state of cake layer in the equilibrium region and the non-equilibrium region, respectively.
    b. What mechanism determines the thickness of the cake later in the equilibrium region and the non-equilibrium region, respectively?
    c. What happens with the retained particles collected in the equilibrium region? How about the retained particles in the non-equilibrium region?
    d. When can we say a crossflow membrane process has reached the steady state from the viewpoint of the state of the cake layer?
16. A crossflow membrane process is operated with a pressure of 5 bars and a shear rate of 100/s. The particles in the suspension have a radius of 0.01 μm. The feed concentration and cake concentration are 0.001 and 0.60, respectively. Calculate the front location of the equilibrium region at 10, 20, 40, 60, 80, and 100 minutes.
17. Determine the time needed to reach steady state in a crossflow membrane process of 2.5 m long with the same parameters as in Question 16.
18. It is observed that the time to reach steady state for a suspension of particles of 0.01 μm is 10 minutes. What would be the time to reach steady state for a suspension of particles of 0.05 μm if all the other parameters remained the same?
19. Why does the cake layer form in a membrane process? What is the ultimate consequence of cake formation in a crossflow membrane process?
20. Can the immature cake layer (before the equilibrium thickness is reached) be a possible reason for the flux paradox observed in colloidal ultrafiltration?

References

1. Field, R.W., Wu, D., Howell, J.A., & Gupta, B.B. (1995). Critical flux concept for microfiltration fouling. Journal of Membrane Science, 100(3), 259–272.
2. Zou, S., Wang, Y.-N., Wicaksana, F., Aung, T., Wong, P.C.Y., Fane, A.G., & Tang, C.Y. (2013). Direct microscopic observation of forward osmosis membrane fouling by microalgae: Critical flux and the role of operational conditions. Journal of Membrane Science, 436, 174–185.
3. Pollice, A., Brookes, A., Jefferson, B., & Judd, S. (2005). Sub-critical flux fouling in membrane bioreactors—A review of recent literature. Desalination, 174(3), 221–230.
4. Wang, Z., Li, J., Liu, H., Ma, J., & Li, Y. (2008). Membrane fouling in a submerged membrane bioreactor (MBR) under sub-critical flux operation: Membrane foulant and gel layer characterization. Journal of Membrane Science, 325, 238–244.

5. Field, R.W., Wu, J.J., & Fane, A.G. (2025). Revisiting the distinct concepts of critical flux and threshold flux. Journal of Membrane Science, 741, 124961.
6. Zhang, M., & Song, L. (2000). Pressure-dependent permeate flux in ultra- and microfiltration. Journal of Environmental Engineering, 126, 667–674.
7. Zhang, M., & Song, L. (2000). Mechanisms and parameters affecting flux decline in cross-flow microfiltration and ultrafiltration of colloids. Environmental Science & Technology, 34(17), 3767–3773.
8. Lee, Y., & Clark, M.M. (1998). Modeling of flux decline during crossflow ultrafiltration of colloidal suspensions. Journal of Membrane Science, 149(2), 181–202.
9. Kirschner, A.Y., Cheng, Y.-H., Paul, D.R., Field, R.W., & Freeman, B.D. (2019). Fouling mechanisms in constant flux crossflow ultrafiltration. Journal of Membrane Science, 574, 65–75.
10. Foley, G. (2006). A review of factors affecting filter cake properties in dead-end microfiltration of microbial suspensions. Journal of Membrane Science, 274, 38–46.
11. Han, Q., Trinh, T.A., & Chew, J.W. (2019). Cake formation of bidisperse suspensions in dead end microfiltration. Journal of Membrane Science, 577, 31–40.
12. Lorenzen, S., Keiding, K., & Christensen, M.L. (2017). The effect of particle surface charge density on filter cake properties during dead-end filtration. Chemical Engineering Science, 163, 155–166.
13. Akhondi, E., Zamani, F., Tng, K.H., Leslie, G., Krantz, W.B., Fane, A.G., & Chew, J.W. (2017). The performance and fouling control of submerged hollow fiber (HF) systems: A review. Applied Sciences, 7.
14. Zamani, F., Ma, A., Han, Q., Ma, Q., Zhang, H., Fane, A.G., & Chew, J.W. (2018). An energy-efficient method for mitigating membrane fouling: A novel embodiment of the inverse fluidized bed. Separation Science and Technology, 53, 683–695.
15. Abid, H.S., Johnson, D.J., Hashaikeh, R., & Hilal, N. (2017). A review of efforts to reduce membrane fouling by control of feed spacer characteristics. Desalination, 420, 384–402.
16. Lohaus, J., Perez, Y.M., & Wessling, M. (2018). What are the microscopic events of colloidal membrane fouling? Journal of Membrane Science, 553, 90–98.
17. Zamani, F., Tanudjaja, H.J., Akhondi, E., Krantz, W.B., Fane, A.G., & Chew, J.W. (2017). Flowfield mitigation of membrane fouling (FMMF) via manipulation of the convective flow in crossflow membrane applications. Journal of Membrane Science, 526, 377–386.
18. Song, L., & Elimelech, M. (1995). Theory of concentration polarization in crossflow filtration. Journal of the Chemical Society, Faraday Transactions, 91(19), 3389–3398.
19. Hegarty, A.F., Miller, J.J.H., O'Riordan, E., & Shishkin, G.I. (1995). Numerical results for advection-dominated heat transfer in a moving fluid with a non-slip boundary condition. International Journal of Numerical Methods for Heat & Fluid Flow, 5(2), 131–140.
20. Zhang, K. (1995). On coupling between the Poincaré equation and the heat equation: Non-slip boundary condition. Journal of Fluid Mechanics, 284, 239–256.
21. Song, L. (1998). A new model for the calculation of the limiting flux in ultrafiltration. Journal of Membrane Science, 144, 173–185.
22. Shen, J.S., & Probstein, R.F. (1977). On the prediction of limiting flux in laminar ultrafiltration of macromolecular solution. Industrial & Engineering Chemistry Fundamentals, 16, 459–465.

23. Trettin, D.R., & Doshi, M.R. (1980). Limiting flux in ultrafiltration of macromolecular solution. Chemical Engineering Communications, 4, 507–522.
24. Denisov, A. (1994). Theory of concentration in crossflow ultrafiltration: Gel-polarization model and osmotic-pressure model. Journal of Membrane Science, 91, 173–187.
25. Song, L. (1998). Flux decline in crossflow microfiltration and ultrafiltration: Mechanisms and modeling of membrane fouling. Journal of Membrane Science, 139(2), 183–200.
26. Cohen, R., & Probstein, R. (1986). Colloidal fouling of reverse osmosis membranes. Journal of Colloid and Interface Science, 114, 194–207.
27. Bacchin, P., Aimar, P., & Sanchez, V. (1995). Model for colloidal fouling of membranes. AIChE Journal, 41, 368–376.
28. Bacchin, P., Aimar, P., & Field, R. (2006). Critical and sustainable fluxes: Theory, experiments and applications. Journal of Membrane Science, 281, 42–69.
29. Davis, R.H. (1992). Modeling of fouling of crossflow microfiltration membranes. Separation and Purification Technology, 21, 75–126.
30. Belfort, G., Davis, R.H., & Zydney, A.L. (1994). The behavior of suspensions and macromolecular solutions in crossflow microfiltration. Journal of Membrane Science, 96, 1–58.
31. Sun, C., Zhang, N., Li, F., Ke, G., Song, L., Liu, X., & Liang, S. (2018). Quantitative analysis of membrane fouling mechanisms involved in microfiltration of humic acid–protein mixtures at different solution conditions. Water, 10(10), 1306.
32. Zhang, Y., Zhang, M.J., Wang, F.Y., Hong, H.C., Wang, A.J., Wang, J., Weng, X.X., & Lin, H.J. (2014). Membrane fouling in a submerged membrane bioreactor: Effect of pH and its implications. Bioresource Technology, 152, 7–14.
33. Cai, H.H., Fan, H., Zhao, L.H., Hong, H.C., Shen, L.G., He, Y.M., Lin, H.J., & Chen, J.R. (2016). Effects of surface charge on interfacial interactions related to membrane fouling in a submerged membrane bioreactor based on thermodynamic analysis. Journal of Colloid and Interface Science, 465, 33–41.
34. Stewart, M.B., Myat, D.T., Kuiper, M., Manning, R.J., Gray, S.R., & Orbell, J.D. (2017). A structural basis for the amphiphilic character of alginates – Implications for membrane fouling. Carbohydrate Polymers, 164, 162–169.
35. Wang, J., Zamani, F., Cahyadi, A., Toh, J.Y., Yang, S., Wu, B., Liu, Y., Fane, A.G., & Chew, J.W. (2016). Correlating the hydrodynamics of fluidized granular activated carbon (GAC) with membrane-fouling mitigation. Journal of Membrane Science, 510, 38–49.
36. Tanudjaja, H.J., & Chew, J.W. (2019). In-situ characterization of cake layer fouling during crossflow microfiltration of oil-in-water emulsion. Separation and Purification Technology, 218, 51–58.
37. Mcdonogh, R.M., Welsch, K., Fane, A.G., & Fell, C.J.D. (1992). Incorporation of the cake pressure profiles in the calculation of the effect of particle charge on the permeability of filter cakes obtained in the filtration of colloids and particulates. Journal of Membrane Science, 72, 197–204.

Part II

Reverse Osmosis Desalination

This part of the book deals with phenomena and issues that are essential to the applications of reverse osmosis technology in water desalination. Mathematical models are developed with emphasis on real-world applications, though some necessary idealizations are used for simplification and elegance. Most of the contents in this part are expected to be relevant to the design and operation of practical reverse osmosis desalination processes.

As one of the most widespread applications of membrane processes, desalination with reverse osmosis membranes is currently the preferred means to produce drinking water of adequately high water quality economically from seawater and brackish water. Reverse osmosis technology is more energy efficient that reduces two-thirds of the energy required by conventional thermal distillation technology. Reverse osmosis membranes are an excellent barrier to all the ions and molecules. The seawater reverse osmosis membrane nowadays can have a rejection rate greater than 99.5% for sodium chloride. It is possible to desalinate seawater to satisfy the drinking water standard by a single pass through the membranes. Other advantages of reverse osmosis desalination technology include small space requirement, simple process configuration, and easy operation.

However, there are challenging issues that require attentions in reverse osmosis desalination processes. One of the major challenges is to be able to describe or model pertinently the reverse osmosis desalination process. Currently, a common way to predict the overall performance of a reverse osmosis process (in terms of average permeate flux or system recovery) is to use a large number of models or equations with fitting coefficients. Such a method is ineffective to study the impact of various parameters on the

DOI: 10.1201/9781003743569-6

membrane process because the links between a particular parameter and the final overall performance indicator become blurred or undefined due to the involvement of too many equations and fitting coefficients. Other challenges include the assessment of concentration polarization on the performance, characterization of fouling strength of feed water, and prediction of fouling development in the reverse osmosis desalination processes.

A concise mechanistic model is first constructed for the performance of a practical reverse osmosis desalination process. Then a generic cost model is introduced for convenient economic analysis of water desalination. Both models form the foundation for process modeling and optimization of reverse osmosis processes. Furthermore, a new method is introduced for better characterization of the fouling potential of feed water to address the most challenging issue in the field. In summary, the following topics will be covered in this part of the book:

Chapter 5. Performance of Reverse Osmosis Desalination Processes

Chapter 6. Process Optimization

Chapter 7. Concentration Polarization Assessment

Chapter 8. Membrane Fouling

5

Performance of Reverse Osmosis Desalination Processes

One of the most widespread applications of membrane processes is reverse osmosis water desalination for drinking water production [1–3]. Desalination has been playing a vital role in water supply for a long time in the Middle East and other water-stressed regions all over the world. The ever-increasing demand for an adequate potable water supply eventually necessitates the technologies that can extract fresh water from saline water, which exists in an abundant amount on the Earth. Therefore, desalination will inevitably be the answer to the worldwide water crisis [4–6]. Reverse osmosis is currently a key technology to secure water supply and will remain that way in the foreseeable future.

Reverse osmosis desalination is a separation process that is affected by a variety of factors, including feed water properties, membrane characteristics, and operational and controlling strategies [7–9]. Significant efforts are made to predict the performance of reverse osmosis processes under various conditions. Most of the efforts are focused on the understanding of fundamental principles on the performance of reverse osmosis processes, such as mass transfer mechanisms and hydrodynamics [10–16]. Reverse osmosis processes are also studied by applications of advanced simulation technologies [17–23].

Over a period of more than half a century, particular features and configurations have developed and evolved in the reverse osmosis processes to better suit practical applications of water desalination. Consequently, practical reverse osmosis desalination processes are quite different from the model membrane processes described in the previous part of this book. For example, a practical reverse osmosis channel usually has a small height and large length so that a high permeate recovery can be obtained. As a result, the hydraulic conditions, e.g., tangential velocity, cannot be maintained constant throughout the entire channel length. In addition, feed water in practical reverse osmosis processes may have a complex chemical composition, which includes ions as the major target species rejected by the reverse osmosis membrane but also other impurities. The other impurities may become foulants of the membrane to cause performance deterioration of the membrane processes over time. Therefore, practical reverse osmosis desalination processes cannot be directly described with mathematical models developed for the model membrane system with an ideal binary solution.

DOI: 10.1201/9781003743569-7

A performance model that is particularly applicable to practical reverse osmosis desalination processes is presented in this chapter. While practical features of reverse osmosis desalination processes are considered, efforts are made to construct the model from the first principles. For this purpose, an idealized long narrow membrane channel is first introduced to mimic the practical reverse osmosis channels used in water desalination. Though some simplifications and idealizations are used, the important features of practical reverse osmosis processes, such as variable salt concentration, permeate velocity, and tangential velocity, are preserved in the ideal long narrow channel. In this way, concise mathematical equations for the practical reverse osmosis desalination systems can be developed.

5.1 Features of Reverse Osmosis Processes in Water Desalination

In the model membrane systems treated in the previous part, properties and flow conditions of the bulk solution remain constant along the whole channel length. The only change in a model membrane process is the particle concentration in a thin boundary layer over the membrane surface, whose thickness is much smaller than the channel height. However, these features do not exist in the practical reverse osmosis desalination applications. Long and narrow channels are commonly used in the practical reverse osmosis systems to achieve high recoveries of permeate from the feed water [24–27]. In such long narrow channels, (1) salt concentration and tangential flow velocity change significantly along the channel length, and (2) the concentration polarization layer can develop up to the whole height of the channel before the exit of the membrane channel. These features must be considered in the development of performance models for accurate simulation of practical reverse osmosis desalination processes.

5.1.1 A Long Narrow Membrane Channel

In the following discussions, a practical reverse osmosis channel is represented by a long narrow membrane channel with a very large length-to-height aspect ratio L/H, as indicated in Figure 5.1. As shown in Figure 5.1, channel height H here is actually half of the distance between the membranes as two walls of a channel. The aspect ratio can be easily greater than 1000 for reverse osmosis channels of spiral wound modules and/or hollow fiber modules. The aspect ratio for the latter is the ratio of fiber length to fiber inner radius.

The aspect ratio may not be the best definition of the long narrow channel because membrane permeability is not considered. When the membrane

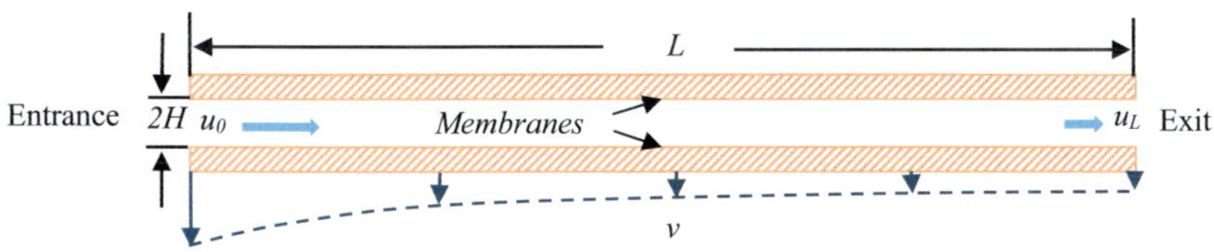

FIGURE 5.1
Schematic of a long narrow membrane channel for water desalination.

permeability is very small, hydraulic conditions such as tangential flow velocity may not change significantly in the channel. A long narrow channel is more appropriately defined by prescribing its recovery to be greater than a specified value.

The recovery in a long narrow channel is calculated with

$$R = \frac{(u_0 - u_L)H}{L} \tag{5.1}$$

where R is the recovery of the membrane channel, u_0 and u_L are the tangential flow velocities at the entrance and exit of the membrane channel, H is the height of the membrane channel, and L is the length of the membrane channel. These parameters are indicated in Figure 5.1. Definitely, recovery is an appropriate indicator for the performance of a membrane channel. Alternatively, performance of a membrane channel can be represented by the average permeate flux, which is related to recovery by

$$\overline{v} = \frac{1}{L}\int_0^L v\,dx = \frac{u_0 H}{L} R \tag{5.2}$$

where $\overline{v}$ is the average permeate flux of a membrane channel.

An important reason to introduce the concept of long narrow membrane channel is that the parameters in a typical practical reverse osmosis membrane channel are no longer constant as in a homogeneous system. Usually, a practical membrane channel has to be treated as a heterogeneous system when the recovery is greater than 10% or another value dependent on the required accuracy in the applications. The major parameters, such as salt concentration, tangential flow velocity, and permeate flux of a membrane channel under that recovery, cannot be regarded as constant anymore.

5.1.2 Equivalent Length of Membrane Channel

Pressure vessels holding eight membrane elements in series are commonly used in seawater desalination. The total length of the channel is roughly 8 m. Most spiral wound membrane elements have a channel height less than

1 mm. Therefore, the aspect ratio of membrane channels used in seawater desalination is greater than 8000. Seawater desalination membrane channels usually work at a recovery of 50% or higher. Tangential flow velocity reduces along the membrane channel to below a half of its initial value by the exit of the membrane channel, while salt concentration increases more than doubled at the same time. Variation of permeate flux along the membrane channel can be even more drastic than tangential flow velocity and salt concentration.

In brackish water desalination, pressure vessels containing six membrane elements in series are commonly used with Christmas tree structure. Figure 5.2 schematically shows such a structure of three stages in a concentrate stream to increase permeate recovery. The channel length in the configuration of the Christmas tree structure cannot be calculated by simply adding up channel lengths in three stages. It can be seen that one pressure vessel in the second stage is shared by two pressure vessels in the first stage, and one pressure vessel in the third stage is shared by four vessels in the first stage. The equivalent length of a membrane channel in the Christmas tree configuration should be a weighted summation of channel lengths in three stages, i.e.,

$$L = L_1 + \frac{1}{2}L_2 + \frac{1}{4}L_3$$

where L is the equivalent length of the membrane channel with a Christmas tree structure, and L_1, L_2, and L_3 are the channel lengths in the first stage, second stage, and third stage, respectively.

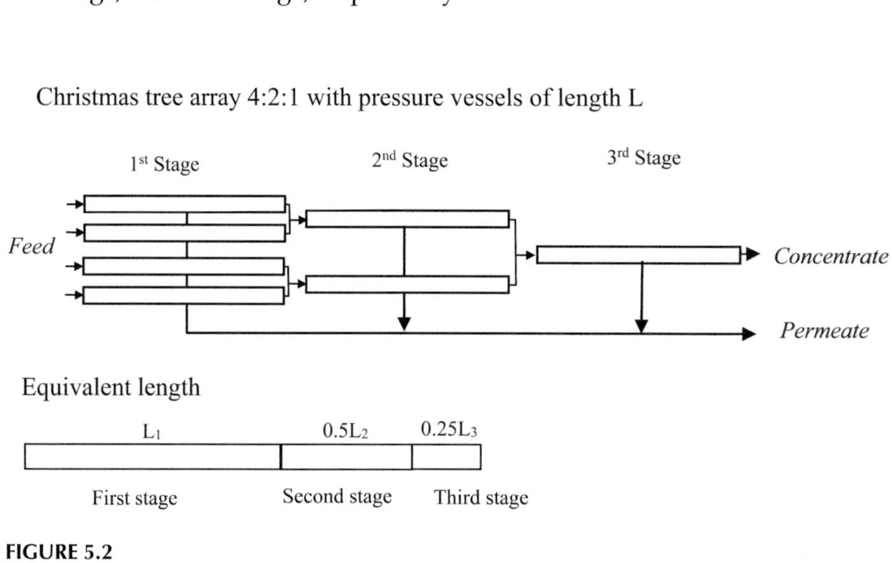

FIGURE 5.2
Equivalent length of membrane channels for water desalination with a Christmas tree structure.

TABLE 5.1

Resistances of Some Brackish Water Reverse Osmosis Membrane Elements[a]

Elements	Area (m^2)	Permeate Flowrate (m^3/day)	Resistance (Pa·s/m)
DOW FILMTEC™ BW30-440	37	40	1.10×10^{11}
DOW FILMTEC™ BW30 PRO-365	34	38	1.07×10^{11}
Hydranautics CPA2	33.9	37.9	1.10×10^{11}
Hydranautics CPA6 MAX	40.9	33.3	1.51×10^{11}
Toray TM720D-440	41	45.8	1.07×10^{11}
Toray TM700D	37	41.6	1.06×10^{11}

[a] Test conditions: Salt concentration of 2000 mg/L NaCl for DOW and Toray membranes but 1500 mg/L for Hydranautics membranes at a pressure of 15.5 bar (225 psi).

5.1.3 Reverse Osmosis Membrane Resistances

By the way, two most important properties of reverse osmosis membranes are their rejection rate to salt and resistance (or their reciprocal permeability) to water. Salt rejection is no longer a big concern currently because seawater membranes with a 99.7% rejection of NaCl are available nowadays. Reverse osmosis membranes with lower salt rejection are chosen sometimes for brackish water desalination because they are sufficient to achieve the treatment goal with lower cost.

Membrane resistance remains a serious concern in the design of a reverse osmosis process because it is highly relevant to the capacity and economy of desalination. For convenience, the resistances of a few brackish water and seawater reverse osmosis membranes are provided in Tables 5.1 and 5.2. (The membranes in the tables are randomly selected without any intention for recommendation or endorsement. Other brands of membranes have roughly the same resistances.) The membrane areas and permeate flowrates are extracted from the information sheets published by the manufacturers or providers.

TABLE 5.2

Resistances of Some Seawater Reverse Osmosis Membrane Elements[a]

Elements	Area (m^2)	Permeate Flowrate (m^3/day)	Resistance ((Pa·s)/m)
DOW FILMTEC™ SW30ULE-440i	41	45.4	2.18×10^{11}
DOW FILMTEC™ SW30HR-380	35	24.6	3.43×10^{11}
Hydranautics SWC5	37	34.1	2.62×10^{11}
Hydranautics SWC4-LD	37	24.6	3.62×10^{11}
Toray TM800V	37	34.1	2.62×10^{11}
Toray TM800M	37	26.5	3.37×10^{11}

[a] Test conditions: Salt concentration of 32,000 mg/L NaCl at a pressure of 55 bar (800 psi).

Referring to these two tables, the typical resistances are 1×10^{11} and 3×10^{11} Pa·s/m for brackish water and seawater membranes, respectively. These two values will be used in the calculations in this part of the book.

Example 5.1

A reverse osmosis brackish water desalination process uses a Christmas tree structure with pressure vessels of 6 m long each, which is just the nominal length that means to hold six membrane elements of standard length (again nominally 1 m long) in series. What are the equivalent lengths of the membrane channels in two-stage and three-stage configurations, respectively?

SOLUTION

In two stages, the equivalent length is

$$L = L_1 + \frac{1}{2}L_2 + \frac{1}{4}L_3 = 6\,m + \frac{1}{2}6\,m + \frac{1}{4}0\,m = 9\,m$$

In three stages

$$L = L_1 + \frac{1}{2}L_2 + \frac{1}{4}L_3 = 6\,m + \frac{1}{2}6\,m + \frac{1}{4}6\,m = 10.5\,m$$

They are 9 and 10.5 m in the two-stage and three-stage configurations, respectively.

5.2 Governing Equations

Because of the large aspect ratio and existence of spacers to promote mixing in the transverse direction, the membrane channel in practical reverse osmosis desalination processes can be described as a one-dimensional problem, i.e., variables or parameters only change along the channel length. In other words, the average value of a variable across the channel height will be used in the problem, which is defined mathematically by

$$\xi(x) = \frac{1}{H}\int_0^H \xi(x,y)\,dy$$

where ξ is a variable or parameter.

5.2.1 Variables in a Long Narrow Membrane Channel

Variables in a long narrow membrane channel can be rigorously described with a set of equations. Because of substantial changes of variables and parameters along the membrane channel, localized equations have to be used. The independent variable x is explicitly included in the dependent variables to emphasize this fact when equations for the variables are introduced.

The first equation is the localized basic filtration equation for permeate flux at any point in the membrane channel:

$$v(x) = \frac{\Delta p(x) - \Delta\pi(x)}{\mathrm{R}_m} \tag{5.3}$$

where $v(x)$ is the permeate flux, $\Delta p(x)$ is the pressure, $\Delta\pi(x)$ is the osmotic pressure, R_m is the membrane resistance, and x is the distance from the entrance of the membrane channel. Permeate flux, applied pressure, and osmotic pressure are functions of x along the membrane channel.

Pressure decreases along the membrane channel due to friction of tangential flow with membrane surfaces [14]:

$$\Delta p(x) = \Delta p_0 - \frac{3K\mu}{H^2}\int_0^x u(x')\,dx' \tag{5.4}$$

where Δp_0 is the applied pressure at the entrance of the membrane channel, and K is the friction coefficient of the channel. $K = 1$ for an empty channel and $K > 1$ when there are spacers in the channel, μ is the viscosity of water, and x' is a dummy integration variable.

The osmotic pressure can be directly related to the salt concentration with

$$\Delta\pi(x) = f_{os}c(x) \tag{5.5}$$

where f_{os} is the osmotic pressure coefficient, and c is the salt concentration. This equation is the localized form of Eq. (1.6).

The above three equations, Eqs. (5.3)–(5.5), contain five variables, which are permeate $v(x)$, pressure $\Delta p(x)$, osmotic pressure $\Delta\pi(x)$, tangential flow velocity $u(x)$, and salt concentration $c(x)$. Two more equations are needed to fully define the filtration problem with this membrane channel. These two equations can be obtained by applying the mass balance principle on salt and water in the membrane channel.

According to the mass balance on salt, salt concentration at any location of the channel can be determined by

$$c(x)u(x)H + (1 - r)\int_0^x c(x')v(x')\,dx' = c_0u_0H \tag{5.6}$$

where r is the rejection rate of salt by the membrane. The first term on the left-hand side of Eq. (5.6) is the amount of salt remaining in the membrane channel at location x. The second term on the left-hand side of Eq. (5.6) is the amount of salt in the permeate up to location x. The term on the right-hand side of Eq. (5.6) is the total amount of salt entering the membrane channel.

Finally, by applying the mass balance principle on water, the tangential flow velocity is determined by

$$u(x)H + \int_0^x v(x')\,dx' = u_0 H \tag{5.7}$$

The two terms on the left-hand side of Eq. (5.7) are the amounts of water remaining in the channel and the amount of permeate, respectively. The term on the right-hand side of the equation is the total amount of water entering the membrane channel.

Now the complete set of equations for variables in the membrane channel are assembled as Eqs. (5.3)–(5.7). When initial values of the five variables are given at the entrance of the membrane channel, this membrane filtration problem is completely defined. It means that values of the five variables can be determined at any point in the membrane channel.

5.2.2 Euler's Numerical Method

Analytical solutions of the set of equations are not available for the general cases. However, a numerical solution procedure of the equations can be constructed. For the sake of numerical solution, the integral equations, Eqs. (5.4)–(5.7), are first converted into differential equations as follows [9,25,28]:

$$\frac{d\Delta p(x)}{dx} = -\frac{3K\mu}{H^2}u(x)$$

$$\frac{dc(x)}{dx}u(x) = r\frac{c(x)v(x)}{H}$$

$$\frac{du(x)}{dx} = -\frac{v(x)}{H}$$

To construct a numerical solution, the channel length is first discretized into n segments with equal step size, as shown in Figure 5.3. The grid points are numbered with 0 at the entrance and n at the exit of the membrane channel.

The distance of grid point i is indicated by x_i with $x_0 = 0$ at the entrance and $x_n = L$ at the exit of the membrane channel, and

$$x_i = x_{i-1} + \Delta x,\ i = 1, 2, \ldots, n$$

where Δx (= L/n) is the step size.

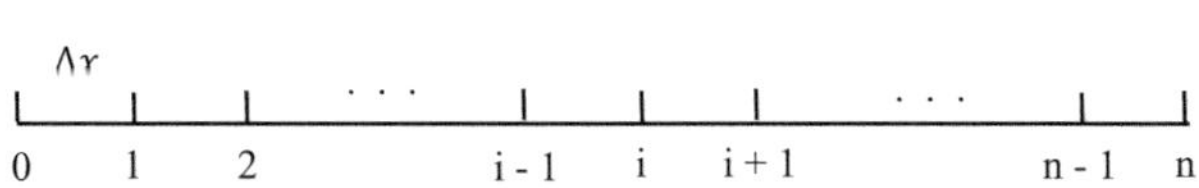

FIGURE 5.3
Discretization of a membrane channel into n segments of equal size.

According to Euler's method [29], the difference forms of the differential equations are

$$\Delta p_{i+1} = \Delta p_i - \frac{3K\mu}{H^2} u_i \Delta x$$

$$c_{i+1} = c_i + r\frac{c_i v_i}{Hu_i}\Delta x$$

$$u_{i+1} = u_i - \frac{v_i}{H}\Delta x$$

There are no difference equations for osmotic pressure and permeate flux. They are determined when the other variables are known on the grid point $i + 1$, i.e.,

$$\Delta\pi_{i+1} = f_{os} c_{i+1}$$

$$v_{i+1} = \frac{\Delta p_{i+1} - \Delta\pi_{i+1}}{R_m}$$

The above equations in discretized form are the numerical model of the membrane channel.

The pressure, salt concentration, and tangential velocity at the entrance of a channel should be specified in any practical reverse osmosis desalination processes. With the knowledge of membrane resistance and osmotic pressure coefficient, the osmotic pressure and permeate flux at the entrance can be determined with Eqs. (5.3) and (5.5). Then values of all five variables on grid point $i = 1, 2, \ldots, n$ along the membrane channel can be calculated with the numerical model progressively.

With the numerical solution, the recovery of the membrane channel is calculated by

$$R = 1 - \frac{u_n}{u_0} \tag{5.8}$$

Average permeate flux can then be calculated from the recovery with Eq. (5.2).

The numerical solution procedure of the governing equations of a long narrow membrane channel is demonstrated with the example below.

Example 5.2

A spacer-free reverse osmosis membrane channel is used to desalinate feed water of salt concentration 2500 mg/L at an applied pressure of 8 bar and temperature of 20°C. The feed velocity is 0.1 m/s. The osmotic pressure coefficient of the feed water is 77 Pa/(mg/L). Other parameters of the membrane channel are equivalent length 9 m, channel height 0.35 mm, salt rejection 95%, and membrane resistance 10^{11} Pa s/m. (1) Construct a numerical solution of this membrane channel and show salt concentration, tangential flow velocity, and permeate flux along the channel graphically; and (2) determine the recovery and average permeate flux from the numerical solution.

SOLUTION

The friction coefficient for a spacer-free channel is $K = 1$. The viscosity of water at 20°C is 10^{-3} Pa·s.

The membrane channel is discretized into 100 equal segments. The step size is

$$\Delta x = \frac{9\,m}{100} = 0.09\,m$$

The initial values of variables at the entrance $i = 0$ are

$$\begin{aligned}
\Delta p_0 &= 8\,bar = 8\times10^5\,Pa \\
c_0 &= 2500\,mg/L \\
u_0 &= 0.1\,m/s \\
\Delta\pi_0 &= f_{os}c_0 = 77\,Pa/(mg/L)\times 2500\,mg/L = 1.93\times10^5\,Pa \\
v_0 &= \frac{\Delta p_0 - \Delta\pi_0}{R_m} = \frac{8\times10^5\,Pa - 1.93\times10^5\,Pa}{10^{11}\,Pa\cdot s/m} = 6.08\times10^{-6}\,m/s
\end{aligned}$$

The values of the five variables at the grid point $i = 1$ can be determined with the numerical model:

$$\begin{aligned}
\Delta p_1 &= \Delta p_0 - \frac{3K\mu}{H^2}u_0\Delta x = 8\times10^5\,Pa - \frac{3\times1\times10^{-3}\,Pa\cdot s}{(0.35\times10^{-3}\,m)^2}\times0.1\,m/s\times0.09\,m \\
&= 7.998\times10^5\,Pa \\
c_1 &= c_0 + r\frac{k_0 v_0}{Hu_0}\Delta x = 2500c_0 + 0.95\frac{2500\,mg/L\times6.08\times10^{-6}\,m/s}{0.35\times10^{-3}\,m\times0.1\,m/s}\times0.09\,m \\
&= 2537\,mg/L \\
u_1 &= u_0 - \frac{v_0}{H}\Delta x = 01\,m/s - \frac{6.08\times10^{-6}\,m/s}{0.35\times10^{-3}\,m}\times009\,m = 0.0984\times10^5\,Pa \\
\Delta x_1 &= f_{os}c_1 = 77\,Pa/(mg/L)\times2537\,mg/L = 1.954\times10^5\,Pa \\
v_1 &= \frac{\Delta p_1 - \Delta\pi_1}{R_m} = \frac{7.998\times10^5\,Pa - 1.954\times10^5\,Pa}{10^{11}\,Pa\cdot s/m} = 6.05\times10^{-6}\,m/s
\end{aligned}$$

With the known values of variables at grid point $i = 1$, the variable values at $i = 2$ can be calculated with the same difference equations. This process is repeated until the variable values at the exit, i.e., $i = n$, are calculated.

The numerical solution is presented graphically below.

From the numerical solution, the tangential velocity at the exit $u_n = 0.022\,m/s$.

The recovery of the membrane channel is

$$R = \frac{u_0 - u_n}{u_0} = \frac{0.1\ m/s - 0.022\ m/s}{0.1\ m/s} = 78.0\%$$

The average permeate flux is

$$\overline{v} = \frac{R u_0 H}{L} = \frac{0.78 \times 0.1 m/s \times 0.35 \times 10^{-3}\ m}{9.0 m} = 0.303 \times 10^{-6}\ m/s$$

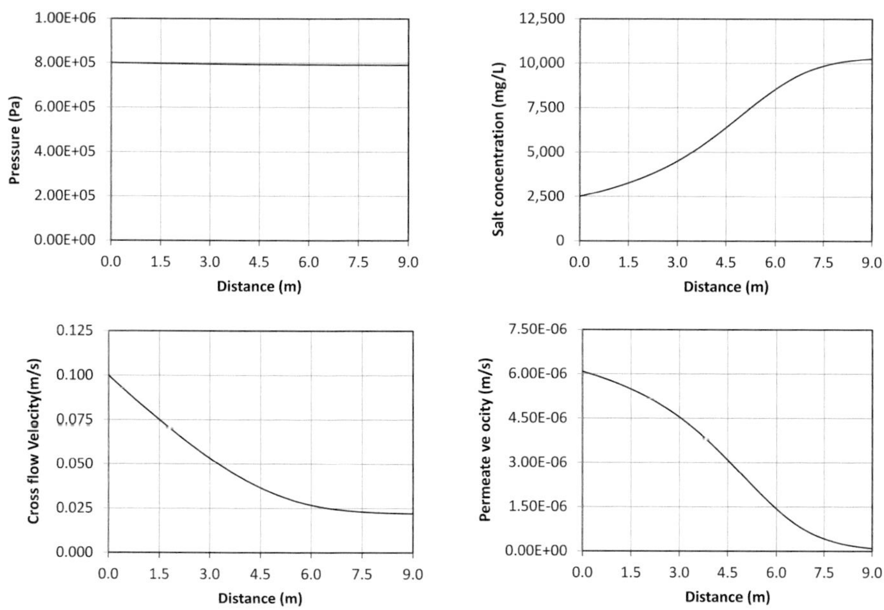

It can be seen from the figures that salt concentration, crossflow velocity, and permeate flux change substantially along the membrane channel. But the driving pressure only changes slightly. This simulation result is consistent with the observations on the practical reverse osmosis desalinations.

Example 5.3

Calculate recoveries of the membrane channel specified in Example 5.2 under the pressures of 3, 4, 5, 6, 8, 11, and 14 bar, respectively.

SOLUTION

Repeat the same numerical procedure used in Example 5.2 for all pressures required in the question. The results are

Δp (bar)	1.93	3	4	5	6	8	11	14
R	0	21.3%	39.1%	51.3%	65.8%	78.0%	85.0%	89.0%

The results are presented graphically in the figure below.

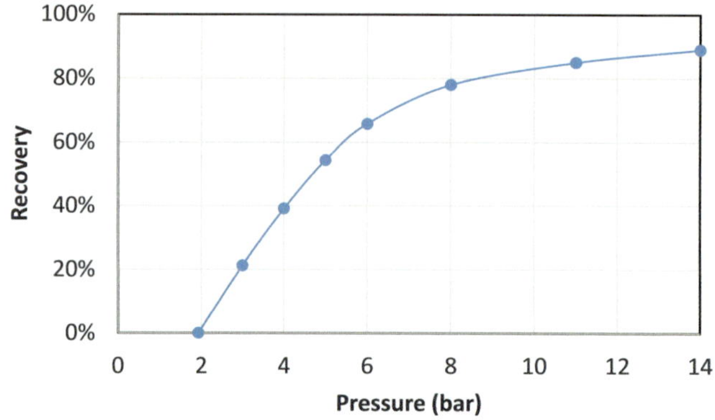

A serious concern about the numerical model is the accuracy of the method. Theoretically, the error of the Euler's numerical method in terms of $\left|u_{a\,i} - u_a(x_i)\right|$ is in the order of $o(\Delta x)$ with $u_a(x_i)$ being the true value at point x_i. That means any accuracy can be obtained as long as a sufficiently small step size Δx is used. More discussions about the accuracy of the Euler's method will be conducted later in Chapter 7 when the numerical solution is assessed by comparison with the true solution of the long narrow channel.

5.3 An Ideal Long Narrow Membrane Channel

An analytical model is always preferred for a system over the numerical one if it is possible. The analytical model can not only show the explicit dependence of a system on parameters but also needs less effort in calculation. These advantages are especially desirable when the controlling mechanisms of a system are investigated.

5.3.1 Governing Equation

In order to pursue an analytical model for the long narrow membrane channel, the concept of an ideal RO channel is first introduced for the sake of rigorousness in model development. A long narrow membrane channel is said to be "ideal" if it has the following features:

1. The driving pressure remains constant all along the membrane channel,
2. The membrane is a perfect barrier for salt, and
3. Salt is completely mixed transversely across the whole channel height.

It can be seen that all three features are closely approximated by the real-world reverse osmosis desalination processes. It is well documented that the decrease of driving pressure along membrane channels is small compared to the applied pressure. The reverse osmosis membrane can have salt rejection greater than 99%. It is also reasonable to assume that salt is well transversely mixed with the existence of spacers in the narrow membrane channel.

The governing equations can be significantly simplified for the ideal membrane. Eq. (5.4) is no longer needed because of constant pressure in the ideal membrane channel. Because of constant pressure and perfect salt rejection, Eqs. (5.3) and (5.6) now become

$$v(x) = \frac{\Delta p - \Delta\pi(x)}{R_m} \tag{5.9}$$

$$c(x)u(x) = c_0 u_0 \tag{5.10}$$

The osmotic pressure Eq. (5.5) and water mass balance Eq. (5.7) remain unchanged. Therefore, an ideal long narrow membrane is completely defined by Eqs. (5.5), (5.7), (5.9), and (5.10) because there are only four unknowns of $v(x)$, $\Delta\pi(x)$, $c(x)$, and $u(x)$ left in the system.

A common method to solve the problem analytically is first to reduce the number of equations and unknowns to one equation with one unknown. Salt concentration is chosen to remain as the final unknown, and the rest need to be eliminated. First, the osmotic pressure is eliminated in Eq. (5.9) with Eq. (5.5) as

$$v(x) = \frac{\Delta p - f_{os}c(x)}{R_m} \tag{5.11}$$

Second, rearranging Eq. (5.10) results in the expression for $u(x)$

$$u(x) = \frac{c_0 u_0}{c(x)} \tag{5.12}$$

Finally, substituting Eqs. (5.11) and (5.12) into Eq. (5.7), we have

$$\frac{c_0 u_0}{c(x)} H + \int_0^x \frac{\Delta p - f_{os}c(x')}{R_m} dx' = u_0 H$$

Rearranging the equation results in an integral equation for salt concentration $c(x)$ in the membrane channel

$$c(x)\left\{u_0 - \frac{1}{HR_m}\int_0^x \left[\Delta p - f_{os}c(\xi)\right]d\xi\right\} = c_0 u_0 \tag{5.13}$$

Taking the derivative of Eq. (5.13), the governing (differential) equation of the long narrow membrane channel is obtained:

$$\frac{dc}{dx} - \left(\frac{\Delta p}{HR_m c_0 u_0} - \frac{f_{os} c}{HR_m c_0 u_0} \right) c^2 = 0 \tag{5.14}$$

Salt concentration c is the only unknown function in Eq. (5.14). The argument or independent variable x is dropped off of function c at this point for convenience without causing unnecessary confusion.

The boundary condition for the differential equation is obviously the salt concentration at the entrance of the membrane channel, i.e.,

$$c = c_0 \quad at \quad x = 0 \tag{5.15}$$

The salt concentration in an ideal long narrow membrane channel is completely defined by Eqs. (5.14) and (5.15).

5.3.2 Solution Procedure

To facilitate the solution procedure for Eqs. (5.14) and (5.15), the dimensionless variables of concentration and distance are first introduced:

$$\begin{aligned} C &= \frac{c}{c_0} \\ X &= \frac{x}{L} \end{aligned} \tag{5.16}$$

The initial condition Eq. (5.15) in dimensionless form becomes

$$C = 1 \quad at \quad X = 0 \tag{5.17}$$

The dimensionless channel length becomes 1.

Also introduced are the dimensionless parameters of osmotic pressure and applied pressure:

$$\begin{aligned} \Delta\Pi_0 &= \frac{f_{os} c_0}{u_0 HR_m / L} = \frac{\Delta\pi_0}{u_0 HR_m / L} \\ \Delta P &= \frac{\Delta p}{u_0 HR_m / L} \end{aligned} \tag{5.18}$$

Substituting Eqs. (5.16) and (5.18) into Eq. (5.14) and rearranging result in

$$\frac{dC}{dX} - \left(\Delta P - \Delta\Pi_0 C \right) C^2 = 0 \tag{5.19}$$

Integrating Eq. (5.19) with boundary condition Eq. (5.17) with the variable separation method

$$\int_{1}^{C_e} \frac{1}{(\Delta P - \Delta\Pi_0 C)C^2} dC = \int_{0}^{1} dX \tag{5.20}$$

where C_e $(= c_e/c_0)$ is the dimensionless salt concentration at the exit of the membrane channel, i.e., in the concentrate stream from the membrane channel. The integration limits of the right-hand side of the equation are 0 and 1.

The function to be integrated on the left-hand side of Eq. (5.20) can be broken up into a few simple terms as

$$\frac{1}{(\Delta P - \Delta\Pi_0 C)C^2} = \frac{a_1}{C^2} + \frac{a_2}{C} + \frac{a_3}{\Delta P - \Delta\Pi_0 C}$$

where a_1, a_2, and a_3 are coefficients to be determined.

Multiplying $(\Delta P - \Delta\Pi_0 C)C^2$ to both sides of the equation and collecting terms of C in the same order, we have

$$1 = a_1 \Delta P + (a_2 \Delta P - a_1 \Delta\Pi_0)C + (a_3 - a_2 \Delta\Pi_0)C^2$$

By comparing terms of the same order of C on both sides of the equation, the coefficients a_1, a_2, and a_3 are determined as

$$a_1 = \frac{1}{\Delta P}$$

$$a_2 = \frac{\Delta\Pi_0}{\Delta P^2}$$

$$a_3 = \frac{\Delta\Pi_0^{\,2}}{\Delta P^2}$$

Now we have

$$\frac{1}{(\Delta P - \Delta\Pi_0 C)C^2} = \frac{1}{\Delta P C^2} + \frac{\Delta\Pi_0}{\Delta P^2 C} + \frac{\Delta\Pi_0^{\,2}}{\Delta P^2 (\Delta P - \Delta\Pi_0 C)} \tag{5.21}$$

Substituting Eq. (5.21) in Eq. (5.20) results in

$$\int_{1}^{C_e} \left[\frac{1}{\Delta P C^2} + \frac{\Delta\Pi_0}{\Delta P^2 C} + \frac{\Delta\Pi_0^{\,2}}{\Delta P^2 (\Delta P - \Delta\Pi C)} \right] dC = 1$$

Carrying out the integration results in

$$\frac{1}{\Delta P}\left(1 - \frac{1}{C_e}\right) + \frac{\Delta\Pi_0}{\Delta P^2} \ln\left[\frac{(\Delta P - \Delta\Pi_0)C_e}{\Delta P - \Delta\Pi_0 C_e}\right] = 1 \tag{5.22}$$

Mass balance on salt in a membrane channel with perfect salt rejection is

$$c_e = \frac{c_0}{1 - R}$$

where R is the recovery of this membrane, and c_e is the salt concentration at the exit of the membrane channel. Then we have

$$C_e = \frac{1}{1-R} \tag{5.23}$$

$$\left(1-\frac{1}{C_e}\right) = 1-\frac{c_0}{c_e} = R \tag{5.24}$$

Substituting Eqs. (5.23) and (5.24) into Eq. (5.22), we have

$$\frac{1}{\Delta P}R + \frac{\Delta\Pi_0}{\Delta P^2}\ln\left[\frac{\left(\Delta P-\Delta\Pi_0\right)\frac{1}{1-R}}{\Delta P-\Delta\Pi_0\frac{1}{1-R}}\right] = 1$$

Converting the logarithm function to an exponential function

$$\left(\Delta P-\Delta\Pi_0\right) = \left(\Delta P - R\Delta P - \Delta\Pi_0\right)exp\left[\frac{\Delta P^2}{\Delta\Pi_0}\left(1-\frac{R}{\Delta P}\right)\right]$$

Rearranging

$$R\Delta P = \left(\Delta P-\Delta\Pi_0\right)\left\{1-exp\left[-\frac{\Delta P^2}{\Delta\Pi_0}\left(1-\frac{R}{\Delta P}\right)\right]\right\}$$

Dividing ΔP on both sides of the equation results in

$$R = \left(1-\frac{\Delta\Pi_0}{\Delta P}\right)\left\{1-exp\left[-\frac{\Delta P}{\Delta\Pi_0}(\Delta P - R)\right]\right\} \tag{5.25}$$

Eq. (5.25) is the solution of the differential equation (Eq. 5.19) with initial condition Eq. (5.17).

5.3.3 State Equation

Substituting Eq. (5.18) into Eq. (5.25) for the dimensionless parameters $\Delta\Pi_0$ and ΔP, the solution for the original problem, Eqs. (5.14) and (5.15), is obtained as

$$R = \left(1-\frac{\Delta\pi_0}{\Delta p}\right)\left\{1-exp\left[-\frac{\Delta p}{\Delta\pi_0}\left(\frac{\Delta p}{u_0 HR_m / L} - R\right)\right]\right\}$$

A new parameter for the long narrow membrane channel is introduced as follows:

$$\Delta p^* = \frac{u_0 HR_m}{L} \tag{5.26}$$

The parameter Δp^* is a lumped parameter of the long narrow membrane channel that includes all parameters relevant to the membrane channel. In other words, the new parameter characterizes a long narrow membrane channel. Because the parameter has the unit of pressure, it is named the *characteristic pressure* of the long narrow membrane channel. The characteristic pressure plays an important role on the performance of a membrane channel, which will be discussed later.

With characteristic pressure Δp^*, the recovery of the long narrow membrane channel is finally obtained [24,30] as

$$R = \left(1 - \frac{\Delta \pi_0}{\Delta p}\right)\left\{1 - exp\left[-\frac{\Delta p}{\Delta \pi_0}\left(\frac{\Delta p}{\Delta p^*} - R\right)\right]\right\} \tag{5.27}$$

Eq. (5.27) is called the *state equation* of the ideal long narrow membrane channel. The state equation links four key variables or parameters of the membrane channel, which are the property of feed water $\Delta \pi_0$, the characteristics of the membrane channel Δp^*, the parameter of operation Δp, and the performance indicator R. Just as the state equation for an ideal gas, the state equation for an ideal long narrow channel reveals the most fundamental relationship of four key variables of the membrane system. The recovery of a long narrow membrane channel can be converted to average permeate flux with Eq. (5.2), which is another indicator for the performance of a membrane channel.

The state equation for the ideal long narrow channel defines the recovery R implicitly because recovery appears on both sides of Eq. (5.27). Further mathematical operations are needed to find R explicitly. It can be done with either the successive substitution method or the graphical method as described below.

5.3.4 Successive Substitution

The recovery can be found from the state equation (Eq. 5.27) by the common successive substitution method. To facilitate description of the successive substitution method, Eq. (5.27) is rewritten as

$$R = f(R)$$

where the symbol "*f*" represents the whole expression on the right side of Eq. (5.27), which is obviously a function of variable R. The successive substitution method consists of the following steps:

1. Select an arbitrary initial guess of $R^{(0)}$,
2. Substitute the current R value into the expression to find out the next round R value: $R^{(n+1)} = f\left(R^{(n)}\right)$, n = 0,1, 2, …,

TABLE 5.3

Parameters for Brackish Water Desalination

Parameter	Symbol	Value	Unit
Membrane resistance	R_m	1.0×10^{11}	Pa·s/m
Channel height	H	3.5×10^{-4}	m
Salt concentration	c_0	2500	mg/L
Osmotic pressure coefficient	f_{os}	77	Pa/(mg/L)
Feed velocity	u_0	0.1	m/s
Channel length	L	9	m

3. Check the result for convergence $\left|R^{(n+1)} - R^{(n)}\right| < \varepsilon$, where ε is a small value preset as the convergence criterion. Go back to step 2 if the criterium is not met. Otherwise, the solution is obtained and stop.

This method is illustrated with a case study of a long narrow RO channel in brackish water desalination as described below. In this case, recoveries of the membrane channel are calculated at pressures of 3, 4, 5, 6, 8, 11, and 14 bar, respectively. The other parameters used in this case study are listed in Table 5.3.

Calculations start with the osmotic pressure and the characteristic pressure:

$$\Delta\pi_0 = f_{os}c_0 = 77\ Pa/(mg/L) \times 2500\ mg/L = 1.93 \times 10^5\ Pa = 1.93\ bar$$

$$\Delta P^* = \frac{u_0 H R_m}{L} = \frac{0.1\ m/s \times 0.35 \times 10^{-3}\ m \times 10^{11}\ Pa/(m/s)}{9\ m}$$

$$= 3.89 \times 10^5\ Pa = 3.89\ bar$$

Then use the successive substitution method to calculate the recovery. For $\Delta p = 3\ bar$, choose randomly an initial guess $R^{(0)} = 0.3$, substituting it into Eq. (5.27), the first estimate of recovery is

$$R^{(1)} = \left(1 - \frac{1.93}{3}\right)\left[1 - e^{-\frac{3}{1.93}\left(\frac{3}{3.89} - 0.3\right)}\right] = 0.186$$

This value is used for successive substitution to estimate the R value in the next round. The converged $R = 0.209$ was obtained at the fifth round of substitution.

The same operations are carried out for all other pressures; the recoveries obtained in the first few rounds of substitutions are presented in Table 5.4.

The recovery of this membrane channel as a function of applied pressure is presented in Figure 5.4. The graph shows that recovery increases smoothly with increasing applied pressure.

The successive substitution method usually can obtain the converged recovery in a few rounds in most cases.

TABLE 5.4

Successive Substitution Results

$\Delta p(bar)$	3	4	5	6	8	11	14
R0	0.300	0.300	0.300	0.300	0.300	0.300	0.300
R1	0.186	0.405	0.567	0.665	0.759	0.825	0.862
R2	0.214	0.377	0.520	0.635	0.756	0.825	0.862
R3	0.208	0.385	0.531	0.639	0.756	0.825	0.862
R4	0.209	0.383	0.528	0.639	0.756	0.825	0.862
R5	0.209	0.383	0.529	0.639	0.756	0.825	0.862
R6	0.209	0.383	0.529	0.639	0.756	0.825	0.862

In some rare cases with a low salt concentration and a short membrane channel simultaneously, Eq. (5.27) with the successive substitution method may not generate the converged recoveries at low applied pressures. In this case, Eq. (5.28) below can be used with the successive substitution method to find the convergent recoveries at the pressures for which Eq. (5.27) does not work.

$$R=\frac{\Delta\pi_0}{\Delta p}ln\left[\left(1-R-\frac{\Delta\pi_0}{\Delta p}\right)\Big/\left(1-\frac{\Delta\pi_0}{\Delta p}\right)\right]+\frac{\Delta p}{\Delta p^*} \tag{5.28}$$

Eq. (5.28) is an alternative form of Eq. (5.27) obtained by solving Eq. (5.27) for the R on the right-hand side of the equation.

This rare situation is described with another case study for a channel length of 6 m and a feed salt concentration of 1000 mg/L. Other parameters

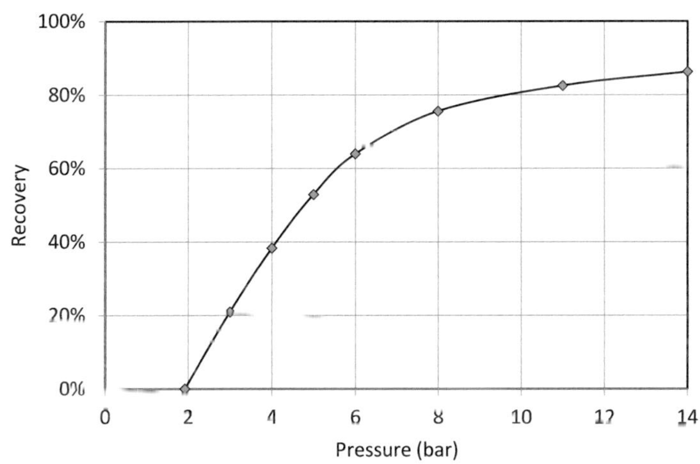

FIGURE 5.4
Recovery of a membrane channel at different applied pressures calculated with parameters given in Table 5.4.

TABLE 5.5

Nonconvergent Successive Substitution Results

$\Delta p(bar)$	3	4	5	6	8	11	14
R0	0.300	0.300	0.300	0.300	0.300	0.300	0.300
R1	0.421	0.699	0.823	0.869	0.904	0.930	0.945
R2	0.227	−0.056	0.167	0.621	0.897	0.930	0.945
R3	0.501	0.790	0.836	0.835	0.897	0.930	0.945
R4	0.038	−0.583	0.106	0.678	0.897	0.930	0.945
R5	0.627	0.806	0.840	0.815	0.897	0.930	0.945
R6	−0.410	−0.704	0.091	0.707	0.897	0.930	0.945
R7	0.723	0.807	0.840	0.801	0.897	0.930	0.945
R8	−0.933	−0.708	0.088	0.724	0.897	0.930	0.945
R9	0.741	0.807	0.840	0.790	0.897	0.930	0.945

are the same as in the previous case study, which are listed in Table 5.3. The recoveries for pressures of 3, 4, and 5 bars do not converge with Eq. (5.27) even for a large number of substitutions. Results of the first ten rounds of substitutions are presented in Table 5.5. By the way, although the recovery for pressure 6 bars has not converged in the first ten substitutions, it eventually converges at 0.762 in step 40.

The recoveries for the consecutive substitutions R8 and R9 are presented in Figure 5.5 as functions of pressure. The recoveries determined in either substitution do not increase monotonically with the applied pressure, which

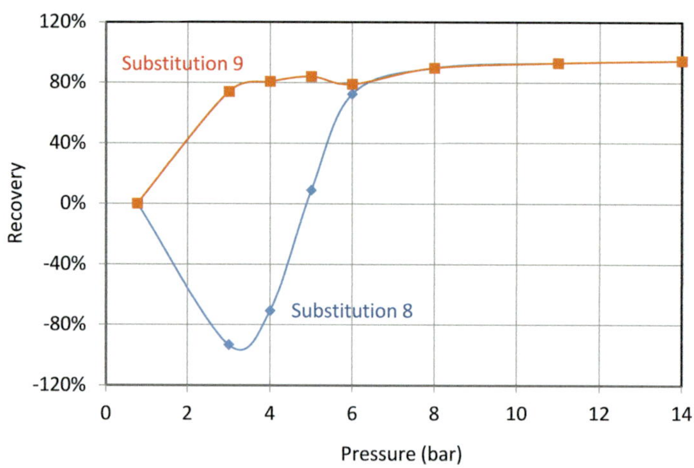

FIGURE 5.5
Unreasonable recoveries at different applied pressures determined in two consecutive substitutions with Eq. (5.27). Substitution numbers are shown in the figure.

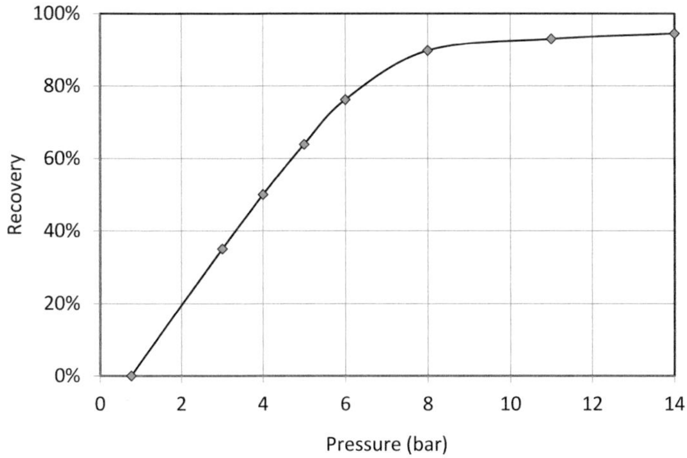

FIGURE 5.6
Reasonable recoveries in the whole pressure range can be obtained with Eqs. (5.27) and (5.28) together.

physically does not make much sense. Even negative recoveries are produced for pressures of 3 and 4 bars in substitution R8.

Eq. (5.28) is then used with the same successive substitution method to calculate recoveries at pressures of 3, 4, and 5 bars. Recoveries at the three pressures all converge within 26 substitutions. The convergent recoveries are 0.351, 0.500, and 0.640 for the pressures 3, 4, and 5 bars, respectively. Combined with the reasonable recoveries calculated with Eq. (5.27) for higher pressures, the recovery as a function of the applied pressure is shown in Figure 5.6. The recovery is seen to increase smoothly and monotonically with the applied pressure.

With numerous trials, it is safe to say that with both Eqs. (5.27) and (5.28), convergent recoveries can always be obtained with the successive substitution method for a membrane channel with any conditions.

5.3.5 Graphical Method

Another method to find the explicit R from Eq. (5.27) is the graphical method. In this method, Eq. (5.27) is first split into two functions of independent variable R:

$$y_1 = R$$
$$y_2 = \left(1 - \frac{\Delta\pi_0}{\Delta p}\right)\left[1 - e^{-\frac{\Delta p}{\Delta\pi_0}\left(\frac{\Delta p}{\Delta p^*} - R\right)}\right] \tag{5.29}$$

where R is an independent variable in the range of 0 to 1, and y_1 and y_2 are functions of R and only have physical meaning with values in the range of 0–1, too.

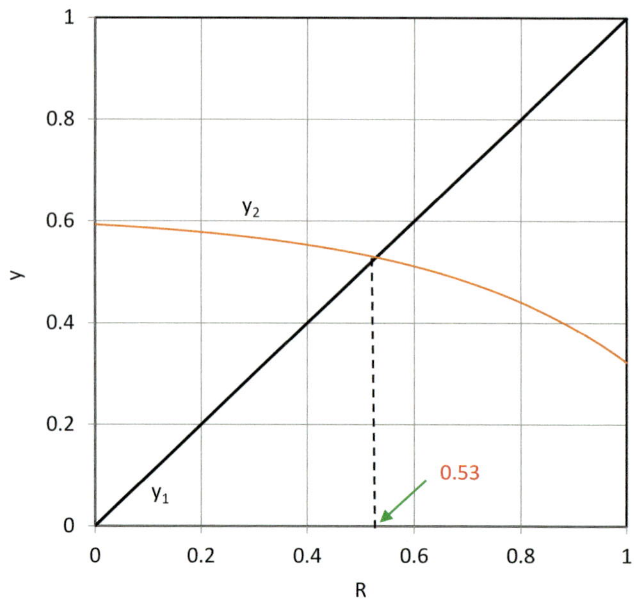

FIGURE 5.7
Determination of recovery with graphical method. The bold diagonal straight line is y_1 and the curve represents y_2. The intersection point is the recovery $R = 0.53$ of the problem.

Second, the two functions are calculated for sufficient R values in the range of 0–1. The two functions for a membrane channel are then plotted in an R–y plane, as shown in Figure 5.7. The recovery of the membrane channel is the intersection point of the two lines.

Function y_1 is the so-called identity function of R, and it is always represented as the diagonal line from point (0,0) to point (1,1) as in Figure 5.7. Function y_2 is a function of not only R but also parameters of a membrane channel, which include osmotic pressure of feed water $\Delta\pi_0$, characteristic pressure of membrane channel Δp^*, and applied pressure Δp. The line of y_2 in the graph represents a membrane channel because it is completely defined by parameters of the membrane channel. The curve for y_2 plotted in Figure 5.7 is for the membrane channel with the parameters given in Table 5.3. The applied pressure used in the development of this curve is 5 bars. This curve of y_2 intersects with the diagonal line of y_1 at $R = 0.53$. Therefore, the recovery of the membrane system is 0.53 or 53%.

The graphical method is robust and always produces a solution for any set of parameters. Another big advantage of the graphical method over the successive substitution method is that it can be used to find solutions for multiple cases in a single figure. The case study, previously used for demonstration of the successive substitution method, is resolved with the graphical method. The results are presented in Figure 5.8. The intersection points (the recoveries) are highlighted with dots in the figure.

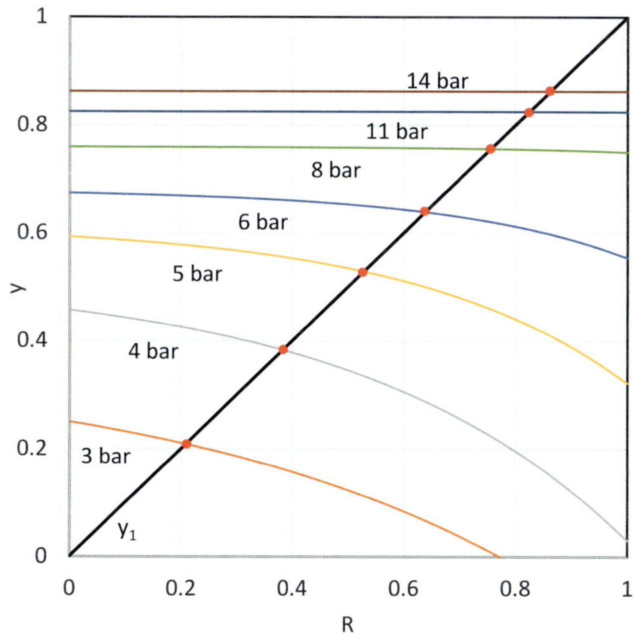

FIGURE 5.8
Graphical solution method for multiple pressures. The intersection point (dot) of each curve with the diagonal line y_1 is the recovery for that pressure.

The recoveries found in Figure 5.8 match well or are compatible with the values obtained with the successive substitution method. It can be verified by comparing the values from Figure 5.8 with the values presented in the last line in Table 5.4.

Example 5.4

The concentrate staging (Christmas tree structure) is studied for its impact on the recovery of a membrane system. The equivalent lengths are 6, 9, and 10.5 m for single stage, two stages, and three stages, respectively, when the ratio of pressure vessels in the Christmas tree structure is 4:2:1. The other parameters are $R_m = 10^{11}$ Pa·s/m, $u_0 = 0.1$ m/s, $H = 0.35 \times 10^{-3}$ m, $c_0 = 2500$ mg/L, and $\Delta p = 5$ bar. Find recoveries of the membrane system with single stage, two stages, and three stages with (1) successive substitution method and (2) graphical method

SOLUTION

Using the osmotic pressure coefficient $f_{os} = 77\ Pa/(mg/L)$, calculate the osmotic pressure of feed water:

$$\Delta\pi_0 = 77\ Pa/(mg/L) \times 2500\ mg/L = 1.93 \times 10^5\ Pa = 1.93\ bar$$

The characteristic pressures are

Single stage:

$$\Delta p_1^* = \frac{u_0 H R_m}{L_1} = \frac{0.1\, m/s \times 0.35 \times 10^{-3}\, m \times 10^{11}\, Pa \cdot s/m}{6\, m}$$

$$= 5.83 \times 10^5\, Pa = 5.83\, bar$$

Two stages:

$$\Delta p_2^* = \frac{u_0 H R_m}{L_2} = \frac{0.1\, m/s \times 0.35 \times 10^{-3}\, m \times 10^{11}\, Pa \cdot s/m}{9\, m}$$

$$= 3.89 \times 10^5\, Pa = 3.89\, bar$$

Three stages:

$$\Delta p_3^* = \frac{u_0 H R_m}{L_3} = \frac{0.1 m/s \times 0.35 \times 10^{-3}\, m \times 10^{11}\, Pa \cdot s/m}{10.5\, m}$$

$$= 3.04 \times 10^5\, Pa = 3.12\, bar$$

1. Using the successive substitution method. Recoveries converge for all three channel lengths within 13 substitutions. The results of substitutions are shown in the table below.

No	6 m	9 m	10.5 m
1	0.5000	0.5000	0.5000
2	0.5484	0.6528	0.6674
3	0.5271	0.6368	0.6593
4	0.5369	0.6389	0.6598
5	0.5325	0.6386	0.6597
6	0.5345	0.6386	0.6597
7	0.5336	0.6386	0.6597
8	0.5340	0.6386	0.6597
9	0.5338	0.6386	0.6597
10	0.5339	0.6386	0.6597
11	0.5338	0.6386	0.6597
12	0.5339	0.6386	0.6597
13	**0.5339**	**0.6386**	**0.6597**

The recoveries are found to be 0.534, 0.639, and 0.660 for single stage, two stages, and three stages, respectively, with the successive substitution method.

2. Using the graphical method.
Both the original graph (a) and the enlarged section (b) are shown graphically. The recoveries can be obtained visually from the graphs,

which are compatible with those found using the successive substitution methods.

Graph (a) on the left side is developed in the whole domain, while graph (b) is the enlarged area of the whole domain that includes all three intersection points. More accurate results can be read from graph (b). The results are 0.534, 0.639, and 0.660 for channel lengths of 6, 9, and 10.5 m, respectively.

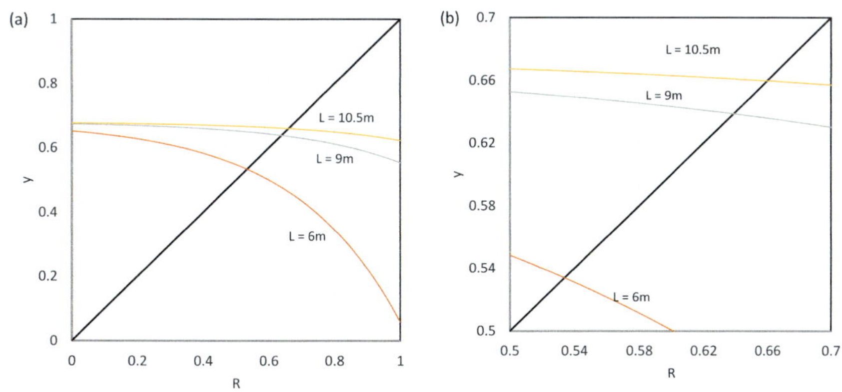

5.4 Two Controlling Mechanisms

A closer and more careful inspection of the recovery curve in Figure 5.4 clearly shows that it can be divided into two stages. Recovery grows roughly linearly with pressure in the first stage where pressure is low. But it levels off in the second stage where pressure is sufficiently high. It is obvious that the recovery, or equivalently average permeate flux, in a membrane channel is controlled by different mechanisms in these two pressure regions. The near linear growth of the recovery in low pressure stage can be easily explained with the dependence of flux on pressure. At first thought, concentration polarization may be the reason for the leveling off behavior of the recovery in high pressure stage. However, it cannot be true because concentration polarization is not considered at all in the derivation of the state equation. Complete mixing in the transverse direction is assumed in the ideal long narrow membrane channel. There must be something new that is responsible for this peculiar behavior of a long narrow channel in the stage with high pressure.

5.4.1 Mass Transfer Controlled Regime

Insights on the mechanisms that control recovery in different pressure ranges can be obtained from the state equation. Let us look at the absolute value of the exponent in the exponential function of Eq. (5.27). There are two extreme cases where it is either much smaller than unity or much greater than unity. The first extreme case is that the exponent is much smaller than unity, i.e.,

$$\frac{\Delta p}{\Delta \pi_0}\left(\frac{\Delta p}{\Delta p^*}-R\right) \ll 1 \tag{5.30}$$

Rearranging Eq. (5.30) results in

$$\frac{\Delta p}{\Delta p^*} \ll \frac{\Delta \pi_0}{\Delta p}+R$$

It can be demonstrated that the following relationship is always true:

$$\frac{\Delta \pi_0}{\Delta p}+R \le 1$$

Therefore, Eq. (5.30) is equivalent to

$$\Delta p \ll \Delta p^*$$

Because of Eq. (5.30), the exponential part of Eq. (5.27) can be expanded into Taylor series. Keeping the first two leading terms, it reduces to

$$exp\left[-\frac{\Delta p}{\Delta \pi_0}\left(\frac{\Delta p}{\Delta p^*}-R\right)\right]=1-\frac{\Delta p}{\Delta \pi_0}\left(\frac{\Delta p}{\Delta p^*}-R\right)$$

Substituting the Taylor expansion into Eq. (5.27) results in

$$R=\left(1-\frac{\Delta \pi_0}{\Delta p}\right)\frac{\Delta p}{\Delta \pi_0}\left(\frac{\Delta p}{\Delta p^*}-R\right)$$

Solving the equation for R leads to

$$R=\frac{\Delta p-\Delta \pi_0}{\Delta p^*} \tag{5.31}$$

Substituting Eq. (5.26) to Eq. (5.31) and converting recovery R to average permeate flux $\overline{v}$ with Eq. (5.2), we have

$$\overline{v}=\frac{\Delta p-\Delta \pi_0}{R_m}$$

It is the basic filtration equation (Eq. 1.11). Therefore, under the condition that the applied pressure is much smaller than the characteristic pressure $\Delta p \ll \Delta p^*$, the recovery of a membrane channel increases linearly with applied pressure. The membrane process under this condition is controlled by mass transfer and affected by the net driving pressure and membrane resistance.

5.4.2 Thermodynamic Restricted Regime

The second extreme case is just the opposite. The absolute value of the exponent in Eq. (5.27) is much greater than unity, i.e.,

$$\frac{\Delta p}{\Delta \pi_0}\left(\frac{\Delta p}{\Delta p^*}-R\right)\gg 1$$

With a similar deduction as for the previous case, the above condition is equivalent to

$$\Delta p \gg \Delta p^*$$

In this case, the exponential function in Eq. (5.27) vanishes, i.e.,

$$e^{-\frac{\Delta p}{\Delta \pi_0}\left(\frac{\Delta p}{\Delta p^*}-R\right)}=0$$

Then, Eq. (5.27) reduces to [26,27,31]

$$R=1-\frac{\Delta \pi_0}{\Delta p} \tag{5.32}$$

Eq. (5.32) shows that the recovery of a membrane channel is controlled only by the thermodynamic parameters, which are the salt concentration and temperature of feed water (through osmotic pressure) and the applied pressure. Under this condition, characteristics of the membrane channel do not affect the recovery at all. Therefore, the membrane channel is said to be controlled by "thermodynamic restriction." This is a new controlling mechanism for the recovery of a long narrow membrane channel that prevails when the applied pressure is much greater than the characteristic pressure of a membrane channel, $\Delta p \gg \Delta p^*$.

5.4.3 Characteristic Pressure

The significance of characteristic pressure to a long narrow membrane channel can be delineated with the help of Figure 5.9. Recovery as a function of applied pressure calculated with the state equation (Eq. 5.27) is plotted in Figure 5.9 together with those calculated with Eq. (5.31) (mass transfer controlled) and Eq. (5.32) (thermodynamically restricted) for the parameters listed in Table 5.3. The characteristic pressure of the membrane channel is also indicated on the figure by a vertical line. The characteristic pressure line just passes the intersection point of the curves for Eq. (5.31) and Eq. (5.32). It can be seen that the linear (mass transfer) equation is a better approximation to the accurate value (predicted by the state equation) in the pressure region smaller than the characteristic pressure. Therefore, the recovery in this pressure range is mainly controlled by the mass transfer mechanism.

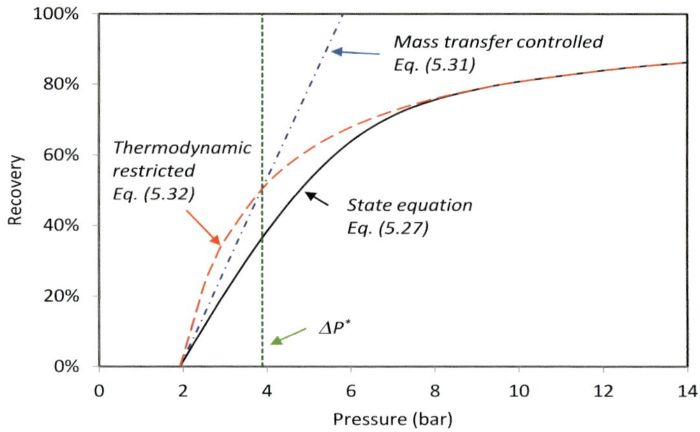

FIGURE 5.9
Recovery is primarily controlled by different mechanisms in different pressure regions. Mass transfer in low pressure regions and thermodynamic restriction in high pressure regions.

On the other hand, thermodynamic restriction is closer to the accurate recovery in the pressure range greater than the characteristic pressure. Recovery predicted with thermodynamic restriction merges to the accurate recovery when the ratio of $\Delta p/\Delta p^*$ is slightly greater than 2. The match of mass transfer recovery with the accurate recovery is not as good as that of thermodynamic restriction. Though the accurate recovery increases almost linearly in this low pressure range, the slope of the straight line is notably smaller than the one of mass transfer. Therefore, Eq. (5.31) cannot be used to predict recovery accurately for any pressure. The state equation should be used to calculate recovery for pressure smaller than 2 Δp^*.

5.4.4 More Discussions About Thermodynamic Restriction

It is not new that the flux or recovery of a membrane process is not affected by the certain characteristics of the membrane process. As discussed in the previous part, limiting flux in crossflow microfiltration and ultrafiltration processes is determined by the carrying capacity of tangential flow that is not affected by pressure and membrane permeability. However, the limiting flux is affected by the length of the membrane channel. Furthermore, more permeate would always be produced by increasing channel length.

The cases under thermodynamic restriction are definitely different from cases of limiting fluxes. In this case, salt concentration at the exit of a membrane channel (concentrate stream) can be calculated with a mass balance relationship:

$$c_e = \frac{c_0}{(1-R)}$$

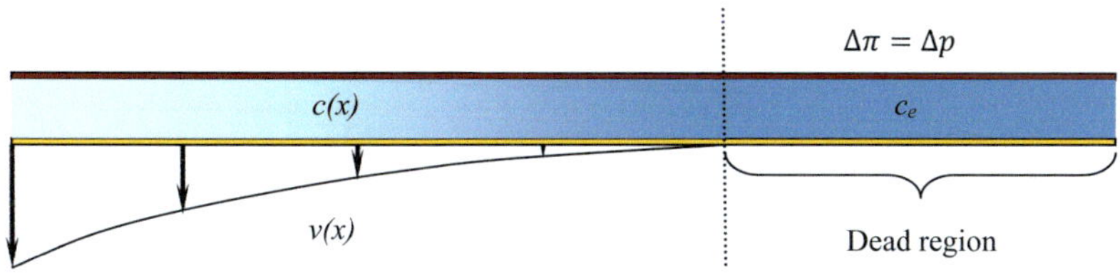

FIGURE 5.10
Thermodynamic restriction is associated with a dead region in the channel where osmotic pressure equals applied pressure.

Then the osmotic pressure at the exit is

$$\Delta\pi = f_{os}c_e = \frac{f_{os}C_0}{(1-R)} = \frac{\Delta\pi_0}{(1-R)} \tag{5.33}$$

When a membrane channel works in a thermodynamic restriction regime, the applied pressure for a given recovery R can be determined by rearranging Eq. (5.32):

$$\Delta p = \frac{\Delta\pi_0}{(1-R)} \tag{5.34}$$

Therefore, osmotic pressure at the exit of the membrane channel is equal to the applied pressure. This means that all water in the membrane channel that can be squeezed out of the membrane channel has been squeezed out.

Actually, osmotic pressure can become equal to the applied pressure before the exit of the membrane channel is reached. The rest of the channel becomes the "dead region" in which no permeate is produced, as shown in Figure 5.10. Salt concentration increases along the channel before the dead region but remains constant in the dead region. In this case, recovery R is only determined by the osmotic pressure of feed water $\Delta\pi_0$ and the applied pressure Δp. All other parameters that characterize the membrane process, i.e., those parameters included in the characteristic pressure, have no impact at all on the recovery of the membrane channel under thermodynamic restriction.

5.4.5 Thermodynamic Restriction Prevailing in Seawater Desalination

The behavior of seawater membrane desalination processes is substantially different from that of brackish water desalination processes because of the much higher salt concentration in seawater. Typical parameters for seawater desalination membrane channels are given in Table 5.6. It is noted that the resistance of seawater reverse osmosis membranes is typically three times that of brackish water reverse osmosis membranes. The osmotic pressure coefficient is relatively constant for seawater due to roughly the

TABLE 5.6

Parameters Used in Seawater Desalination

Parameter	Symbol	Value	Unit
Membrane resistance	R_m	3.00×10^{11}	Pa·s/m
Channel height	H	3.50×10^{-4}	m
Channel length	L	8	m
Salt concentration	c_0	35,000	mg/L
Osmotic pressure coefficient	f_{os}	73.5	Pa/(mg/L)
Feed tangential velocity	u_0	0.1	m/s

same chemical composition, while it changes with compositions of brackish water.

The osmotic pressure of feed water and characteristic pressure of the membrane channel with parameters in Table 5.6 are

$$\Delta\pi_0 = 73.5\,Pa/(mg/L) \times 35{,}000\,mg/L = 2.57 \times 10^6\,Pa = 25.7\,bar$$

$$\Delta p_1^* = \frac{u_0 H R_m}{L} = \frac{0.1\,m/s \times 0.35 \times 10^{-3}\,m \times 3 \times 10^{11}\,Pa \cdot s/m}{8\,m}$$

$$= 1.31 \times 10^6\,Pa = 13.1\,bar$$

It is interesting to note that the osmotic pressure of seawater almost doubles the characteristic pressure of the seawater membrane desalination channel! Because the applied pressure should be much greater than the osmotic pressure of feed water, the seawater desalination channel certainly works under thermodynamic restriction. This observation is checked with the following example.

Example 5.5

Calculate the recoveries of a seawater desalination membrane channel as functions of applied pressure assuming thermodynamic restriction. Comparing the results with the true value determined with the state equation of an ideal long narrow membrane channel. The parameters of the membrane channel are the same as in Table 5.6.

SOLUTION

Assuming thermodynamic restriction, the recovery of the membrane channel can be calculated by Eq. (5.32). The osmotic pressure and characteristic pressure are known to be 25.7 bars and 13.1 bars, respectively.

A sample calculation for recovery at a pressure of 30 bar is

$$R = 1 - \frac{\Delta\pi_0}{\Delta p} = 1 - \frac{25.7\,bar}{30\,bar} = 14.3\%$$

A sample calculation of the true recovery by Eq. (5.27) with successive substitution at a pressure of 30 bar is:

R0	0.500
R1	0.125
R2	0.131
R3	0.131

The recovery is $R = 13.1\%$

The results for a series of applied pressures are presented in the table below.

Pressure (bar)	**30**	**35**	**40**	**45**	**50**	**60**	**70**
T. R. (%)	14.3	26.5	35.7	42.8	48.6	57.1	63.3
True (%)	13.1	25.5	35.1	42.6	48.5	57.1	63.2
Relative Error (%)	8.8	3.9	1.5	0.5	0.2	0.0	0.0

The results show that the relative errors of the thermodynamic restriction model with the true value is smaller than 1% when the applied pressure is greater than 45 bar. In other words, the simple equation (Eq. 5.32) can accurately predict the recovery of seawater desalination membrane processes for all practical applied pressures (>50 bar). There is no need to use the complex state equation Eq. (5.27) with tedious numerical or graphical methods to calculate the recovery of seawater desalination membrane processes.

The results in the table are graphically presented in the figure below for visual appreciation of the fitness of goodness of Eq. (5.32) to Eq. (5.27) in the whole pressure range.

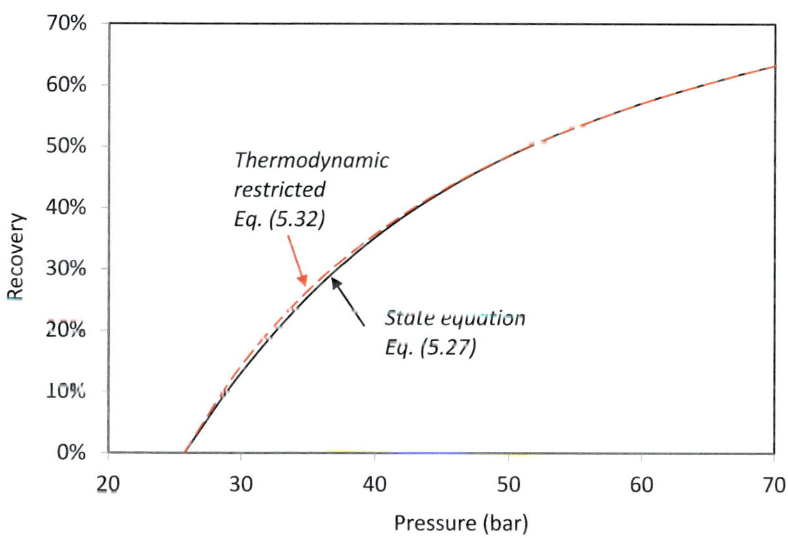

5.5 Problems

1. Discuss the features of a reverse osmosis channel used in water desalination. What is the principal consideration of such a configuration?
2. Why is the configuration of two- and three-concentration staging commonly used in desalination operation?
3. Calculate the equivalent channel length of a reverse osmosis process of three stages in a concentrate stream. The same channels of 5 m are used in all stages.
4. A seawater reverse osmosis membrane has a resistance of 3.3×10^{11} Pa·s/m. Convert the resistance to the unit of bar·h·m²/L, and psi·day·ft²/gal, respectively.
5. The height and length of a membrane channel are 0.3 mm and 6 m with a membrane resistance of 1.2×10^{11} Pa·s/m. The feed and exit tangential velocities of the membrane channel is 0.1 and 0.045 m/s, respectively. Calculate the recovery and average permeate flux of this membrane channel.
6. Calculate numerically the changes of driving pressure, salt concentration, tangential flow velocity, and permeate flux along a reverse osmosis channel with parameters given in the table below. The numerical solution of this problem is required for two-step numbers of 50 and 100, respectively. Show the results graphically.

Parameters of the Reverse Osmosis Membrane Channel

Parameter	Symbol	Value	Unit
Membrane resistance	R_m	1.10×10^{11}	Pa·s/m
Channel height	H	3.50×10^{-4}	m
Channel length	L	9	m
Friction coefficient	K	5	
Salt concentration	c_0	2000	mg/L
Applied pressure	$\Delta\pi$	6	bar
Osmotic pressure coefficient	f_{os}	80	Pa/(mg/L)
Feed tangential velocity	u_0	0.11	m/s

7. Determine the recovery and average permeate velocity from the numerical solutions obtained in Question 6. Compare the results for two step numbers used in the numerical solution.
8. Calculate numerically the changes of salt concentration, tangential flow velocity, and permeate flux along a reverse osmosis channel

with parameters given in the table below. The pressure drop along the channel is assumed negligible. Determine the recovery and average permeate flux of the membrane.

Parameters of the Reverse Osmosis Membrane Channel

Parameter	Symbol	Value	Unit
Membrane resistance	R_m	1.00×10^{11}	Pa·s/m
Channel height	H	3.50×10^{-4}	m
Channel length	L	6	m
Step number	N	100	
Salt concentration	c_0	3000	mg/L
Applied pressure	Δp	10	bar
Osmotic pressure coefficient	f_{os}	80	Pa/(mg/L)
Feed tangential velocity	u_0	0.10	m/s

9. What is the characteristic pressure of the membrane channel in Question 8?
10. Using the state equation for the long narrow membrane channel, calculate the recovery of the membrane channel in Question 8 with both successive substitution and graphical methods. Compare the results with that determined numerically in Question 8.
11. An RO membrane channel of 6 m long is used in the treatment of a feed water of 5000 mg/L salt. The height of the membrane channel is 0.35 mm, the clean membrane resistance is 1.0×10^{11} Pa·s/m, and the feed cross flow velocity is 0.13 m/s. Calculate the recoveries of the RO process for the driving pressures of 8, 10, 12, 15, and 20 bar, with the graphical method.
12. For the same channel as in Question 11, calculate the recoveries of the RO process for the driving pressures of 8, 10, 12, 15, and 20 bar, with the successive substitution method.
13. The state equation

$$R = \left(1 - \frac{\Delta\pi_0}{\Delta p}\right)\left\{1 - exp\left[-\frac{\Delta p}{\Delta\pi_0}\left(\frac{\Delta p}{\Delta p^*} - R\right)\right]\right\}$$

is commonly used in the successive substitution method to find out the explicit recovery R. However, in some rare cases with low salt concentration at low applied pressure, this equation does not converge. In that case, the alternative equation below can be used for the pressures for which the successive substitution method does not converge.

$$R=\frac{\Delta\pi_0}{\Delta p}ln\left[\left(1-R-\frac{\Delta\pi_0}{\Delta p}\right)\Big/\left(1-\frac{\Delta\pi_0}{\Delta p}\right)\right]+\frac{\Delta p}{\Delta p^*}$$

Try to find out the recoveries at pressure of 3.0, 4.0, and 5.0 bar for the membrane channel of the parameters in the table below.

Parameters of the Reverse Osmosis Membrane Channel

Parameter	Symbol	Value	Unit
Membrane resistance	R_m	1.0×10^{11}	Pa·s/m
Channel height	H	3.50×10^{-4}	m
Channel length	L	6	m
Salt concentration	c_0	1200	mg/L
Osmotic pressure coefficient	f_{os}	75	Pa/(mg/L)
Feed tangential velocity	u_0	0.11	m/s

14. The osmotic pressure profiles in two membrane channels are shown in the graphs A and B. Indicate the recovery or average permeate flux is controlled by which mechanisms.

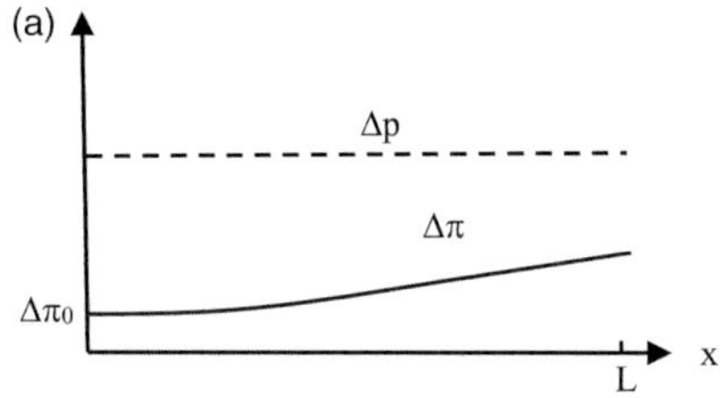

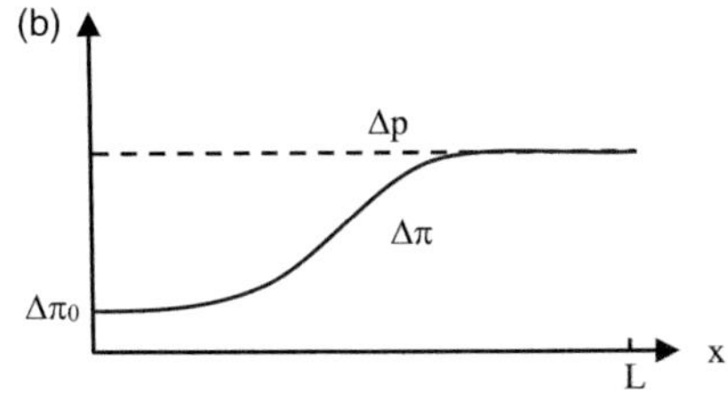

15. There are two tangential velocity profiles in the figure below. Which profile indicates the process is controlled by thermodynamic restriction?

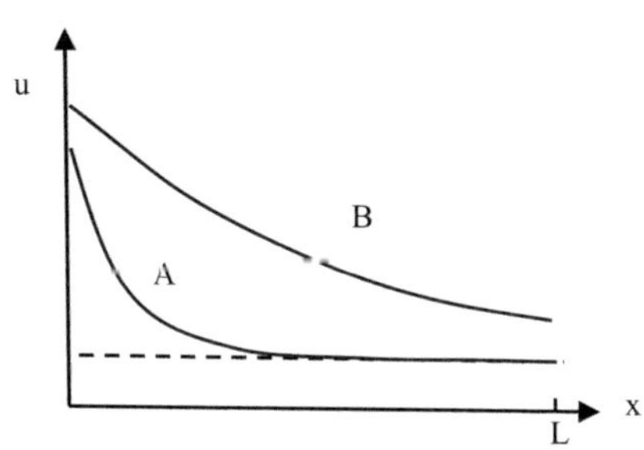

16. The permeate flux profile is presented in the figure below. What mechanism controls the membrane process?

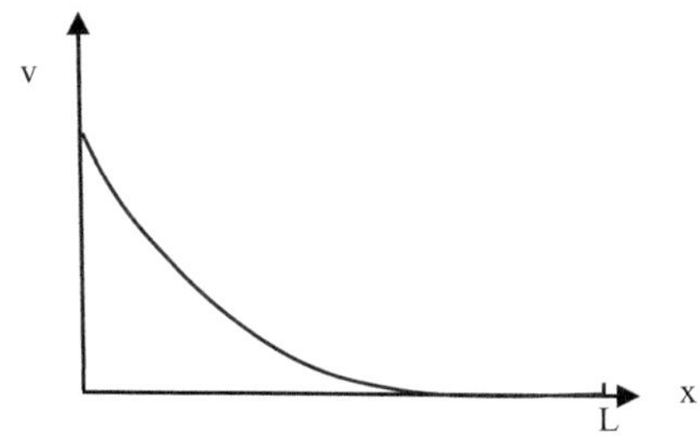

17. The parameters of a reverse osmosis channel for seawater desalination are given in the table below. Calculate recoveries with the state equation and the equation for thermodynamic restriction, respectively, for pressures of 30, 40, 50, 60, and 70 bars. Compare the results.

Parameters of the Reverse Osmosis Membrane Channel

Parameter	Symbol	Value	Unit
Membrane resistance	R_m	3.00×10^{11}	Pa·s/m
Channel height	H	3.50×10^{-4}	m
Channel length	L	8	m
Salt concentration	c_0	35,000	mg/L
Osmotic pressure coefficient	f_{os}	73.5	Pa/(mg/L)
Feed tangential velocity	u_0	0.12	m/s

18. When a membrane channel is working in the regime of thermodynamic restriction, can you increase the recovery of the desalination process by increasing channel length with all the other parameters unchanged?
19. What will happen to increase the applied pressure on the membrane channel that is operating under thermodynamic restriction?
20. Do you agree with the assertion that the performance of seawater desalination membrane channels is controlled by thermodynamic restriction? Explain your answer.

References

1. Wilf, M., & Klinko, K. (2001). Optimization of seawater RO systems designs. Desalination, 138, 299–306.
2. Bremere, I., Kennedy, M., Stikker, A., & Schippers, J. (2001). How water scarcity will affect the growth in the desalination market in the coming 25 years. Desalination, 138, 7–15.

3. Hajeeh, M., & Chaudhuri, D. (2000). Reliability and availability assessment of reverse osmosis. Desalination, 130, 185–192.
4. Qasim, M., Badrelzaman, M., Darwish, N.N., Darwish, N.A., & Hilal, N. (2019). Reverse osmosis desalination: A state-of the-art review. Desalination, 459, 59–104.
5. Shenvi, S.S., Isloor, A.M., & Ismail, A.F. (2015). A review on RO membrane technology: Developments and challenges. Desalination, 368, 10–26.
6. Prihasto, N., Liu, Q.F., & Kim, S.H. (2009). Pre-treatment strategies for seawater desalination by reverse osmosis system. Desalination, 249, 308–316.
7. Tayeh, Y.A. (2024). A comprehensive review of reverse osmosis desalination: Technology, water sources, membrane processes, fouling, and cleaning. Desalination and Water Treatment, 320(4), 100882.
8. Geraldes, V., Pereira, N.E., & Pinho, M.N. (2005). Simulation and optimization of medium-sized seawater reverse osmosis processes with spiral-wound modules. Industrial & Engineering Chemistry Research, 44, 1897–1905.
9. Zhou, W., Song, L., & Guan, T.G. (2006). A numerical study on concentration polarization and system performance of spiral wound RO membrane modules. Journal of Membrane Science, 271, 38–46.
10. Al-Obaidi, M.A., Kara-Zaitri, C., & Mujtaba, I.M. (2017). Modeling of a spiral-wound reverse osmosis process and parameter estimation. Desalination and Water Treatment, 69, 93–101.
11. Altaee, A. (2012). A computational model to estimate the performance of 8 inches RO membranes in pressure vessel. Journal of Membrane and Separation Technology, 1, 60–71.
12. Bahoosh, M., Kashi, E., Shokrollahzadeh, S., & Rostami, K. (2019). Comparison of the performance of different reverse osmosis membrane modules by CFD modeling. Iranian Journal of Chemical Engineering, 16, 101–116.
13. Absar, B., Kadi, L., & Belhamiti, O. (2008). Mathematical modeling of reverse osmosis process by the orthogonal collocation on finite element method. Asian Journal of Applied Sciences, 1, 1–18.
14. Bouchard, C.R., Carreau, P.J., Matsuura, T., & Sourirajan, S. (1994). Modeling of ultrafiltration: Predictions of concentration polarization effects. Journal of Membrane Science, 97, 215–229.
15. Ghernaout, D. (2017). Reverse osmosis process membranes modeling – A historical overview. Journal of Civil, Construction and Environmental Engineering, 2(4), 112–122.
16. Fang, Y., & Duranceau, S.J. (2013). Modeling mass transfer using surface morphology in full-scale reverse osmosis membrane processes. Desalination and Water Treatment, 51, 6459–6471.
17. Alardhi, S.M., Ahmed, M.L., Breig, S.J., Jaber, A.A., Fiyadh, S.S., Salman, A.D., AlJaberi, F.Y., Duc Nguyen, D., Van, B., & Le, P. (2024). Artificial neural network and response surface methodology for modeling reverse osmosis process in wastewater treatment. Journal of Industrial and Engineering Chemistry, 133, 599–613.
18. Absar, B., Kadi, S.E., & Belhamiti, O. (2010). Reverse osmosis modeling with the orthogonal collocation on finite element method. Desalination and Water Treatment, 21, 23–32.
19. Mielczarek, K., Bohdziewicz, J., & Kwarciak-Kozłowska, A. (2011). Modeling performance in reverse osmosis wastewater treatment post-trial coke in the integrated systems based on the assumptions of the relaxation model. Civil and Environmental Engineering Reports, 1, 19–32.

20. Avlonitis, S.A., Pappas, M., & Moutesidis, K. (2007). A unified model for the detailed investigation of membrane modules and RO plants performance. Desalination, 203, 218–228.
21. Salami, E.S., Ehetshami, M., Karimi-Jashni, A., Salari, M., Nikbakht Sheibani, S., & Ehteshami, A.A. (2016). A mathematical method and artificial neural network modeling to simulate osmosis membrane's performance. Modeling Earth Systems and Environment, 2, 1–11.
22. Sobana, S., & Panda, R.C. (2014). Modeling and control of reverse osmosis desalination process using centralized and decentralized techniques. Desalination, 344, 243–251.
23. Tekle, B., Alazzam, A., Ibrahim, A., Malkawi, G., Fares, A., Qureshi, N., & Al Hamadat, A. (2022). Using artificial intelligence for reverse osmosis desalination. 2022 8th International Conference on Information Technology Trends (ITT), 56–60.
24. Song, L., & Tay, K.G. (2006). Performance prediction of a long crossflow reverse osmosis membrane channel. Journal of Membrane Science, 281, 163–169.
25. Song, L., Hong, S., Hu, J., Ong, S.L., & Ng, W.J. (2002). Simulations of full-scale reverse osmosis membrane process. Journal of Environmental Engineering, 128, 960–966.
26. Song, L., Hu, J., Ong, S.L., Ng, W.J., Elimelech, M., & Wilf, M. (2003). Performance limitation of the full-scale reverse osmosis process. Journal of Membrane Science, 214, 239–244.
27. Tay, K.G., Song, L., Ong, S.L., & Ng, W.J. (2005). Nonlinear behavior of permeate flux in full-scale reverse osmosis processes. Journal of Environmental Engineering, 131, 1481–1487.
28. Song, L. (2010). Concentration polarization in a narrow reverse osmosis membrane channel. AIChE Journal, 56, 143–149.
29. Brauer, F., & Nohel, J.A. (1967). Ordinary Differential Equations: A First Course. W. A. Benjamin, New York, p. 340.
30. Song, L. (2023). A concise analytical model for the ideal reverse osmosis desalination processes. Water Environment Research, 95, e10939.
31. Song, L., Hu, J.Y., Ong, S.L., Ng, W.J., Elimelech, M., & Wilf, M. (2003). Emergence of thermodynamic restriction and its implications for full-scale reverse osmosis processes. Desalination, 155(3), 213–228.

6

Process Optimization

The reverse osmosis membrane itself is expensive, and the desalination process is energy intensive that incurs high cost, too. It is imperative to optimize the design and operation of reverse osmosis processes to reduce desalination cost for more widespread economical applications of this technology [1–7].

The cost effectiveness of a reverse osmosis process is affected by many factors, including membrane cost, energy (mostly electricity) cost, feed water quality, concentrate handling, and operation requirement (recovery or average permeate flux). A serious challenge in process optimization is to construct an optimization model that contains major components essential to represent the reverse osmosis desalination processes. At the same time, the model should not include trivial features that only occur in some particular cases to avoid unnecessary complexity and tediousness of optimization processes.

It is also of great interest and benefit to have an optimization procedure that is clearly delineated, easily implementable, and without the need for enormous computational power to complete.

6.1 A Generic Cost Model for Membrane Desalination

There are different ways to study the cost of membrane processes in desalination. A popular way is to build a specific model that includes numerous particular equipment and devices [8–15]. There are multiple parameters in the model that need to be optimized to minimize desalination cost. Parameters in the model may include types and sizes of particular devices, pretreatment options, operation strategies, performance indicators, and more [16–20]. In optimization of a membrane system, the cost model needs to run in tandem with a performance model that maybe is of the same complexity [21–29]. Such a complex cost model is inefficient in process optimization for two major reasons. First, it is difficult or too tedious to construct an accurate cost model because there is a wide variety of equipment and devices to choose from. Second, multiple variable optimizations usually require considerable computational power, and therefore it is not convenient for practicing engineers or operators of desalination processes to use.

DOI: 10.1201/9781003743569-8

Another way to handle the cost for membrane desalination is developing a generic model that focuses on the costs associated with the functions or stages of the membrane process operation rather than the individually itemized equipment and device [30,31]. The total cost of a reverse osmosis desalination system can be divided into four cost components associated with stages of operation. Two major cost components are the membrane and power in the membrane filtration stage, and the other two cost components are associated with pretreatment stage and concentrate handling stage. The main focus of optimization is the tradeoff between membrane cost and energy cost. Costs associated with pretreatment and concentrate handling phases come into the picture only through the prices for treatments of unit volume of feed water or concentrate stream. The generic model does not only substantially reduce the difficulty and working load of optimization but also makes the optimization process easy to follow and adapt by the practicing engineers and operators.

A generic cost model with components of a membrane desalination plant in three stages is schematically described in Figure 6.1. According to model, the total cost of desalination can be associated with three stages: feed water pretreatment, membrane filtration, and concentrated handling. The key component is the membrane process in the middle of the diagram. The capacity (daily permeate production rate) of Q_p is a requirement for the desalination process. The performance indicator of the membrane process recovery R, which is equivalent to specifying the average permeate flux, is the principal design parameter to optimize the membrane desalination system. The flowrate of feed water that needs pretreatment and the flowrate of concentrate stream that needs handling and discharge are determined by the recovery of the membrane process as, respectively,

$$Q_f = Q_p/R$$
$$Q_c = (1-R)Q_p/R$$

where Q_f, Q_p, and Q_c are the flowrates of feed water, permeate, and concentrate streams, respectively, and R is the recovery of this membrane process.

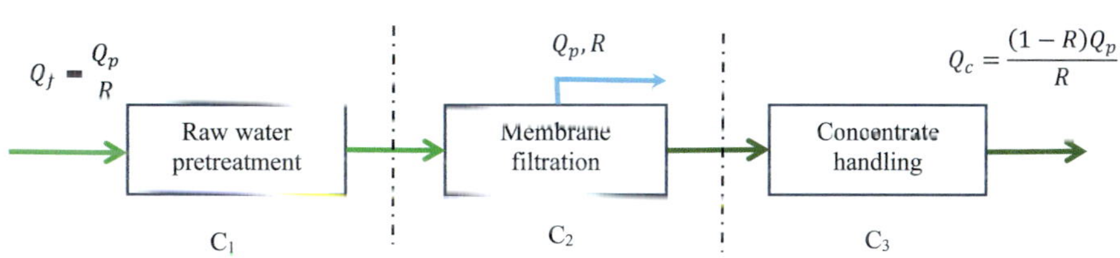

FIGURE 6.1
A generic cost model for reverse osmosis desalination process.

For a given capacity Q_p, the cost components in the three stages are affected by the principal design parameter R. Total cost of the desalination process is the sum of the three cost components, i.e.,

$$C_t = C_1 + C_2 + C_3 \tag{6.1}$$

where C_t is the total cost, C_1 is the cost for feed water pretreatment, C_2 is the cost for membrane process, and C_3 is the cost for concentrate handling. It needs to point out particularly that the total cost and cost components in Eq. (6.1) are for production of permeate at flowrate Q_p. Therefore, it is not the total cost of the desalination plant for whole service life. It is rather a "money flow" into the desalination plant per unit time. When the capacity is specified as daily permeate flowrate, Eq. (6.1) will be the daily cost of desalination plant. The use of daily cost here is convenient to convert to the specific cost of permeate, which is the most commonly used term for desalination cost later and will be discussed in more details.

6.1.1 Price of Membrane

In the generic cost model, the cost component C_2 for membrane process can be further divided into membrane cost and energy cost. The most challenging part is to quantify membrane cost because it has to be amortized over its service time.

Before determination of membrane cost in a desalination process, the price for using a unit area membrane per unit time is first needed. Pressure vessels should be included in the membrane cost because they are always required to hold the membrane elements. Because the service times of membrane elements and pressure vessels are different, the cost for pressure vessels should be included in the membrane cost proportionally. For a pressure vessel holding several membrane elements, the membrane price is then determined as

$$c_m = \frac{1}{S_e}\left(\frac{p_e}{T_e} + \frac{p_v}{nT_v}\right) \tag{6.2}$$

where c_m is the membrane price per unit area per unit time, p_e and p_v are the prices of a membrane element and a pressure vessel, respectively, T_e and T_v are the service times of membrane element and pressure vessel, respectively, n is the number of elements holding in a pressure vessel, and S_e is the membrane area in a membrane element.

Example 6.1

A standard BW8040 membrane element is available at the price of \$500. The membrane element has a membrane area of 40 m^2 and a warranty service time of five years. Price of a pressure vessel that holds six BW8040 membrane elements is $1500. The service time of the pressure vessel is 20 years. Calculate the costs for using a square meter membrane for a year or a day.

SOLUTION

With the information available, the cost to use the membrane can be directly calculated with Eq. (6.2):

$$c_m = c_m = \frac{1}{S_e}\left(\frac{p_e}{T_e} + \frac{p_v}{nT_v}\right) = \frac{1}{40m^2}\left(\frac{\$500}{5y} + \frac{\$1500}{6\times 20y}\right)$$
$$= \$2.81/\left(m^2 \cdot y\right) = 0.771\ ¢/\left(m^2 \cdot d\right)$$

The cost to use 1 m^2 membrane is 2.81 dollars per year or 0.771 cents per day.

6.1.2 Cost of Membrane Filtration

Energy cost in a membrane unit can be determined straightly from energy needed to pump feed water into the membrane unit at the required flowrate and pressure:

$$C_e = c_e Q_f \Delta p \tag{6.3}$$

where C_e is the energy cost, c_e is the price per unit of energy, Q_f is the flowrate of feed water, and Δp is the applied pressure. Attention should be given to the unit of energy. If the basic metric units of m^3/s and Pa are used for flowrate and pressure, respectively, the energy unit is Joule. Because the unit Joule is too small, kilowatt hour is commonly used in practical desalination applications, whose conversion factor to Joule is 1 kWh = 3.60×10^6 J.

To determine the membrane cost, the area of membrane needed to produce the required permeate flowrate is first determined as follows:

$$S = \frac{Q_p}{\overline{v}}$$

where S is the total membrane area and $\overline{v}$ is the average permeate flux of the membrane unit. The membrane cost per unit time is

$$C_m = c_m S = c_m \frac{Q_p}{\overline{v}}$$

where C_m is the membrane cost and c_m is the cost per unit area of membrane.

Therefore, the cost of membrane filtration stage is the sum of energy cost and membrane cost:

$$C_2 = C_e + C_m = c_e Q_f \Delta p + c_m \frac{Q_p}{\overline{v}} \tag{6.4}$$

Example 6.2

A reverse osmosis desalination plant has a capacity of 10,000 m^3/day with a recovery of 75%. The membrane unit is operated at an average permeate flux of 5×10^{-6} m/s under the applied pressure of 10 bar. Prices for electricity and membrane are \$0.09/kWh and \$0.01/m^2·d.What is the cost of membrane filtration stage of this desalination plant?

SOLUTION

The feed flowrate is

$$Q_f = \frac{Q_p}{R} = \frac{10,000\, m^3/d}{0.75} = 1.33 \times 10^4\, m^3/d$$

The electricity cost is

$$C_e = c_e Q_f \Delta p = \frac{\$0.09/kWh}{3.6 \times 10^6\, J/kWh} \times 1.33 \times 10^4\, m^3/d \times 10\, bar \times 10^5\, Pa/bar \times 1\, J/\left(Pa \cdot m^3\right)$$

$$= \$333/d$$

The membrane cost is

$$C_e = c_m \frac{Q_p}{\overline{v}} = \$0.01/\left(m^2 \cdot d\right) \times \frac{10,000\, m^3/d}{5 \times 10^{-6}\, m/s \times 86,400\, s/d} = \$231/d$$

The total cost to run this membrane filtration unit is

$$C_2 = C_e + C_m = \$333/d + \$231/d = \$564/d$$

6.1.3 Costs for Feed Water Pretreatment and Concentrate Handling

The cost for feed water pretreatment can be calculated by referring to a known case in the following form:

$$C_1 = M_0 \left(\frac{Q_f}{Q_0} \right)^{S_1}$$

where Q_0 is the flowrate of a reference pretreatment process, M_0 is the cost to run the reference pretreatment process per unit time, Q_f is the flowrate of feed water of the plant under consideration, and S_1 (≤ 1) is the scale effect parameter for the pretreatment unit. The reference pretreatment process can be an existing facility running under similar conditions if it is available. Otherwise, it can be a pilot testing facility to generate the required information, or just a virtual one on computer built from past experience. The economics of scale means the cost per unit of product decreases with the increasing scale of production. The scale effect parameter is smaller than unity ($S_1 < 1$) when there is a scale effect. If scale effect is not obvious, the scale effect parameter is unity $(S_1 = 1)$. In this case, the cost for pretreatment becomes

$$C_1 = M_0 \frac{Q_f}{Q_0} = c_1 Q_f = c_1 Q_p / R \tag{6.5}$$

where $c_1 (= M_0/Q_0)$ is the pretreatment price of a unit volume of feed water.

Similarly, the cost for concentrate handling with and without scale effect can be written as, respectively,

$$C_3 = M_1 \left(\frac{Q_c}{Q_1} \right)^{S_2}$$

and

$$C_3 = c_3 Q_c = c_3 (1 - R) Q_p / R \tag{6.6}$$

where Q_1 is the flowrate of a reference concentrate handling facility, M_1 is the cost to run the facility per unit time, Q_c is the flowrate of concentrate stream in the plant under consideration, S_2 (≤ 1) is the scale effect parameter for the concentrate handling unit, and c_3 is the handling price per unit volume of concentrate. Eq. (6.6) is for the cases without scale effect.

Example 6.3

The sum of equivalent daily cost of capital investment and operational cost of an existing pretreatment facility is \$100/day. The flowrate of feed water treated in the facility is 10,000 m^3/day. (1) What is the price to treat one unit volume of feed water in the existing facility? (2) Estimate the cost for a new pretreatment facility of five times the capacity of the existing one if the scale effect parameter is 0.75. (3) What is the pretreatment cost if there is no scale effect?

SOLUTION

1. The pretreatment price for a cubic meter feed water is

$$c_1 = \frac{M_0}{Q_0} = \frac{\$100/d}{10{,}000\ m^3/d} = \$0.01/m^3$$

2. The cost for pretreatment in new facility with scale effect is

$$C_1 = M_0\left(\frac{Q_f}{Q_0}\right)^{S_1} = \$100/d\left(\frac{50,000\ m^3/d}{10,000\ m^3/d}\right)^{0.75} = \$334.4/day$$

3. The cost for pretreatment in the new plant without scale effect is calculated with Eq. (6.5):

$$C_1 = c_1 Q_f = \$0.01/m^3 \times 50,000\ m^3/d = \$500/d$$

6.1.4 Total Cost of Membrane Desalination

The total cost for the membrane desalination process can be obtained by the substitution of Eqs. (6.4), (6.5), and (6.6) into Eq. (6.1):

$$C = c_1\frac{Q_p}{R} + c_e\frac{Q_p}{R}\Delta p + c_m\frac{Q_p}{\bar{v}} + c_3\frac{(1-R)Q_p}{R} \tag{6.7}$$

Dividing both sides of Eq. (6.7) by Q_p results in

$$\frac{C}{Q_p} = c_1\frac{1}{R} + c_e\frac{1}{R}\Delta p + c_m\frac{1}{\bar{v}} + c_3\frac{(1-R)}{R}$$

Now the terms in cost equation change from money flows for a given desalination capacity to the specific cost of permeate, which is the cost per unit volume of permeate produced. Let us introduce a new variable for the specific cost of permeate:

$$c_s = \frac{C}{Q_p}$$

Furthermore, the average permeate flux is related to the recovery by Eq. (5.2):

$$\bar{v} = \frac{u_0 H}{L}R$$

Then the total cost model becomes

$$c_s = c_1\frac{1}{R} + c_e\frac{1}{R}\Delta p + c_m\frac{L}{u_0 HR} + c_3\frac{(1-R)}{R} \tag{6.8}$$

Eq. (6.8) is the final form of the cost model for reverse osmosis desalination process. As a generic model, Eq. (6.8) is generally applicable to all cases while the cost coefficients c_1, c_e, c_m, and c_3 represent localities of a reverse osmosis process at a particular location and should be determined prior to process optimization. Of course different values can be used for cost coefficients to compare different alternatives. For example, the prices for feed water pretreatment and concentrate handling can vary with different options.

Usually, the height of the membrane channel H and the feed velocity u_0 are not the variables to be optimized. The membrane elements used in the process have a fixed channel height. Similarly, the feed flow velocity in most cases also has a fixed value recommended by manufacturer of the membrane elements. Length of membrane channel L can be a variable in the optimization process. However, because of the use of membrane elements, only discrete values can be used for channel length. It means that the length of membrane channel can only take integer times of the length of a membrane element. If the common 6m element pressure vessel is used, the equivalent lengths of the membrane channel can only take the values of 6, 9, and 10.5 m for single-stage, two-stage, and three-stage array structures when the length of a membrane element is nominally taken as 1 m.

Recovery R is the only free variable in Eq. (6.8) on which all cost components depend. Though the applied pressure Δp appears in Eq. (6.8), it will be shown in the next section that it is not a free variable but is fully determined by the recovery of a membrane process. Therefore, Eq. (6.8) is a single variable cost model for a reverse osmosis desalination process, with which the optimal recovery is searched to minimize the specific cost for permeate production

6.2 A Reverse Problem

Applied pressure is required to determine the specific cost of permeate with Eq. (6.8). In the previous chapter, it is shown that the applied pressure is implicitly related to recovery by the state equation Eq. (5.27). The state equation can be rewritten in the form below:

$$\frac{\Delta p}{\Delta \pi_0} = \frac{1}{R}\left(\frac{\Delta p}{\Delta \pi_0} - 1\right)\left[1 - e^{-\frac{\Delta p}{\Delta \pi_0}\left(\frac{\Delta p}{\Delta \pi_0}\frac{\Delta \pi_0}{\Delta p} - R\right)}\right] \tag{6.9}$$

In Eq. (6.9), the recovery now becomes an independent variable while the applied pressure Δp is defined implicitly as a function of the recovery R. The explicit Δp can be found from the equation with a graphical method similar to the one for R in the previous chapter. Actually, Eq. (6.9) is a reverse problem of Eq. (5.27) for the recovery R.

To find the explicit Δp, the applied pressure, Eq. (6.9) is first split into two functions of free variable $\Delta p/\Delta \pi_0$,

$$y_1 = \frac{\Delta p}{\Delta \pi_0} \tag{6.10}$$

$$y_2 = \frac{1}{R}\left(\frac{\Delta p}{\Delta \pi_0} - 1\right)\left[1 - e^{-\frac{\Delta p}{\Delta \pi_0}\left(\frac{\Delta p}{\Delta \pi_0}\frac{\Delta \pi_0}{\Delta p^*} - R\right)}\right] \tag{6.11}$$

where y_1 and y_2 are functions of the dimensionless pressure $\Delta p/\Delta\pi_0$. The use of the dimensionless form of $\Delta p/\Delta\pi_0$ is simply more convenient. The two functions are then plotted in a $(\Delta p/\Delta\pi_0)$y-plane. The intersection of the two lines is the dimensionless pressure for the given recovery.

The graphical method is demonstrated below with a membrane channel in brackish water desalination. The question in this case is to find the applied pressure for a recovery of 60%. Parameters of the membrane channel were previously given in Table 5.3.

Osmotic pressure of the feed water and characteristic pressure of the membrane channel are

$$\Delta\pi_0 = f_{os}c_0 = 77\ Pa/(mg/L) \times 2500\ mg/L = 1.93 \times 10^5\ Pa = 1.93\ bar$$

and

$$\Delta p^* = \frac{u_0 H R_m}{L} = \frac{0.1\ m/s \times 0.35 \times 10^{-3}\ m \times 10^{11}\ Pa/(m/s)}{9\ m}$$

$$= 3.89 \times 10^5\ Pa = 3.89\ bar$$

$$\frac{\Delta\pi_0}{\Delta p^*} = \frac{1.93\ bar}{3.89\ bar} = 0.495$$

Calculate y_1 and y_2 as function of $\Delta p/\Delta\pi_0$ and the results are presented in Table 6.1.

A graph is developed from the data in Table 6.1 and presented in Figure 6.2.

TABLE 6.1

y_1 and y_2 for Various Dimensionless Pressures

$\Delta p/\Delta\pi_0$	y_1	y_2
0.00	0.00	0.00
1.00	1.00	0.00
1.50	1.50	0.16
2.00	2.00	0.90
2.50	2.50	1.99
3.00	3.00	3.10
3.50	3.50	4.09
4.00	4.00	4.98
4.50	4.50	5.83
5.00	5.00	6.67
5.50	5.50	7.50
6.00	6.00	8.33

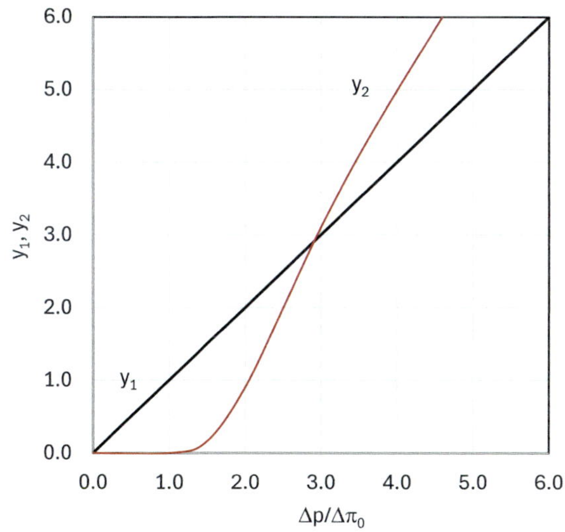

FIGURE 6.2
Graphical solution method for pressure in the reverse problem. The intersection point of y_1 and y_2 is the dimensionless pressure determined with the graphical method.

It can be seen in Figure 6.2 that two lines of y_1 and y_2 intersect at 2.9. Therefore, the dimensionless pressure is

$$\frac{\Delta p}{\Delta \pi_0} = 2.9$$

Then the applied pressure in this case is

$$\Delta p = 2.9\Delta \pi_0 = 2.9 \times 1.93\ bar = 5.58\ bar$$

The graphical method can be used to determine pressures for multiple recoveries. This is described with the example below.

Example 6.4

Using the graphical method to determine the applied pressures of a membrane process at recoveries of 10, 20, 30, 40, 50, 60, and 70%, respectively. Parameters used in the calculation are the same as given in Table 5.3.

SOLUTION

From the case study, it is known that

$$\Delta \pi_0 = 1.93\ bar$$
$$\Delta p^* = 3.89\ bar$$
$$\frac{\Delta \pi_0}{\Delta p^*} = 0.495$$

The function y_2 for all seven recoveries are plotted in one graph with y_1 in Figure 6.3.

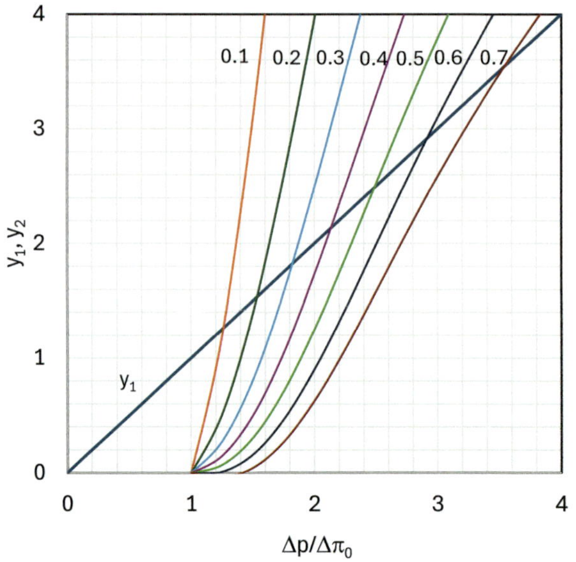

FIGURE 6.3
Pressures determined with graphical method for multiple recoveries as indicated on the curves.

The dimensionless pressure $\Delta p/\Delta\pi_0$ can be read from the figure, from which the applied pressure is calculated. More accurate values can be obtained by enlarging the area around the intersection points. The results are presented in table below.

The Applied Pressure for Given Recoveries

R	0.1	0.2	0.3	0.4	0.5	0.6	0.7
$\Delta p/\Delta\pi_0$	1.26	1.53	1.82	2.13	2.48	2.91	3.54
Δp (bar)	2.43	2.95	3.50	4.10	4.77	5.60	6.81

6.3 Cost Minimization

The most important and challenging step for process optimization is to construct a cost model that relates cost to process performance. With Eq. (6.8), performance optimization of a reverse osmosis desalination process can be simply expressed as

$$\min_{0\le R\le 1}\left[c_1\frac{1}{R}+c_e\frac{1}{R}\Delta p+c_m\frac{L}{u_0 HR}+c_3\frac{(1-R)}{R}\right] \tag{6.12}$$

Eq. (6.12) is an optimization of a single variable. As shown before, the applied pressure is solely determined by the recovery. Searching procedure for the optimal recovery as the sole optimization variable is simple and straight. Another advantage of Eq. (6.12) is that the total cost is explicitly split into components for pretreatment, energy, membrane, and concentration handling.

For any given recovery, the applied pressure can be found with the state equation. The specific cost can be readily determined for that recovery. By plotting the specific cost over the recovery in a reasonable range, the optimal recovery that minimizes the specific cost can be visually identified from the graph. This optimization procedure is illustrated by a few cases.

6.3.1 Case 1: Design of a Brackish Desalination Plant

This case is to design a two-stage reverse osmosis desalination process of capacity 1000 m^3/day with the lowest cost. The cost coefficients for pretreatment, energy, membrane, and concentrate handling are listed in Table 6.2. The parameters of the membrane channel are given in Table 5.3. Though the question seems to ask only for the optimal recovery of the membrane process, membrane area, applied pressure, and all cost components can actually be determined at the same time.

Before process optimization, some issues should be clarified. First, parameters of membrane channel are not design parameters because the membrane elements are manufactured with fixed sizes, as given in Table 5.3. The equivalent length of membrane channel here is 9 m because 6 m vessels are assumed in two stages, with the length of an element being nominally set as 1.00 meter for convenience. The actual length of elements should be used in the real-world problems.

Next, units should be consistent for all variables and parameters in Eq. (6.8). Because the basic metric units of meter and second are used for length and time in the cost model, the cost coefficients for energy and membrane need to be modified accordingly,

$$c_e = \$0.10/kWh \times \frac{1\,kWh}{3.6\times10^6\,Pa\cdot m^3} = \$2.78\times10^{-8}/\left(Pa\cdot m^3\right)$$

$$c_m = \$0.01/\left(m^2\cdot d\right)\times\frac{1\,d}{86,400\,s} = \$1.16\times10^{-7}/\left(m^2\cdot s\right)$$

TABLE 6.2

Price Coefficients for Cost Components

Cost Component	Pretreatment	Energy	Membrane	Concentrate
Symbol	c_1 (\$/m^3)	c_e (\$/kWh)	c_m (\$/(m^2·d))	c_3 (\$/m^3)
Value	0.05	0.10	0.01	0.03

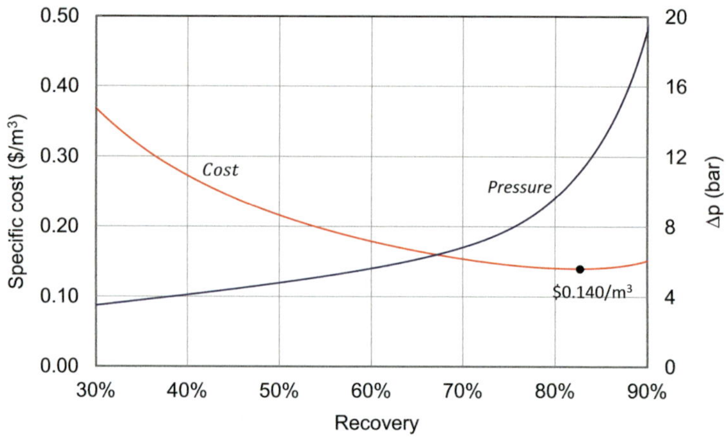

FIGURE 6.4
Specific cost of permeate as the function of recovery. It can be visually determined that the lowest specific cost occurs at the recovery of 83% with the value of $0.140/m^3$.

Now we can start process optimization. The applied pressure as a function of the recovery is determined first using Eq. (6.9) with a graphical method. Then the specific cost of permeate is calculated with Eq. (6.8). The results are graphically presented in Figure 6.4.

It can be seen from the figure that specific cost is substantially affected by recovery. The specific cost reduces by about half when the recovery increases from 40 to 80%. The optimal recovery to minimize the specific cost to 14.0 ¢/m^3 is 83%. The corresponding applied pressure is 11.2 bar (162 psi) that can also be found in the graph. With the known optimal recovery, the specific cost can be easily split into 6.0, 3.8, 3.6, and 0.6 ¢/m^3 on pretreatment, energy, membrane, and concentration handling.

Furthermore, the important design parameters of this membrane process can be determined from the optimal recovery as follows. The average permeate flux is calculated from the recovery by Eq. (5.2):

$$\bar{v} = \frac{u_0 H}{L} R = \frac{0.1\, m/s \times 0.35 \times 10^{-3}\, m}{9\, m} \times 83\%$$
$$= 3.22 \times 10^{-6}\, m/s$$

The membrane area needed is calculated by Eq. (1.10):

$$S = \frac{Q_p}{\bar{v}} = \frac{1000\, m^3/d \times \dfrac{1\, d}{86,400\, s}}{3.22 \times 10^{-6}\, m/s} = 3594\, m^2$$

Assume the membrane area in an 8040 membrane element is 40 m², the number of elements is

$$n = \frac{3594\ m^2}{400\ m^2/element} \approx 90\ element$$

When the pressure vessels holding 6 elements are used, 15 vessels are needed, with 10 vessels in 1st stage and 5 vessels in 2nd stage.

An important feature of process optimization with the generic model is the least involvement of particular equipment and devices. In this current case only the prices of membrane elements and pressure vessels are needed.

6.3.2 Case 2: Processes Optimization of Different Stages

Reverse osmosis processes are commonly used in two or three stages for brackish water desalination to increase recovery. The pressure vessel of a nominal length of 6 m (holding six membrane elements) is almost the default choice. Optimization will be conducted for reverse osmosis processes with single-stage, two-stage, and three-stage membrane arrays. The number of stages will be assessed for the most desirable results.

The parameter that characterizes stages is the equivalent length of membrane channel, which are L_1 = 6 m, L_2 = 9 m, L_3 = 10.5 m for single stage, two stages, and three stages, respectively. Capacity of the desalination plant is 1000 m³/day. Other parameters of the membrane processes are the same as in Table 5.3. The solution procedure to optimize the processes is the same as described in Case 1.

Specific costs for permeate as functions of recovery are presented in Figure 6.5. The optimal recovery occurs at about 83% unanimously for membrane processes with different stages. It is interesting to note that the process with single stage has the lowest specific cost. More details on the optimal design parameters can be found in Table 6.3.

It can be seen from Table 6.3 that the processes with different stages optimize at about the same applied pressure and recovery. The membrane area varied significantly. There are only 60 pressure vessels in the membrane process of single stage. The process with two stages has 60 pressure vessels in the first stage and 30 pressure vessels in the second stage. There are an additional 15 more pressure vessels in the third stage in the process with three stages.

Because the three membrane processes work at about the same recovery, it is obvious that the membranes in the second stage of the two-stage process and in second and third stages of the three-stage process do not produce any permeate. It means that the membrane processes with different stages are all under thermodynamic restriction. Permeate has been

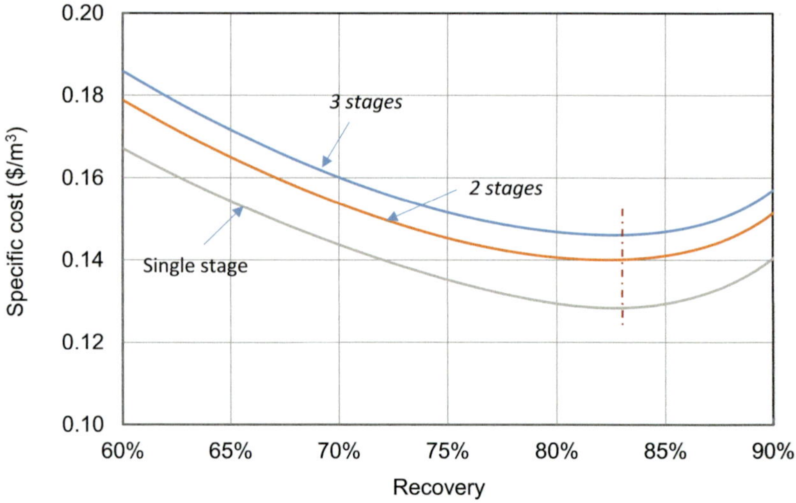

FIGURE 6.5
Specific costs of permeate in reverse osmosis desalination processes with single-stage, two-stage, and three-stage arrays, respectively. Parameters of the membrane processes are the same as in Table 5.3 except for channel length. Equivalent channel lengths are 6, 9, and 10.5 m for single, two, and three stages, respectively.

produced even before the end of pressure vessel in the first stage. The membrane elements in the second and third stages are not used and redundant. These redundant membrane elements only increase the cost of desalination. Definitely, the membrane process with only one stage is the best configuration in this case.

The cost breakdowns of the membrane desalination processes are listed in Table 6.4. It can be seen in the table that only membrane cost increases with the number of stages. Other cost components remain the same for membrane processes with different stages. The membrane process with single stage gives the lowest specific cost of 12.9 ¢, compared to 14.0 and 14.6 ¢ for membrane processes with two stages and three stages, respectively.

TABLE 6.3

Optimal Design Parameters for Different Stages

Parameter	Single Stage	Two Stages	Three Stages
Recovery (%)	82.6	82.8	82.8
Pressure (bar)	11.2	11.2	11.2
Membrane area (m^2)	2402	3594	4194
Number of elements	60	90	105
Array structure	60	60:30	60:30:15

TABLE 6.4

Specific Cost Components and Total Cost (¢/m^3) for Different Stages

Cost Component	Single Stage	Two Stages	Three Stages
Pretreatment	6.1	6.0	6.0
Energy	3.8	3.8	3.8
Membrane	2.4	3.6	4.2
Concentrate	0.6	0.6	0.6
Total	12.9	14.0	14.6

6.3.3 Case 3: Processes Optimization with Short Channels

The results from process optimization with different stages raise a serious question about the current practice of the concentrate staging with 6 m long pressure vessels. Optimization results show that one stage is the best configuration under the optimized operating conditions. Membrane elements in the second and third stages do not produce any permeate. If we want the membrane elements in second and third stages to contribute, the whole membrane processes have to work at suboptimal conditions, which means lower recoveries. From Figure 6.5, it can be seen that the specific cost would increase substantially at lower recoveries.

How about using shorter pressure vessels in the concentrate staging membrane processes? Optimizations are carried out for pressure vessels of 4 and 3 m long in the Christmas tree array with the other parameters given in Table 5.3. The results are presented in Figures 6.6 and 6.7, respectively.

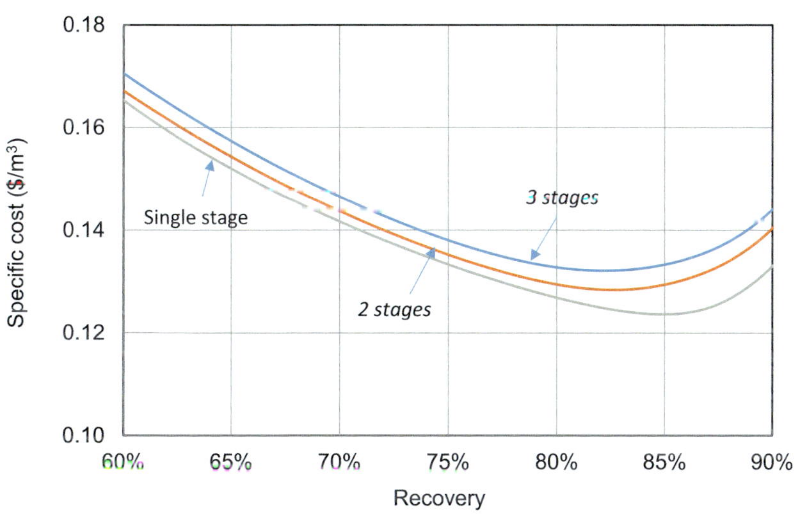

FIGURE 6.6
Specific cost of permeate in reverse osmosis processes with single stage, two-stage, and three-stage arrays. The vessel length is 4 m. Other parameters are given in Table 5.3.

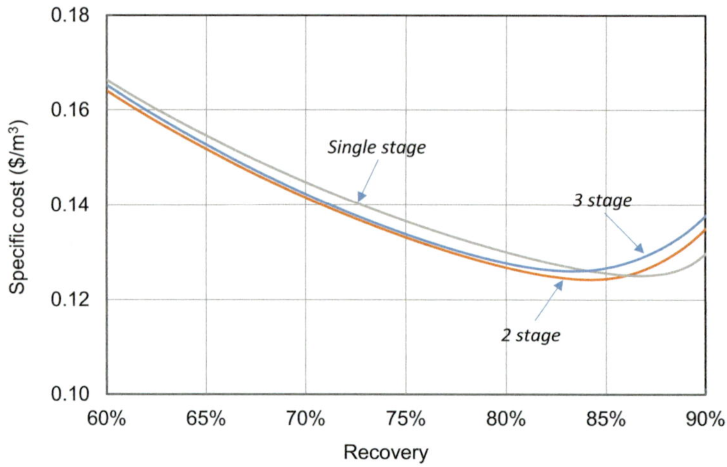

FIGURE 6.7
Specific cost of permeate in reverse osmosis processes with single stage, two-stage, and three-stage arrays. The vessel length is 3 m. Other parameters are given in Table 5.3.

The optimal design parameters and cost components are jointly presented in Tables 6.5 and 6.6, respectively.

Figure 6.6 shows that the membrane processes with 4 m long pressure vessels have the similar pattern with those with 6 m pressure vessels in that the specific cost is always the smallest for the process with single stage and the largest for that with three stages. However, the minimum specific costs occur at different recoveries. The largest recovery occurs for single stage, and the smallest one occurs for three stages. Overall, the specific costs are smaller for all three configurations with 4-m pressure vessels than these with 6-m vessels.

The behaviors of specific costs for membrane processes with 3 m–long pressure vessels are different from these with longer pressure vessels. As shown in Figure 6.7, the process with two stages has the lowest specific cost in almost the whole range of recovery. On the contrary, the process with single

TABLE 6.5
Optimal Design Parameters for Different Pressure Vessel Lengths

	Pressure Vessel Length L = 4 m			Pressure Vessel Length L = 3 m		
Parameter	**Single Stage**	**Two Stages**	**Three Stages**	**Single Stage**	**Two Stages**	**Three Stages**
Recovery (%)	84.5	82.6	82.1	87.0	84.5	82.9
Pressure (bar)	13.3	11.2	10.8	16.3	12.8	11.6
Membrane area (m^2)	1565	2402	2820	1140	1761	2094
Number of elements	40	60	70	30	45	53
Array structure	40	40:20	40:20:10	30	30:15	30:15:8

TABLE 6.6

Specific Cost Components and Total Cost (¢/m^3) for Different Pressure Vessel Lengths

	Pressure Vessel Length L = 4 m			Pressure Vessel Length L = 3 m		
Cost Component	**Single Stage**	**Two Stages**	**Three Stages**	**Single Stage**	**Two Stages**	**Three Stages**
Pretreatment	5.9	5.9	5.9	5.7	5.9	6.0
Energy	4.4	4.4	4.4	5.2	4.2	3.9
Membrane	1.6	2.3	2.7	1.1	1.8	2.1
Concentrate	0.6	0.6	0.6	0.4	0.6	0.6
Total	12.4	13.2	13.6	12.5	12.4	12.6

stage has the highest specific cost over the most range of recovery. Again the minimum specific costs occur at different recoveries, which decrease in the order of single stage, two stages, and three stages.

More detailed information about the optimized parameters and specific energy components are listed in Tables 6.5 and 6.6. From the optimization results, it can summarize that the process with three stages should not be used with any length of pressure vessels because it is always the most expensive configuration. Besides, process complexity increases with the number of stages. Processes with single stage of 4-m pressure vessels and two stages of 3-m pressure vessels both produce the same lowest specific cost of 12.4 ¢/m^3 and work at the same recovery of 84.5%. Under same specific cost and recovery, the process with 3-m pressure vessels in two stages should be chosen over the one with 4-m pressure vessels in single stage. The flowrate in pressure vessels is more even in processes with two stages than that in the processes with single stage. In the process with single stage, the flowrate by the end of pressure vessel is 15.5% of the feed flowrate. However, it is 31% of the feed flowrate by the end of pressure vessel in the second stage because it combines two flows from the first stage.

In conclusion, if membrane process parameters are given as in Table 5.3, the best design of reverse osmosis process is the one with 3-m pressure vessels in two stages. The optimal design parameters are highlighted in Table 6.5. Specific cost components and total cost are highlighted in Table 5.6. The specific cost drops by more than 11% in the process with 3-m pressure vessels than that with the common 6-m pressure vessels when they are used in two-stage configurations in both processes.

6.4 Factors Affecting Process Optimization

Process optimization is affected by many factors involved in the operation of membrane processes. For instance, higher prices associated with feed water pretreatment, concentrate handling, and membrane tend to push the optimization toward higher recovery. On the other hand, higher electricity price and higher salt concentration work against recovery in process optimization.

In the optimizations that have been done before, pretreatment cost is always the largest contributor to the total cost regardless of the stages and vessel length. Then energy, membrane, and concentration contribute in decreasing order. Reasons for the highest pretreatment cost component include the high pretreatment price coefficient and low salt concentration used in the calculations. Different results can be obtained with different conditions. A few examples are used below to show the extent that these factors affect the recovery and specific cost of membrane processes.

6.4.1 Prices for Pretreatment

First, the impact of prices for pretreatment and concentrate handling is considered as additional cost to the membrane process because both are somehow isolated from the membrane processes but heavily depend on the local conditions, such as feed water quality, concentrate handling options, and regulations. The prices can change significantly with locations. On the contrary, the membrane process is relatively less location dependent and without big variations in prices.

In the following process optimizations, the capacity of desalination plant is 1000 m^3/day. Values in Table 5.3 will be used for parameters of the membrane process. Particularly, the equivalent length of 9 m is used for the membrane channel, which means that the membrane process uses 6-m pressure vessels in two stages. The electricity price, membrane price, and concentrate handling price listed in Table 6.2 will be used. Instead of the value in Table 6.2, five pretreatment prices of 0, 2.5, 5.0, 7.5, and 10 ¢/m^3 are compared in process optimization.

Specific costs with the five pretreatment prices are presented in Figure 6.8. The figure shows that the specific cost is strongly affected by the pretreatment prices. The higher pretreatment price leads to generally higher specific cost over the whole range of recovery and pushes the optimal recovery to higher value.

The optimal design parameters for various pretreatment prices are listed in Table 6.7. It can be seen that there is a significant increase in applied pressure (44%) when the pretreatment price change from 0 to 10 ¢. The recovery increases from 78.4 to 85.0% and the membrane area decreases from about 3800 to 3500 m^2 for the same change of pretreatment price.

The weights of each cost component changing with pretreatment prices are listed in Table 6.8. It can be seen that pretreatment is always one important cost component in brackish water desalination unless there is no need for pretreatment. When the pretreatment price is 2.5 ¢/m^3, pretreatment cost is comparable to these for energy and membrane. It becomes the predominant

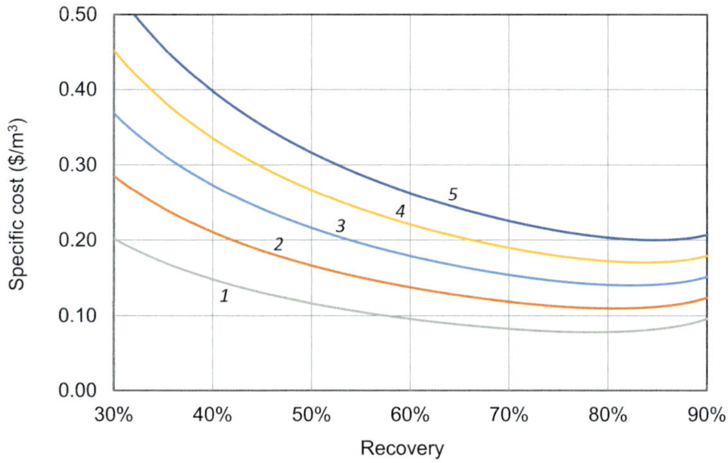

FIGURE 6.8
Specific costs in reverse osmosis brackish desalination processes with different pretreatment prices: (1) 0 ¢/m^3, (2) 2.5 ¢/m^3, (3) 5.0 ¢/m^3, (4) 7.5 ¢/m^3, and (5) 10 ¢/m^3. Parameters of membrane channel are given in Table 5.3.

cost component when the pretreatment price is greater than 5.0 ¢/m^3. Energy and membrane costs have about the same weights that decrease with the increasing pretreatment price. Concentrate handling remains the smallest cost component in all cases and become trivial when the pretreatment cost is not zero.

TABLE 6.7
Optimal Design Parameters for Different Pretreatment Prices

Pretreatment price (¢/m^3)	0.0	2.5	5.0	7.5	10.0
Optimal recovery (%)	78.4	81.0	82.8	83.9	85.0
Optimal pressure (bar)	8.9	9.9	11.2	12	12.8
Membrane area (m^2)	3796	3674	3594	3547	3501

TABLE 6.8
Optimal Design Parameters for Different Pretreatment Prices

Pretreatment price (¢/m^3)	0.0	2.5	5.0	7.5	10.0
Specific cost (¢/m^3)	7.8	10.9	14.0	17.0	20.0
Pretreatment (%)	0.0	28.4	43.1	52.5	58.9
Energy (%)	40.5	31.3	26.8	23.3	20.9
Membrane (%)	48.9	33.8	25.6	20.8	17.5
Concentrate handling (%)	10.6	6.5	4.4	3.1	2.6

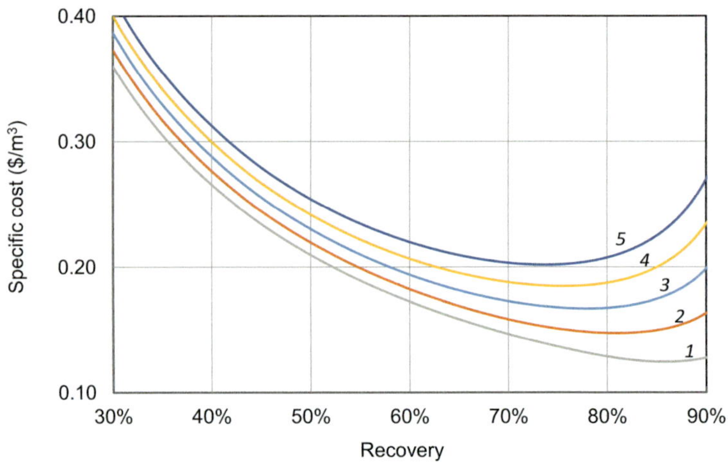

FIGURE 6.9
The specific cost in reverse osmosis brackish desalination processes with different feed concentration: (1) 1500 mg/L, (2) 3000 mg/L, (3) 4500 mg/L, (4) 6000 mg/L, and (5) 7500 mg/L. Parameters of membrane channel are given in Table 5.3.

6.4.2 Feed Water Salinity

Another important parameter, salt concentration of feed water, is investigated for its impact on process optimization. Salt concentrations of 1500, 3000, 4500, 6000, and 7500 mg/L are used in this investigation. Capacity of the desalination plant is 1000 m^3/day. Other parameters used are the same as in Table 5.3. Specific costs are presented as functions of recovery in Figure 6.9. It is obvious that specific cost increases but the optimal recovery decreases with increasing feed water salinity. Detailed information about the optimal design parameters and cost components are presented in Tables 6.9 and 6.10, respectively.

Table 6.9 shows that applied pressure again changes most significantly with feed concentration. It increases from 8.2 to 22.3 bars for feed salt concentration from 1500 to 7500 mg/L, which is about 171% increase in applied pressure. However, the corresponding changes in the recovery and membrane area are much more moderate with about 13.6 and 15.7%, respectively, for the same change in feed salinity.

TABLE 6.9
Optimal Design Parameters for Different Feed Concentrations

Feed concentration (mg/L)	1500	3000	4500	6000	7500
Optimal recovery (%)	85.9	80.9	77.4	75.8	74.2
Optimal pressure (bar)	8.2	12.1	15.4	19.1	22.3
Membrane area (m^2)	3465	3679	3845	3926	4011

TABLE 6.10

Total Specific Cost and Component Percentages for Different Feed Concentrations

Feed concentration (mg/L)	1500	3000	4500	6000	7500
Specific cost (¢/m³)	12.4	14.7	16.7	18.5	20.1
Pretreatment (%)	46.8	42.0	38.7	35.7	33.5
Energy (%)	21.4	28.2	33.1	37.9	41.4
Membrane (%)	27.9	25.0	23.0	21.2	19.9
Concentrate handling (%)	4.0	4.8	5.2	5.2	5.2

It can be seen from Table 6.10 that pretreatment is the largest cost component for feed concentrations up to feed salt concentration of 4500 mg/L and remains one of the major cost components for higher feed salt concentrations. Energy is the second largest cost component for feed concentrations of 3000 and 4500 mg/L and overtakes pretreatment as the largest cost component for feed concentrations of 6000 and 7500 mg/L. Cost associated with concentrate handling is the smallest cost component, which is about 5% of the total cost with all feed salt concentrations.

From the optimization studies, it can be seen that pretreatment is the largest cost component in brackish water membrane desalinations processes in the common range of salt concentration (<4500 mg/L). Development of more efficient and affordable pretreatment methods is the most effective way to reduce the cost of brackish water desalination. On the other side, development of more permeable membranes can only reduce desalination cost marginally because it is not the largest cost component after all.

6.5 Seawater Desalination

Seawater desalination membrane processes are different from brackish desalination membrane processes in that they all work in the regime of thermodynamic restriction. Therefore, there is no need of the graphical method to determine the applied pressure as in the brackish desalination cases. The pressure for a given recovery in seawater desalination can be simply determined with Eq. (5.32) as

$$\Delta p = \frac{\Delta \pi_0}{(1-R)}$$

Substituting it into Eq. (6.8), the cost function becomes

$$c_s = c_1 \frac{1}{R} + c_e \frac{\Delta \pi_0}{R(1-R)} + c_m \frac{L}{u_0 HR} + c_3 \frac{(1-R)}{R} \tag{6.13}$$

Eq. (6.13) is an explicit function of a single variable R. Therefore, the recovery that minimizes specific cost can be analytically determined as below.

Taking derivative of Eq. (6.13) with respect to R,

$$\frac{dc_s}{dR} = -\left(c_1 + c_e \Delta\pi_0 + c_m \frac{L}{u_0 H} + c_3 \right) \frac{1}{R^2} + c_e \Delta\pi_0 \frac{1}{(1-R)^2}$$

Let the derivative equal to zero:

$$-\left(c_1 + c_e \Delta\pi_0 + c_m \frac{L}{u_0 H} + c_3 \right) \frac{1}{R^2} + c_e \Delta\pi_0 \frac{1}{(1-R)^2} = 0$$

Rearranging results in

$$\left[1 + \frac{1}{c_e \Delta\pi_0} \left(c_1 + c_m \frac{L}{u_0 H} + c_3 \right) \right] (1-R)^2 - R^2 = 0 \quad (6.14)$$

Let,

$$\omega = \frac{1}{c_e \Delta\pi_0} \left(c_1 + c_m \frac{L}{u_0 H} + c_3 \right) \quad (6.15)$$

where ω is a dimensionless parameter that characterizes the cost structure for seawater desalination with membrane processes. Therefore, it is named as *the compound cost coefficient* of a membrane seawater desalination process.

Substituting Eq. (6.15) into Eq. (6.14), we have

$$(1+\omega)(1-R)^2 - R^2 = 0$$

Rearranging results in

$$\omega R^2 - 2(1+\omega)R + (1+\omega) = 0$$

This is a quadratic equation of R, the roots of this equation are

$$R = \frac{2(1+\omega) \pm \sqrt{4(1+\omega)^2 - 4\omega(1+\omega)}}{2\omega} = \frac{1+\omega \pm \sqrt{1+\omega}}{\omega}$$

Because $R < 1$, minus sign before the square root operation must be chosen. The final result is

$$R = \frac{\omega + 1 - \sqrt{1+\omega}}{\omega} = 1 + \frac{1}{\omega}\left(1 - \sqrt{1+\omega}\right) \quad (6.16)$$

With Eqs. (6.13) and (6.16), the optimization of seawater desalination membrane processes is simple and straightforward. The following case study is carried out to illustrate the use of these two equations.

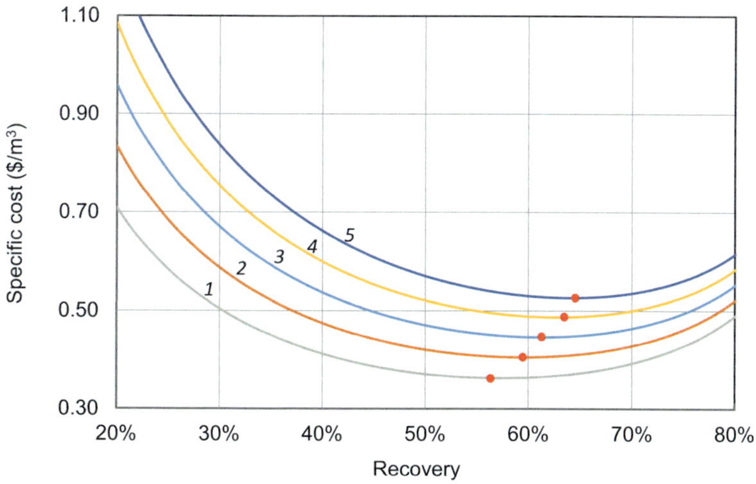

FIGURE 6.10
The specific costs in reverse osmosis seawater desalination processes with different pretreatment alternatives: (1) 0 ¢/m^3, (2) 2.5 ¢/m^3, (3) 5.0 ¢/m^3, (4) 7.5 ¢/m^3, and (5) 10 ¢/m^3. Parameters of membrane channel are given in Table 5.6.

The case study is to find out the minimal specific cost for a seawater desalination membrane process with five alternative pretreatment methods. The capacity of the desalination plant is 1000 m^3/d. The parameters of the membrane process are given in Table 5.6. The electricity price, membrane price, and concentrate handling price are 10 ¢/kWh, 0.01 ¢/(m^2·d), and 0.03 ¢/m^3, respectively. The five pretreatment alternative prices used in process optimization are 0, 2.5, 5.0, 7.5, and 10 ¢/m^3.

The specific costs for five pretreatment alternatives are calculated as functions of recovery with Eq. (6.13). The results are presented in Figure 6.10. The minimal specific costs are visually identified and indicated on the curves with small dots in the figure. The optimal design parameters thus found are listed in Table 6.11.

The osmotic pressure of seawater is

$$\Delta\pi_0 = f_{os}c_0 = 73.9\,Pa/(mg/L)\times 35{,}000\,mg/L = 2.59\times 10^6\,Pa = 25.9\,bar$$

The optimal recoveries in Table 6.11 can be directly calculated with Eqs. (6.15) and (6.16). It is demonstrated below.

TABLE 6.11

Optimal Design Parameters for Seawater Desalination

Pretreatment price (¢/m^3)	0.0	2.5	5.0	7.5	10.0
Optimal recovery (%)	57.0	59.4	61.4	63.2	64.1
Optimal pressure (bar)	60.1	63.7	67.0	70.3	72.1
Membrane area (m^2)	4641	4454	4312	4186	4127

TABLE 6.12

Optimal ω and Recovery in Seawater Desalination

Pretreatment price (¢/m^3)	0.0	2.5	5.0	7.5	10.0
ω	0.786	1.134	1.482	1.830	2.178
Recovery (%)	57.2	59.4	61.2	62.7	64.1

For pretreatment alternative 1 (no treatment) $c_1 = 0$

$$\omega = \frac{1}{c_e \Delta\pi_0}\left(c_1 + c_m \frac{L}{u_0 H} + c_3\right)$$

$$= \frac{0 + \$1.16\times10^{-7}/(m^2\cdot s)\dfrac{8m}{0.1\, m/s \times 0.35\times10^{-3}\, m} + \$0.03/m^3}{\$2.78\times10^{-8}/(Pa\cdot m^3)\times 2.59\times10^6\, Pa}$$

$$= 0.786$$

$$R = 1 + \frac{1}{\omega}\left(1-\sqrt{1+\omega}\right) = 1 + \frac{1}{0.786}\left(1-\sqrt{1+0.786}\right) = 57.2\%$$

The calculations are repeated for the other four pretreatment alternatives. The results are shown in Table 6.12.

Compare the recoveries in Tables 6.11 and 6.12, the differences are smaller than 1%. Actually, the recoveries determined analytically are the exact values. The graphically determined recoveries have some errors due to the limited accuracy of visual identification of the values.

Example 6.5

Energy consumption in seawater desalination membrane processes can be significantly reduced by installation of energy recovery devices (ERDs) in the concentrate stream. (1) Derive the equation of specific energy for seawater desalination membrane process with an ERD to recover the energy in concentrate stream with 100% efficiency. (2) What is the equation for optimal recovery for such a membrane process?

SOLUTION

When there is an ERD to recover energy in the concentrate stream, the feed flowrate $Q_f\left(=\dfrac{Q_f}{R}\right)$ in second term for energy cost in Eq. (6.7) should be replaced by permeate flowrate Q_p. The total cost equation becomes

$$C = c_1\frac{Q_p}{R} + c_e Q_p \Delta p + c_m\frac{Q_p}{\overline{v}} + c_3\frac{(1-R)Q_p}{R} \tag{s1}$$

Following the same procedure as derive Eq. (6.8), the specific cost for membrane processes with the ERD is

$$c_s = c_1 \frac{1}{R} + c_e \Delta p + c_m \frac{L}{u_0 HR} + c_3 \frac{(1-R)}{R} \tag{s2}$$

The applied pressure in seawater membrane processes is linked to the recovery by Eq. (5.32). Then we have

$$c_s = c_1 \frac{1}{R} + c_e \frac{\Delta \pi_0}{(1-R)} + c_m \frac{L}{u_0 HR} + c_3 \frac{(1-R)}{R} \tag{s3}$$

Eq. (s3) is the equation of specific cost for membrane desalination processes with the ERD with 100% efficiency.

Taking derivative of Eq. (s3) and let it equal to zero

$$\frac{dc_s}{dR} = -\left(c_1 + c_m \frac{L}{u_0 H} + c_3 \right) \frac{1}{R^2} + c_e \Delta \pi_0 \frac{1}{(1-R)^2} = 0$$

Rearranging

$$\left[\frac{1}{c_e \Delta \pi_0} \left(c_1 + c_m \frac{L}{u_0 H} + c_3 \right) \right] (1-R)^2 - R^2 = 0$$

Let

$$\omega = \frac{1}{c_e \Delta \pi_0} \left(c_1 + c_m \frac{L}{u_0 H} + c_3 \right)$$

Then we have

$$\omega (1-R)^2 - R^2 = 0$$

Rearranging results in

$$\omega - 2\omega R + \omega R^2 - R^2 = 0$$

$$(\omega - 1) R^2 - 2\omega R + \alpha = 0$$

This is a quadratic equation of R, the roots of this equation are

$$R = \frac{2\omega \pm \sqrt{4\omega^2 - 4(\omega - 1)\omega}}{2(\omega - 1)} = \frac{\omega \pm \sqrt{\omega}}{\omega - 1}$$

Because R < 1, minus sign before the square root operation must be chosen. The final result is

$$R = \frac{\omega - \sqrt{\omega}}{\omega - 1} = \frac{\sqrt{\omega}}{\sqrt{\omega} + 1} \tag{s4}$$

Example 6.6

A seawater desalination process has parameters as listed in Table 5.6. Pretreatment price, electricity price, membrane price, and concentrate handling price are 5 ¢/m^3, 10 ¢/kWh, 0.01 ¢/(m^2·d), and 0.03 ¢/m^3, respectively. (1) Determine the optimal recovery to minimize specific cost, and (2) determine the minimal specific cost.

SOLUTION

(1) The optimal recovery

$$\Delta\pi_0 = f_{os}c_0 = 73.5\ Pa/(mg/L)\times 35{,}000\ mg/L = 2.59\times 10^6\ Pa = 25.7\ bar$$

$$\omega = \frac{c_1 + c_m \dfrac{L}{u_0 H} + c_3}{c_e \Delta\pi_0}$$

$$= \frac{\$0.05/m^3 + \$1.16\times 10^{-7}/(m^2\cdot s)\dfrac{8\ m}{0.1\ m/s\times 0.35\times 10^{-3}\ m} + \$0.03/m^3}{\$2.78\times 10^{-8}/(Pa\cdot m^3)\times 2.57\times 10^6\ Pa}$$

$$= 1.48$$

$$R = \frac{\sqrt{\omega}}{\sqrt{\omega}+1} = \frac{\sqrt{1.48}}{\sqrt{1.48}+1} = 54.9\%$$

(2) The minimal specific cost

$$c_s = c_1\frac{1}{R} + c_e\frac{\Delta\pi_0}{(1-R)} + c_m\frac{L}{u_0 HR} + c_3\frac{(1-R)}{R}$$

$$= \frac{\$0.05/m^3}{54.9\%} + \$2.78\times 10^{-8}/(Pa\cdot m^3)\frac{2.59\times 10^6\ Pa}{(1-54.9\%)}$$

$$+1.16\times 10^{-7}\frac{8\ m}{0.1\ m/s\times 0.35\times 10^{-3}\ m\times 54.9\%} + \frac{\$0.03/m^3(1-54.9\%)}{54.9\%}$$

$$= \$0.32/m^3$$

6.6 Problems

1. What are the major cost components for membrane desalination?
2. A membrane desalination plant with a capacity of $1000 m^3/d$ (permeate) is operated at a recovery of 70%. What is the volume of feed water to the plant daily? What is the volume of concentrate discharged from the plant daily?
3. Convert the energy of 1 kWh to the unit of Joule.
4. What is the energy needed to pump feedwater of 1500 m^3 at 6 bars into membrane process if the pump efficiency is assumed to be 100%?
5. What is the energy cost in Question 4 if the electricity price is 10 ¢/kWh?

6. A standard SWRO membrane element is available at the price of \$800. The membrane element has a membrane area of 40 m^2 and a warranty service time of five years. The price of a pressure vessel that holds eight membrane elements is \$2500. The service time of the pressure vessel is 25 years. Calculate the costs for using a square meter membrane per year or per day.
7. A membrane process of capacity 1000 m^3/day is designed to work at the pressure of 10 bars with an average permeate flux of 5×10^{-6} m/s and a recovery of 75%. The membrane price and electricity prices are estimated at \$3.00 per square meter per year and \$0.10/kWh. Determine
   a. the energy cost,
   b. the membrane cost, and
   c. the total cost with membrane filtration.
8. In Question 7, if the prices for pretreatment and concentrate handling are 1 ¢ and 0.5 ¢/m^3, determine the costs associated with pretreatment and concentrate handling, respectively.
9. With the results from Questions 7 and 8, determine
   a. the total cost and
   b. the specific cost of the membrane process.
10. The height and length of a membrane channel are 0.0035 and 9 m, respectively. Calculate the average of the membrane channel for the recovery of 75% when the feed velocity at the entrance of the membrane channel is 0.12 m/s.
11. Use graphical method to find the pressure for a brackish desalination membrane channel to have a recovery of 75%. The parameters of the membrane channel are listed below:

Parameters of a brackish desalination membrane channel

Parameter	Symbol	Value	Unit
Membrane resistance	R_m	1.0×10^{11}	Pa·s/m
Channel height	H	3.5×10^{-4}	m
Salt concentration	c_0	2000	mg/L
Osmotic pressure coefficient	f_{os}	80	Pa/(mg/L)
Feed velocity	u_0	0.1	m/s
Channel length	L	9	m

12. Determine the specific energy of a membrane process that works at a pressure of 10 bar and a recovery of 70%.
13. Calculate the specific cost of desalination of brackish water with a salinity of 5000 mg/L as a function of the recovery and identify the

optimal recovery that minimizes the specific cost. The other parameters are given below.

Parameter	Symbol	Value	Unit
Membrane resistance	R_m	1.1×10^{11}	Pa·s/m
Channel height	H	3.5×10^{-4}	m
Osmotic pressure coefficient	f_{os}	80	Pa/(mg/L)
Feed velocity	u_0	0.12	m/s
Stage number		2	
Pressure vessel length	L	6	m
Electricity price	c_e	10	¢/kWh
Pretreatment price	c_1	3	¢/m^3
Membrane price	c_m	1	¢/(m^2·day)
Concentrate handling price	c_3	1	¢/m^3

14. With the information obtained in Question 13, Determine
    a. the optimal recovery,
    b. calculate the average permeate flux, and
    c. total membrane area when the capacity of the desalination plant is 1000 m^3/day.
15. Calculate the specific cost for a membrane desalination process for brackish water of 5000 mg/L salinity. The parameters are the same as given in Question 13 except for pressure vessel length, which is 4 m in this question. Find the optimal recovery, average permeate flux, and total membrane area in this membrane process for a capacity of 1000 m^3/day.
16. Find the optimal recovery and the minimal specific cost for a seawater desalination channel with the parameters given below.

Parameter	Symbol	Value	Unit
Channel height	H	3.5×10^{-4}	m
Membrane resistance	R_m	3.0×10^{11}	Pa·s/m
Osmotic pressure coefficient	f_{os}	74	Pa/(mg/L)
Feed velocity	u_0	0.10	m/s
Channel length	L	8	m
Salt concentration	c_0	40,000	mg/L
Electricity price	c_e	10	¢/kWh
Pretreatment price	c_1	3	¢/m^3
Membrane price	c_m	1.5	¢/(m^2·day)
Concentrate handling price	c_3	0.5	¢/m^3

17. For the same membrane channel as in Question 15, what are the optimal recovery and minimal specific cost if the ERD is installed with 100% efficiency to recover the energy in concentrate stream?
18. A seawater desalination membrane channel has the parameters as listed in the table below.

Parameter	Symbol	Value	Unit
Channel height	H	3.5×10^{-4}	m
Feed velocity	u_0	0.10	m/s
Channel length	L	8	m
Salt concentration	c_0	35,000	mg/L
Osmotic pressure coefficient	f_{os}	73.5	Pa/(mg/L)
Electricity price	c_e	10	¢/kWh
Pretreatment price	c_1	3.0	¢/m^3
Membrane price	c_m	1.5	¢/(m^2·day)
Concentrate handling price	c_3	1.0	¢/m^3

a. Calculate the compound cost coefficient of this membrane seawater desalination process,
b. Determine the optimal recovery to minimize the specific cost, and
c. Calculate the minimum specific cost.

19. Energy is a major cost component in seawater desalination. If the electricity price reduces from 10 to 5 ¢/kWh due to either invention of the cheap energy or subsidization from government, determine (a) the compound cost coefficient, (b) the optimal recovery, and (c) the minimum specific cost of a membrane process with other parameters as given in Question 18.
20. Calculate (a) the compound cost coefficient, (b) the optimal recovery, and (c) the minimum specific cost of a membrane process when the membrane price reduces to 1.0 ¢/(m^2·day) while other parameters remain the same as given in Question 18.
21. Calculate (a) the compound cost coefficient, (b) the optimal recovery, and (c) the minimum specific cost of a membrane process when the pretreatment price reduces to 2.0 ¢/m^3 while other parameters remain the same as given in Question 18.
22. The ERD is installed in a seawater desalination membrane channel that has the same parameters as in Question 18. The device is assumed to recover the energy in the concentrate stream with 100% efficiency.

a. Determine the optimal recovery to minimize the specific cost, and
b. Calculate the minimum specific cost.

23. Recalculate Question 19 with an ERD installed to recover energy in concentrate stream completely.
24. Recalculate Question 20 with an ERD installed to recover energy in concentrate stream completely.
25. Recalculate Question 21 with an ERD installed to recover energy in concentrate stream completely.

References

1. Ahunbay, M.G., Tantekin-Ersolmaz, Ş.B., & Krantz, W.B. (2018). Energy optimization of a multistage reverse osmosis process for seawater desalination. Desalination, 429, 1–11.
2. Amusat, O.O., Atia, A.A., Dudchenko, A.V., & Bartholomew, T.V. (2024). Modeling framework for cost optimization of process-scale desalination systems with mineral scaling and precipitation. ACS Es&t Engineering, 4, 1028–1047.
3. Aridi, R., Al Mawla, M., Harika, E., Lemenand, T., Khaled, M., & Gad El-Rab, M. (2024). A new algorithmic method for reverse osmosis desalination analysis: Design optimization and parametric study. Eng, 5(3), 1183–1208.
4. Alspach, B., Abada, B., & Chellam, S. (2023). The future of reverse osmosis optimization, part 1: Intelligent water. Journal – American Water Works Association, 115, 42–47.
5. Bartholomew, T.V., Siefert, N.S., & Mauter, M.S. (2018). Cost optimization of osmotically assisted reverse osmosis. Environmental Science & Technology, 52(20), 11813–11821.
6. Mirghaderi, F., Rahmanian, N., Patel, R., Manca, D., & Mujtaba, I. (2017). Simulation and optimization of a continuous reverse osmosis desalination process for making fresh water. Chemical Engineering Transactions, 61, 1783–1788.
7. Cho, H.U., Cho, K.H., Kang, S., & Kim, Y.M. (2016). Optimization of operating variables in a pilot-scale reverse osmosis membrane process for reclamation of tunnel construction wastewater. Desalination and Water Treatment, 57, 12082–12089.
8. Cai, Q., Lee, B.C.Y., Ong, S.L., & Hu, J. (2021). Application of a multiobjective artificial neural network (ANN) in industrial reverse osmosis concentrate treatment with a fluidized bed Fenton process: Performance prediction and process optimization. ACS ES&T Water, 1(4), 847–858.
9. Haefner, M.W., & Haji, M.N. (2022). Integrated pumped hydro reverse osmosis system optimization with enhanced reverse osmosis modeling. Energy Proceedings, 23(2021), 1–9.
10. Reverberi, A.P., Meshalkin, V.P., Cerrato, C., & Savina, Y.O. (2011). Dynamics of a reverse osmosis unit with application to pulsating regimes for process optimization. Theoretical Foundations of Chemical Engineering, 45, 190–197.
11. Jeong, K., Park, M., & Chong, T.H. (2019). Numerical model-based analysis of energy-efficient reverse osmosis (EERO) process: Performance simulation and optimization. Desalination, 453, 19–21.

12. Saini, M., Kumar, N., Sinwar, D., & Kumar, A. (2024). Stochastic modeling and availability optimization of reverse osmosis water purification system using metaheuristic algorithms. Journal of Quality in Maintenance Engineering, 30(4), 584–604.
13. McMordie Stoughton, K.L., Duan, X., & Wendel, E.M. (2013). Reverse Osmosis Optimization (Tech. Rep. No. PNNL-22652). Pacific Northwest National Laboratory (PNNL), Richland, WA.
14. Haidar, Z.A., Al-Saud, M., Orfi, J., & Al-Ansary, H. (2021). Reverse osmosis desalination plants energy consumption management and optimization for improving power systems voltage stability with PV generation resources. Energies, 14(22), 7739.
15. Lim, S.J., Ki, S.J., Lim, J.-L., Lee, K.-H., Kim, J., Moon, J.-G., & Kim, J. (2022). Optimization of the design configuration and operation strategy of single-pass seawater reverse osmosis. Membranes, 12, 1145.
16. Zhu, Z., Tan, G., Lei, D., Yang, Q., Tan, X., Liang, N., & Ma, D. (2021). Omniphobic membrane with process optimization for advancing flux and durability toward concentrating reverse-osmosis concentrated seawater with membrane distillation. Journal of Membrane Science, 639, 119763.
17. Najdawi, F.Z., & Neptune, K.T. (2022). Optimizing reverse osmosis membrane parameters through the use of the solution-diffusion model: A review. Engineering, 14, 9–32.
18. Seo, J., Kim, Y.-M., Chae, S.H., Lim, S.J., Park, H., & Kim, J. (2019). An optimization strategy for a forward osmosis-reverse osmosis hybrid process for wastewater reuse and seawater desalination: A modeling study. Desalination, 463, 40–49.
19. Jeong, K., Park, M., Ki, S.J., & Kim, J. (2017). A systematic optimization of internally staged design (ISD) for a full-scale reverse osmosis process. Journal of Membrane Science, 540, 285–296.
20. Leon, F., Ramos, A., & Perez-Baez, S.O. (2021). Optimization of energy efficiency, operation costs, carbon footprint and ecological footprint with reverse osmosis membranes in seawater desalination plants. Membranes, 11, 781.
21. Abba, S.I., Usman, J., Abdulazeez, I., Lawal, D.U., Baig, N., Usman, A.G., & Aljundi, I.H. (2023). Integrated modeling of hybrid nanofiltration/reverse osmosis desalination plant using deep learning-based crow search optimization algorithm. Water, 15(19), 3515.
22. Hadadian, Z., Zahmatkesh, S., Ansari, M., Haghighi, A., & Moghimipour, E. (2021). Mathematical and experimental modeling of reverse osmosis (RO) process. Korean Journal of Chemical Engineering, 38, 366–379.
23. Aghilesh, K., Chaturvedi, A., Ali, J., Singh, R., Agarwal, S., & Garg, M.C. (2021). Response surface methodology-based modeling and optimization of chromium removal using spiral-wound reverse-osmosis membrane setup. International Journal of Environmental Science and Technology, 19, 5999–6010.
24. Madaeni, S.S., Shiri, M., & Kurdian, A.R. (2015). Modeling, optimization, and control of reverse osmosis water treatment in Kazeroon power plant using neural network. Chemical Engineering Communications, 202, 14–16.
25. Mahadeva, R., Kumar, M., Manik, G., & Patole, S.P. (2021) Modeling, simulation, and optimization of the membrane performance of seawater reverse osmosis desalination plant using neural network and fuzzy based soft computing techniques. Desalination and Water Treatment, 229, 17–30.
26. Emamjome, A., Mahdi Zahedi, M., & Ziyaadini, M. (2019). Economic analysis for process optimization of Chabahar maritime university reverse osmosis desalination plant: A case study. Applied Water Science, 9, 1–8.

27. Virapan, S., Saravanane, R., & Murugaiyan, V. (2016). Reverse osmosis desalination performance using artificial neural network approach with optimization. Asian Journal of Water, Environment and Pollution, 13, 95–102.
28. Yousra, J., & Souad, A. (2022). Reverse osmosis process optimization for acetic acid rejection, by coupling RSM to PSO. 2022 2nd International Conference on Innovative Research in Applied Science, Engineering and Technology (IRASET), 1–6.
29. Avlonitis, S.A., Pappas, M., & Moutesidis, K. (2007). A unified model for the detailed investigation of membrane modules and RO plants performance. Desalination, 203, 218–228.
30. Liang, S., Liu, C., & Song, L. (2009). Two-step optimization of pressure and recovery of reverse osmosis desalination process. Environmental Science & Technology, 43, 3272–3277.
31. Song, L. (2024). Optimization of seawater desalination processes with the ideal reverse osmosis equation. Desalination, 576, 117376.

7

Concentration Polarization Assessment

Retained salt is assumed to be completely mixed in the transverse direction across the entire height in the long narrow membrane channel in Chapter 5. A single value is used to indicate salt concentration at a location along the membrane channel. In other words, concentration polarization is not considered in the state equation to calculate recovery (equivalently average permeate flux) of a membrane channel. However, there is always a concern about the impact of concentration polarization on the performance of the long narrow channel.

Concentration polarization in a long narrow channel is also very different from that in the model membrane system discussed in Chapter 3 [1–9]. In the model membrane system, concentration polarization is assumed to develop in membrane channels that have sufficiently large heights (theoretically the semi-plane over the membrane surface). The thickness of the concentration polarization layer is always much smaller than the channel height. Therefore, there is always a bulk solution above the concentration polarization layer in which salt concentration is unaffected by concentration polarization. Furthermore, hydraulic conditions (mainly tangential velocity) in the channel are not affected by the loss of water as permeate. The transverse velocity in the concentration polarization layer can be assumed to be constant and equal to the permeate velocity (flux).

These conditions for concentration polarization in the model membrane system for sure cannot be used in a long narrow channel [10,11]. In the long narrow channel, the concentration polarization layer can develop to the full channel height. Because of the required substantial recovery in practical desalination applications, the tangential velocity and salt concentration no longer remain constant along the channel length. Transverse flow velocity cannot be assumed constant across the concentration polarization layer, as will be shown later.

In this chapter, mathematical models for concentration polarization will be constructed with conditions that are pertinent to the long narrow channel. Then concentration polarization will be simulated under various conditions. Finally, the impact of concentration polarization on the performance of the long narrow membrane channel will be assessed.

7.1 Flow Field in a Long Narrow Membrane Channel

Compared to concentration polarization in the semi-plane over a membrane, the constant tangential velocity and transverse velocity cannot be

DOI: 10.1201/9781003743569-9

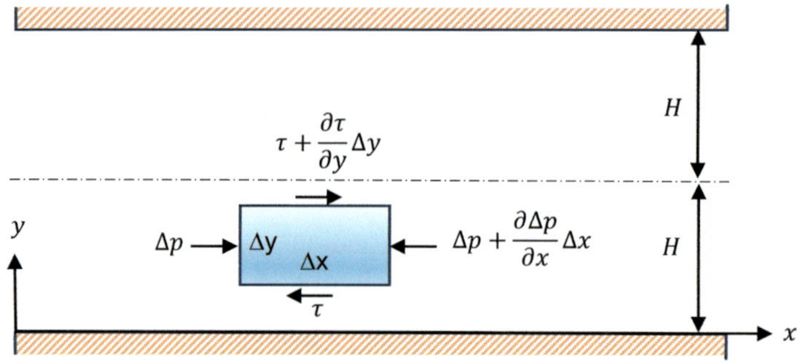

FIGURE 7.1
Force balance on a fluid particle in the membrane channel at steady state.

used in a long, narrow channel. Both tangential and transverse flow velocities vary significantly along the channel length or across the channel height [12–15]. It is imperative to determine the flow field in the channel because it is a major force for concentration polarization to develop over the membrane surface.

7.1.1 Tangential Flow Velocity

Tangential flow is assumed to be fully developed at any point x in the membrane channel. Figure 7.1 shows a fluid particle inside the channel with sides of Δx and Δy, respectively. There are four forces acting on the fluid particle: two pressure forces $\Delta p \Delta y$ and $-\left(\Delta p + \frac{\partial \Delta p}{\partial x}\Delta x\right)\Delta y$ on the left and right sides, and two shear forces $-\tau\Delta x$ and $\left(\tau + \frac{\partial \tau}{\partial y}\Delta y\right)\Delta x$ on the top and the bottom, respectively. At steady state, the total force is equal to zero, i.e.,

$$\Delta p \Delta y - \left(\Delta p + \frac{\partial \Delta p}{\partial x}\Delta x\right)\Delta y - \tau\Delta x + \left(\tau + \frac{\partial \tau}{\partial y}\Delta y\right)\Delta x = 0$$

where Δp is the pressure in the fluid, and τ is the shear stress.

Eliminating the same terms, the equation becomes

$$\frac{\partial \Delta p}{\partial x} = \frac{\partial \tau}{\partial y} \tag{7.1}$$

The shear stress τ in Eq. (7.1) is related to the velocity gradient with

$$\tau = \mu \frac{\partial u}{\partial y} \tag{7.2}$$

where μ is the viscosity of the fluid.

Substituting Eq. (7.2) into Eq. (7.1) results in

$$\frac{\partial \Delta p}{\partial x} = \mu \frac{\partial^2 u}{\partial y^2} \tag{7.3}$$

Eq. (7.3) is the governing equation for tangential flow velocity in the membrane channel.

No-slip condition at the channel walls (membrane surfaces) can be used for the boundary conditions of this problem. In the coordinates and parameters shown in Figure 7.1, the boundary conditions can be written as

$$u = 0 \ at \ y = 0$$

and

$$u = 0 \ at \ y = 2H$$

Integrating Eq. (7.3) for u twice gives

$$u = \frac{1}{2\mu} \frac{\partial \Delta p}{\partial x} y^2 + a_1 y + a_2 \tag{7.4}$$

where a_1 and a_2 are the integration constants.

Using the two boundary conditions, the two integration constants are determined as

$$a_2 = 0$$

$$a_1 = -\frac{1}{\mu} \frac{\partial \Delta p}{\partial x} H$$

Substituting a_1 and a_2 into Eq. (7.4) results in

$$u = \frac{1}{\mu} \frac{\partial \Delta p}{\partial x} \left(\frac{1}{2} y^2 - Hy \right)$$

The average tangential velocity can be calculated by integrating u over the channel height:

$$u_a = \frac{1}{H} \int_0^H u \, dy = -\frac{1}{3\mu} \frac{\partial \Delta p}{\partial x} H^2 \tag{7.5}$$

where u_a is the average tangential velocity. Combining Eq. (7.5) with the equation above, the final expression for tangential velocity (with the coordinates x and y explicitly included) is

$$u(x, y) = \frac{3}{2} u_a(x) \frac{y}{H} \left(2 - \frac{y}{H} \right) \tag{7.6}$$

Eq. (7.6) is different from the equation for the tangential velocity between two impermeable plates. Because of the loss of water as permeate along the membrane channel, the average tangential velocity is a function of x, so does the general tangential velocity. Both are invariants along the x direction in cases of two impermeable plates.

7.1.2 Transverse Flow Velocity

Because there is a permeate flow through the membrane, the fluid velocity in the transverse direction of the channel needs to be delineated. Transverse flow velocity in the membrane channel can be derived from the continuity equation:

$$\frac{\partial u}{\partial x}+\frac{\partial v}{\partial y}=0 \tag{7.7}$$

Rearranging Eq. (7.7) results in

$$\frac{\partial v}{\partial y}=-\frac{\partial u}{\partial x}$$

Substituting Eq. (7.6) into the equation, we have

$$\frac{\partial v}{\partial y}=-\frac{3}{2}\frac{\partial u_a(x)}{\partial x}\frac{y}{H}\left(2-\frac{y}{H}\right) \tag{7.8}$$

Because of the symmetry of the channel with respect to the central line, transverse velocity vanishes at the central line of the channel, i.e.,

$$v=0 \ at \ y=H \tag{7.9}$$

Eq. (7.8) can be integrated from $y = 0$ to H with Eq. (7.9) as the boundary condition at H:

$$v(x,y)=v_w(x)\left[1-\frac{1}{2}\left(\frac{y}{H}\right)^2\left(3-\frac{y}{H}\right)\right] \tag{7.10}$$

where $v_w(x)$ is the transverse velocity on the channel wall, which is the permeate flux through the membrane. With the mass balance principle, the transverse velocity on the membrane surface can be determined by

$$v_w(x)=-H\frac{du_a(x)}{dx} \tag{7.11}$$

Since only one independent variable x is involved, the partial differential symbol is replaced by an ordinary differential symbol in Eq. (7.11). The dimensionless transverse velocity is shown in Figure 7.2.

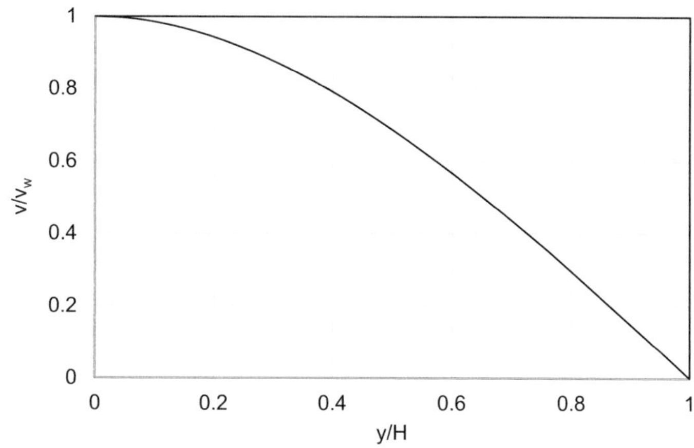

FIGURE 7.2
Dimensionless transverse velocity profile from membrane surface to the center of a channel.

Up to now, the flow field in the membrane channel is fully described by Eqs. (7.6) and (7.10). By the way, the pressure drop along the channel can be calculated from the average tangential velocity by integrating Eq. (7.5):

$$\Delta p(x) = \Delta p_0 - \frac{3\mu}{H^2} \int_0^x u_a(x')\,dx' \tag{7.12}$$

where Δp_0 is the pressure applied at the entrance of the membrane channel. Eq. (7.12) is actually Eq. (5.4) with $K = 1$.

7.2 Quantification of Concentration Polarization

Development of concentration polarization in a long narrow channel is different from that in the model membrane processes discussed in Chapter 3, where many parameters can be treated as constant, such as tangential flow velocity and the bulk concentration above the concentration polarization layer. Most parameters in the long narrow membrane channel can no longer be assumed constant. In fact, tangential flow velocity and salt concentration change significantly along the membrane channel. The impact of variable parameters on concentration polarization must be considered.

7.2.1 Polarization Factor

Before studying concentration polarization, an appropriate indicator for concentration polarization in the long narrow channel is needed. When

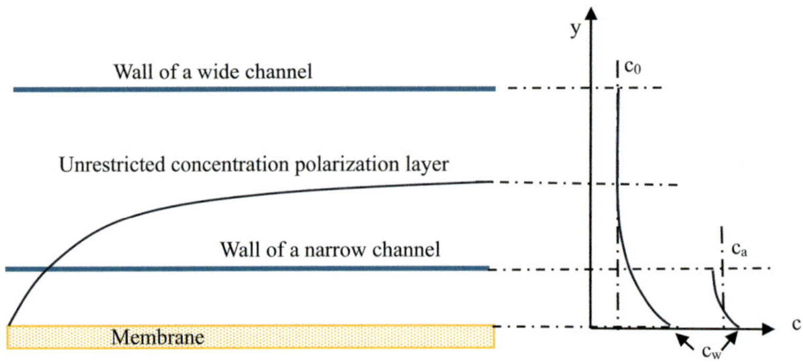

FIGURE 7.3
Thickness of unrestricted concentration polarization layer with upper walls of wide and narrow membrane channels, respectively. Transverse concentration distributions in the two channels are indicated on the right side.

concentration polarization was discussed in Chapter 3, the ratio of wall concentration to feed concentration c_w/c_0, which is commonly named concentration polarization modulus or factor, was used as the indicator for the extent or strength of polarization. It is an easy and appropriate indicator for concentration polarization in that context. The model membrane system has a channel height that is much greater than the thickness of the concentration polarization layer. As schematically illustrated in Figure 7.3, the "wall of a wide channel" is way above the concentration polarization layer in the channel. There are two profiles on the right side of the figure for concentration distribution along the height of the membrane channel. As indicated by the one with wider distribution, there is always a bulk solution above the concentration polarization layer with salt concentration c_0 intact. It is natural to use it as a reference to measure salt concentration on the membrane surface. The ratio $c_w/c_0 = 1$ means that there is no concentration polarization. But $c_w/c_0 > 1$ means that there is concentration polarization because of a higher wall concentration.

The situation in a narrow membrane channel is completely different. Also illustrated schematically in a channel with the "wall of a narrow channel" in Figure 7.3, the concentration polarization layer can develop to full channel height shortly into the channel and cannot further expand. The concentration profile in the narrow channel is shown on the right side of the figure as the curve with narrow distribution. The concentration in the entire channel height is greater than the feed concentration c_0 except for a short distance from the entrance. The ratio $c_w/c_0 > 1$ does not necessarily mean uneven transverse distribution of salt concentration, which is the definition of concentration polarization. For example, when a brackish desalination membrane channel works at a recovery of 80%, the average salt concentration by the exit of the channel is five times as high as the feed salt concentration, i.e., $c_a = 5c_0$. Even though there is no concentration polarization and salt distributes evenly in the transverse direction, i.e., $c_w = c_a$, the ratio c_w/c_0 is still 5! Therefore, the ratio

c_w/c_0 is no longer an appropriate indicator for concentration polarization in the long narrow membrane channel.

In a narrow channel where concentration polarization can develop to full height, the ratio of wall concentration to average concentration c_w/c_a can be a pertinent indicator for concentration polarization in the long narrow membrane channel. The ratio $c_w/c_a = 1$ means no concentration polarization where salt is evenly distributed throughout the channel height. On the other hand, the ratio $c_w/c_a > 1$ means concentration polarization occurs where wall concentration is truly higher than the concentration measured away from the membrane surface. Therefore, the ratio c_w/c_a will be used as the indicator of concentration polarization in the long narrow channel. This is the new concentration polarization factor suitable for use as an indicator for concentration polarization in long narrow membrane channels.

7.2.2 Concentration Polarization with Constant Transverse Velocity

Concentration polarization in a long narrow channel is governed by mass balance on retained salt. At any location x in the channel, the convective flux of retained salt to the membrane surface is always balanced by the diffusive flux of retained salt from the membrane surface due to the induced concentration gradient:

$$v(c - c_0) + D\frac{d(c - c_0)}{dy} = 0 \tag{7.13}$$

where c and c_0 are the salt concentration and the feed salt concentration, respectively, v is the transverse velocity, and D is the diffusion coefficient. The use of $c - c_0$ (retained salt) rather than c (all salt) indicates concentration polarization is only related to the retained salt, while the feed salt concentration c_0 is just the background.

The amount of retained salt at any location in the channel is

$$M = \int_0^H (c - c_0)\,dy = (c_a - c_0)H \tag{7.14}$$

where M is the amount of retained salt per unit membrane area, and c_a is the average salt concentration across the channel height. Transverse distribution of salt in the channel is completely defined by Eqs. (7.13) and (7.14).

The general solution of Eq. (7.13) can be found with the variable separation method as follows:

$$\int_{c_w}^{c} \frac{d(c - c_0)}{c - c_0} = -\int_0^y \frac{v}{D}\,dy \tag{7.15}$$

where c_w is the salt concentration on the wall of the channel (membrane surface).

In cases where v is constant, the above equation can be integrated directly to result in

$$c - c_0 = (c_w - c_0)e^{-\frac{v}{D}y} \tag{7.16}$$

Wall concentration c_w is unknown yet. It can be determined by substituting the above equation into Eq. (7.15):

$$\int_0^H (c_w - c_0)e^{-\frac{v}{D}y}\,dy = (c_a - c_0)H$$

Carrying out the integration on the left-hand side of the equation, we have

$$c_w - c_0 = (c_a - c_0)\frac{vH}{D}\frac{1}{\left(1 - e^{-\frac{v}{D}H}\right)} \tag{7.17}$$

The dimensionless parameter vH/D is Peclet number. The term $(c_a - c_0)$ on the right-hand side of Eq. (7.17) is related to the amount of retained salt by Eq. (7.14) as

$$(c_a - c_0) = \frac{M}{H}$$

The remaining part of the term on the right-hand side of Eq. (7.17) is assigned to the name of the polarization number (PN), i.e.,

$$PN = \frac{vH}{D}\frac{1}{1 - e^{-\frac{v}{D}H}} \tag{7.18}$$

where PN is the polarization number at a point in the membrane channel. PN combines all parameters that are relevant to concentration polarization, including permeate velocity, salt diffusion coefficient, and channel height. Now Eq. (7.17) can be rewritten as

$$c_w - c_0 = PN(c_a - c_0) \tag{7.19}$$

Eq. (7.19) is of fundamental importance to concentration polarization. It clearly indicates that the concentration polarization is caused by the accumulation of retained salt: no retained salt ($c_a - c_0 = 0$), no concentration polarization ($c_w - c_0 = 0$)!

Substituting Eq. (7.19) into Eq. (7.16) and rearranging result in

$$c = c_0 + PN(c_a - c_0)e^{-\frac{v}{D}y} \tag{7.20}$$

The salt concentration along the channel height can be calculated with this equation.

By the way, since sodium chloride is the major salt to remove in the desalination process, the diffusion coefficient of sodium chloride as a function of temperature is provided below for convenience [16]:

$$D = (0.44 + 0.0423T) \times 10^{-9}\ m^2/s \tag{7.21}$$

where D is the diffusion coefficient of sodium chloride, and T is temperature in degree Celsius.

Example 7.1

The parameters of a membrane process are c_0 = 2500 mg/L, $v = 4 \times 10^{-6}$ m/s, and T = 20°C. The amount of retained salt at the point of concern in the channel is M = 175 mg/m². Find the transverse distribution of salt concentration in the membrane channel for half channel heights of 0.35×10^{-3}, 0.7×10^{-3}, 1.40×10^{-3}, and 2.80×10^{-3} m, respectively.

SOLUTION

The diffusion coefficient of sodium chloride at 20°C is

$$D = (0.44 + 0.0423 \times 20) \times 10^{-9}\ m^2/s = 1.29 \times 10^{-9}\ m^2/s$$

Sample calculation for half channel height of 0.35×10^{-3} m; the average salt concentration is first calculated:

$$c_a = c_0 + \frac{M}{H} = 2500\ mg/L + \frac{175 \times mg/m^2}{0.35 \times 10^{-3}\ m} = 3000\ mg/L$$

The PN is calculated with Eq. (7.18):

$$PN = \frac{\frac{v}{D}H}{1 - exp\left(-\frac{v}{D}H\right)} = \frac{\frac{4 \times 10^{-6}\ m/s}{1.29 \times 10^{-9}\ m^2/s} \times 0.35 \times 10^{-3}\ m}{1 - exp\left(-\frac{4 \times 10^{-6}\ m/s}{1.29 \times 10^{-9}\ m^2/s} \times 0.35 \times 10^{-3}\ m\right)} = 1.64$$

The half channel height H is divided into ten equal segments for the calculation of concentration distribution. At the first point y = 0.10 H, the salt concentration at is calculated:

$$\begin{aligned} c &= c_0 + PN \times (c_a - c_0)\exp\left(-\frac{v}{D}y\right) \\ &= 2500\ mg/L + 1.64 \times (3000\ mg/L - 2500\ mg/L) \\ &\quad \times \exp\left(-\frac{4 \times 10^{-6}\ m/s \times 0.10 \times 0.35 \times 10^{-3}\ m}{1.28 \times 10^{-9}\ m^2/s}\right) \\ &= 3236\ mg/L \end{aligned}$$

Average salt concentration for all half channel heights is calculated and presented in the table below.

Average Salt Concentration for Different Channel Heights				
H (mm)	3.50	7.00	10.5	14.0
c_a (mg/L)	3000	2750	2667	2625

The transverse distributions of salt concentration in channels of different heights are presented in the table below.

Transverse Distributions of Salt Concentrations in Channels

	H (mm)			
y/H	**3.50**	**7.00**	**10.5**	**14.0**
0.00	3319	3113	3064	3050
0.10	3236	2993	2908	2856
0.20	3160	2897	2794	2731
0.30	3092	2819	2713	2649
0.40	3031	2757	2653	2597
0.50	2976	2707	2611	2563
0.60	2927	2667	2580	2541
0.70	2883	2634	2558	2526
0.80	2844	2608	2542	2517
0.90	2809	2587	2530	2511
1.00	2777	2570	2522	2507

The concentrations in the above table are also plotted versus dimensionless channel height y/H in the figure below.

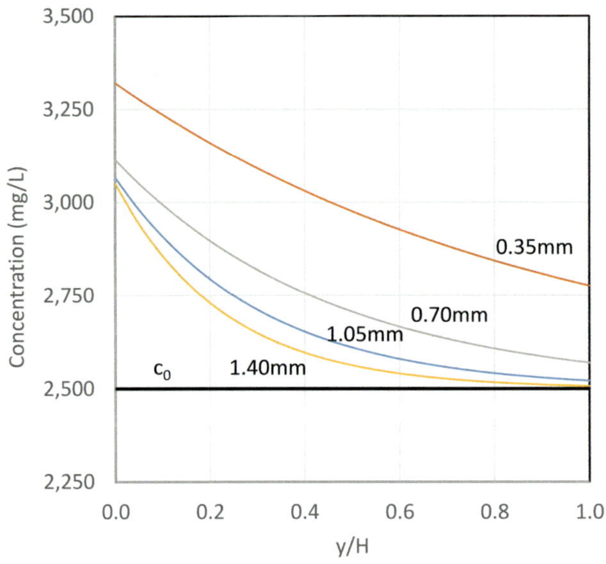

The figure shows the vertical distributions of salt concentration in membrane channels with different heights (indicated in the figure). y/H = 1 is the central line of membrane channels.

It is clear from the table and figure that concentration polarization develops to the whole height of the membrane channel even for the half channel height as big as 1.40 mm.

7.2.3 Concentration Polarization with Variable Transverse Velocity

When transverse flow velocity is not constant but a function of y, Eq. (7.15) cannot be integrated simply as was done with the constant transverse flow velocity. Eq. (7.11) is needed to substitute for v in Eq. (7.15) to transform it into the following form:

$$\int_{c_w}^{c} \frac{d(c-c_0)}{c-c_0} = -\frac{v_w(x)}{D} \int_0^y \left[1-\frac{1}{2}\left(3\left(\frac{y}{H}\right)^2-\left(\frac{y}{H}\right)^3\right)\right] dy$$

Carrying out the integrations, we have

$$c-c_0=(c_w-c_0)exp\left\{-\frac{v_w(x)H}{D}\left[\frac{y}{H}-\frac{1}{2}\left(\frac{y}{H}\right)^3+\frac{1}{8}\left(\frac{y}{H}\right)^4\right]\right\} \tag{7.22}$$

The wall concentration c_w in Eq. (7.22) is undetermined yet. Substituting Eq. (7.22) into Eq. (7.14), we obtain

$$\int_0^H (c_w-c_0)exp\left\{-\frac{v_w(x)H}{D}\left[\frac{y}{H}-\frac{1}{2}\left(\frac{y}{H}\right)^3+\frac{1}{8}\left(\frac{y}{H}\right)^4\right]\right\} dy=(c_a-c_0)H$$

The term (c_w-c_0) is constant and can be taken out of the integration operation. Rearranging the equation results in

$$(c_w-c_0)=\frac{H}{\int_0^H exp\left\{-\frac{v_w(x)H}{D}\left[\frac{y}{H}-\frac{1}{2}\left(\frac{y}{H}\right)^3+\frac{1}{8}\left(\frac{y}{H}\right)^4\right]\right\} dy}(c_a-c_0) \tag{7.23}$$

The integration term in the denominator is needed for the calculation of wall salt concentration.

Compared to Eq. (7.19) for the cases with constant transverse velocity, the expression in Eq. (7.23) between the terms of $(c_w - c_0)$ and $(c_a - c_0)$ is exactly the PN for the cases with a variable transverse velocity [10], i.e.,

$$PN = \frac{1}{\int_0^H exp\left\{-\left(\frac{v_w(x)H}{D}\right)\left[\frac{y}{H} - \frac{1}{2}\left(\frac{y}{H}\right)^3 + \frac{1}{8}\left(\frac{y}{H}\right)^4\right]\right\} d\left(\frac{y}{H}\right)} \tag{7.24}$$

When the PN is known, wall concentration can be calculated with Eq. (7.19).

The polarization number (PN) cannot be determined directly with Eq. (7.24) because the integration in it is not integrable analytically. However, the PN can be determined numerically in advance for a relevant range of parameter $\frac{v_w(x)H}{D}$. Usually, the reverse osmosis desalination processes have the permeate velocity $v_w < 10^{-5}\ m/s$, channel height $H < 10^{-3}\ m$, and diffusion coefficient $D > 10^{-9}\ m^2/s$. Therefore, a magnitude analysis gives $\frac{v_w H}{D} < 10$. The PNs in the $\frac{v_w H}{D}$ range of 0–10 are calculated with the integration term in Eq. (7.24) determined numerically. The results are listed in Table 7.1.

The PN values can be fitted perfectly ($R^2 = 1$) in MS Excel with a third-order polynomial trendline (Figure 7.4). The resultant polynomial equation is

$$PN = -3.59 \times 10^{-3} Pe^3 + 8.04 \times 10^{-2} Pe^2 + 3.74 \times 10^{-1} Pe + 1.00 \tag{7.25}$$

where Pe is the Peclet number of the membrane channel at location x, defined by

$$Pe = \frac{v_w(x)H}{D} \tag{7.26}$$

TABLE 7.1

Polarization Number

$v_w H/D$	PN
0	1
1	1.45
2	2.03
3	2.74
4	3.55
5	4.44
6	5.37
7	6.33
8	7.29
9	8.25
10	9.20

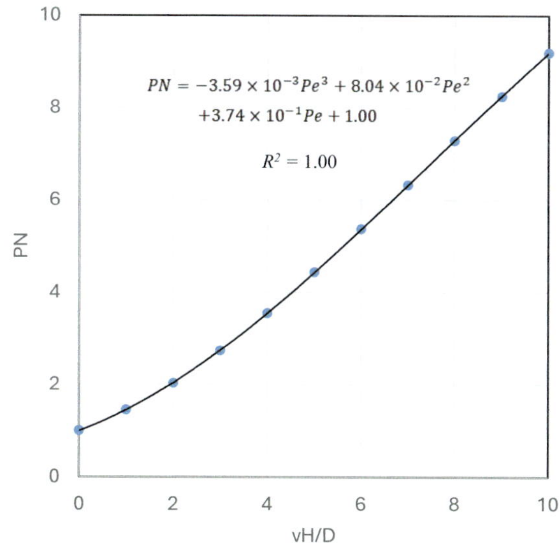

FIGURE 7.4
Polarization number as a function of Peclet number with a third-order polynomial trendline.

Eq. (7.25) is more convenient to use than Eq. (7.24) because numerical integration of the denominator is needed every time for the latter.

Example 7.2

In the calculation of wall concentration with variable transverse velocity described by Eq. (7.11), a constant transverse velocity of the same wall velocity v_w is used. What is the error in terms of $(c_w - c_0)/(c_a - c_0)$?

SOLUTION

The terms of $(c_w - c_0)/(c_a - c_0)$ are actually the PN that are calculated with Eqs. (7.18) and (7.24), respectively, for the constant and variable transverse velocities. The errors are then determined by comparing PN values determined in both cases. The PNs with variable transverse velocity can be taken directly from Table 7.1. The PNs for constant transverse velocity can be calculated using Eq. (7.18).

For vH/D = 0, the PN cannot be calculated directly with the equation. However, it can be found out by taking the limit of Eq. (9.18) with vH/D approaches to 0. Thus determined PN = 1.

For vH/D = 1,

$$PN = \frac{\frac{vH}{D}}{1 - e^{-\frac{v}{D}H}} = \frac{1}{1 - e^{-1}} = 1.58$$

The PNs are calculated for constant transverse velocity up to vH/D = 10, and the results are listed in the table below with those for variable transverse velocity taken from Table 7.1. The relative errors of these two PNs are also provided in the table below.

Comparison of Polarization Numbers with Constant or Variable Transverse Velocity

vH/D	Constant	Variable	Error (%)
0	1.00	1.00	0.0
1	1.58	1.45	9.3
2	2.31	2.03	13.8
3	3.16	2.74	15.2
4	4.07	3.55	14.8
5	5.03	4.44	13.5
6	6.01	5.37	12.0
7	7.01	6.33	10.8
8	8.00	7.29	9.8
9	9.00	8.25	9.1
10	10.00	9.20	8.7

The data in the above table are also graphically presented in the figure below. It can be seen from the table and the figure that the PN determined with constant transverse velocity is greater than the corresponding one with variable transverse velocity. The relative error of these two PNs can be as high as 15% at the vH/D value of about 3.

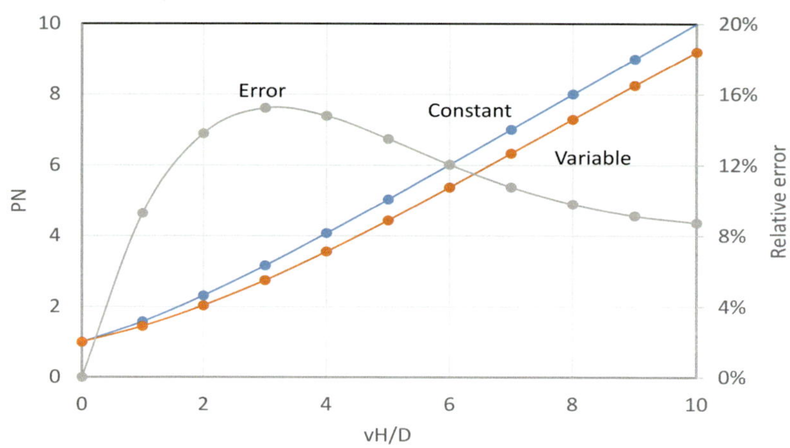

7.3 Concentration Polarization Modeling

The impact of concentration polarization on permeate flux in a membrane channel can be investigated by including Eq. (7.23) in the group of governing equations initially constructed in Chapter 5 for the membrane system. The group of governing equations are again solved numerically with Euler's difference method.

7.3.1 Governing Equations

The full set of governing equations become

$$v(x)=\frac{\Delta p(x)-\Delta\pi(x)}{R_m} \tag{5.3}$$

$$\Delta p(x)=\Delta p_0-\frac{3K\mu}{H^2}\int_0^x u(x')\,dx' \tag{5.4}$$

$$\Delta\pi(x)=f_{os}c_w(x) \tag{5.5}$$

$$c_a(x)u(x)+\frac{(1-r)}{H}\int_0^x c_a(x')v(x')\,dx'=c_0u_0 \tag{5.6}$$

$$u(x)H+\int_0^x v(x')\,dx'=u_0H \tag{5.7}$$

$$Pe=\frac{v(x)H}{D} \tag{7.26}$$

$$PF=-3.59\times10^{-3}Pe^3+8.04\times10^{-2}Pe^2+3.74\times10^{-1}Pe+1.00 \tag{7.25}$$

$$c_w(x)-c_0+PN\left(c_a(x)-c_0\right) \tag{7.23}$$

Eqs. (5.3)–(5.7) retain their original equation numbers for easy reference of their formulations. Two modifications made in these equations are to replace $c(x)$ with $c_w(x)$ in Eq. (5.5) and to replace $c(x)$ with $c_a(x)$ in Eq. (5.6). The reason for these replacements is that salt concentration is variable in the transverse direction. The wall concentration is needed to determine osmotic pressure on the membrane surface, while the mass balance is on the average concentration. In order to calculate wall concentration with Eq. (9.23), two more equations, Eqs. (9.25) and (9.26), are also included. Now eight equations are needed to fully define the membrane system with concentration polarization being considered.

7.3.2 A Numerical Model

Similar to cases without considering concentration polarization, the set of governing equations can be easily solved numerically [10,17,18]. The pressure, average concentration, and tangential velocity are determined by Euler's difference method as

$$\Delta p_{i+1} = \Delta p_i - \frac{3K\mu}{H^2} u_i \Delta x$$

$$c_{a,i+1} = c_{a,i} + r\frac{c_{a,i} v_i}{Hu_i} \Delta x$$

$$u_{i+1} = u_i - \frac{v_i}{H} \Delta x$$

The other variables at step $i + 1$ are calculated with the values at previous step i as

$$\Delta\pi_{i+1} = f_{os} c_{w,i}$$

$$v_{i+1} = \frac{\Delta p_i - \Delta\pi_i}{R_m}$$

$$Pe_{i+1} = \frac{v_i H}{D}$$

$$PN_{i+1} = f\left(Pe_i\right)$$

$$c_{w,i+1} = c_0 + PN_i\left(c_{a,i} - c_0\right)$$

The equations above are the numerical model of the membrane system. In the calculation of *Pe*, *f* is the shorthand for Eq. (7.25). The first three difference equations follow strictly the Euler's method. There are some flexibilities for the calculations of the other five variables. Now all the numerical equations above use the values at the previous step i in the calculations. Actually, the newly obtained values at the current step i+1 can be used in the calculation of others.

For a well-defined membrane desalination problem, in addition to membrane channel characteristics, the applied pressure Δp_0, feed salt concentration c_0, and tangential velocity u_0 at the entrance of the membrane channel must be provided. Otherwise, the problem is not well defined. The initial conditions at the entrance $i = 0$ can be set up as

$$\Delta p_0 = \Delta p_0$$
$$u_0 = u_0$$
$$c_{a,0} = c_0$$
$$c_{w,0} = c_0$$
$$\Delta \pi_0 = f_{os} c_0$$
$$v_0 = \frac{\Delta p - \Delta \pi_0}{R_m}$$
$$Pe_0 = \frac{v_0 H}{D}$$
$$PN_0 = f(Pe_0)$$

With the initial conditions on step $i = 0$, the numerical model can be used to calculate values of all eight variables on the next node $i = 1$. This procedure is repeated until the values on the last step $i = n$ on the exit of the membrane channel are determined.

With this numerical solution, recovery and the average permeate velocity of the membrane channel can be calculated by Eqs. (5.8) and (5.2), respectively:

$$R = 1 - \frac{u_n}{u_0}$$
$$\bar{v} = \frac{R u_0 H}{L}$$

All numerical calculations can be conveniently implemented on a Microsoft Excel spreadsheet.

7.3.3 Accuracy of Numerical Solution

The accuracy of Euler's difference method is assessed by comparing its results with these calculated with the state equation of a long narrow membrane channel. The numerical solutions are obtained under exactly the same conditions with which the state equation is derived, which means concentration polarization is not considered. It would not compromise the purpose because the numerical solution is checked with the exact analytical solution for the same equations under the same conditions. The recoveries with Euler's method were obtained in Chapter 5. Recoveries obtained with the

TABLE 7.2

Comparison of Recoveries by Euler's Method to the Analytical Solution

Pressure (bar)	State Equation[a] (%)	Number of Steps (%)				
		25	50	100	200	400
3	**20.9**	21.1	21.0	21.0	20.9	20.9
4	**38.3**	38.7	38.5	38.4	38.4	38.3
5	**52.9**	53.5	53.2	53.0	53.0	52.9
6	**63.9**	64.5	64.2	64.0	63.9	63.9
8	**75.6**	75.8	75.7	75.7	75.6	75.6
11	**82.5**	82.5	82.5	82.5	82.5	82.5
14	**86.2**	86.3	86.2	86.2	86.2	86.2

[a] Copied from Table 5.4.

state equation for the same case are considered the true solutions because no further assumptions or simplifications are used. The true solutions are listed in Table 7.2 as the first column next to the pressure. Recoveries listed in the next columns are calculated with Euler's method with step numbers n = 25, 50, 100, 200, and 400, respectively. Results show that the accuracy of Euler's method is very high. Even with the smallest step number n = 25, the numerical solutions reproduce the true recoveries well at every pressure. The numerical recoveries become identical with the true solutions when the step number increases to 400.

The relative errors of the numerical solutions to the true solution are calculated and presented in Table 7.3. The biggest relative error is 1.10% that occurs for a pressure of 5 bar when step number n = 25 is used in the numerical solution. It can also be seen that the error generally reduces to half when

TABLE 7.3

Relative Errors of Euler's Method to the True Solution

Pressure (bar)	Number of Steps (%)				
	25	50	100	200	400
3	0.90	0.45	0.22	0.11	0.06
4	1.03	0.51	0.25	0.13	0.06
5	1.10	0.54	0.27	0.14	0.07
6	0.99	0.49	0.24	0.12	0.06
8	0.26	0.14	0.07	0.04	0.02
11	0.00	0.00	0.00	0.00	0.00
14	0.00	0.00	0.00	0.00	0.00

the step number doubles, as theoretically predicted for Euler's method. Step number n = 100 will be used in the following numerical investigations of concentration polarization. The relative errors with this step number are around a quarter of 1% or smaller.

7.4 Impact of Concentration Polarization on Performance

Concentration polarization in a long narrow membrane channel is investigated with this numerical model under conditions similar to those of practical applications. In this investigation, tangential velocity, permeate velocity, average salt concentration, and wall salt concentration along the whole channel will be produced. The development of concentration polarization and its impact on the performance of the membrane channel are discussed.

7.4.1 Brackish Water Desalination

The performance of brackish water membrane desalination is investigated with the numerical model for parameters given in Table 7.4. The channel height H of 0.35 mm, which is measured by the membrane surface to the central line of the channel, corresponds to one of the most commonly used feed spacers in reverse osmosis elements that has a thickness of 28 mil (~0.7 mm) [19]. The equivalent channel length of 9 m is the most commonly used configuration in brackish water desalination, which is two stages of pressure vessels each holding six membrane elements in series. The osmotic pressure coefficient varies in a range with the salt composition in feed water. A value

TABLE 7.4

Parameters Used in Brackish Water Desalination

Parameter	Symbol	Value	Unit
Membrane resistance	R_m	1.00×10^{11}	Pa·s/m
Channel height	H	3.50×10^{-4}	m
Channel length	L	9	m
Salt concentration	c_0	2500	mg/L
Osmotic coefficient	f_{os}	77	Pa/(mg/L)
Feed tangential velocity	u_0	0.1	m/s
Pressure	Δp	11	bar
Diffusion coefficient	D	1.29×10^{-9}	m^2/s

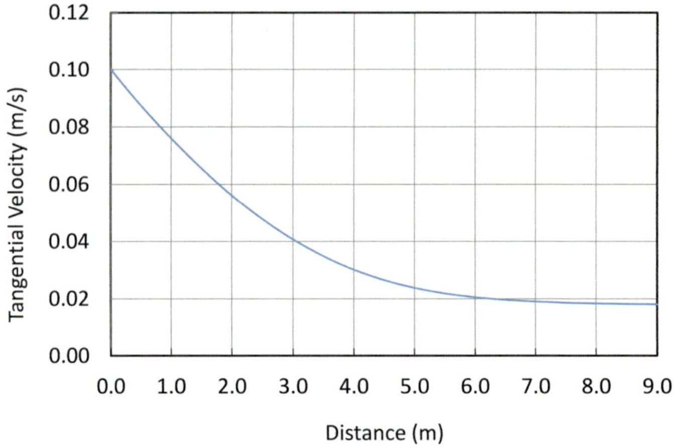

FIGURE 7.5
Tangential velocity along a membrane channel for brackish water desalination calculated with parameters given in Table 7.4.

of 77 Pa/(mg/L) is used for brackish water in this part of the book. All the other parameters in the table are common to practically reverse osmosis processes in brackish water desalination.

With the numerical model, distributions of all eight parameters along the membrane channel can be obtained simultaneously. Figure 7.5 shows that the tangential velocity decreases about 80% along the first 6.0 m in the membrane channel. It declines along the rest of the channel much more slowly. The numerical solution shows that the tangential velocity at the exit of the membrane channel is 1.81×10^{-2} m/s, with which the recovery is determined with Eq. (5.8) to be 81.9%.

Distribution of permeate velocity (flux) along the channel is presented in Figure 7.6. Permeate velocity is seen to decrease all the way along the membrane channel and approaches zero by the exit of the channel, which indicates that this membrane channel works almost under thermodynamic restriction.

Areas under the permeate velocity curve are proportional to the rates of permeate production. Under this situation, permeate is mainly produced in the first stage ($x < 6$ m). Though permeate flux is not zero in the second stage, the contributions from the second stage (x > 6 m) to the total permeate production are really small (~3.04%), as shown by the shadowed area in the figure.

Wall concentration c_w and average concentration c_a are presented in Figure 7.7, as well as their ratio c_w/c_a as a polarization factor. Concentration polarization is visualized with a higher wall salt concentration than the average salt concentration along the channel. The polarization factor c_w/c_a maximizes at the location beyond $x = 2\ m$ with the peak value of about 1.38. It increases from the entrance to the maximum and then decreases along the channel.

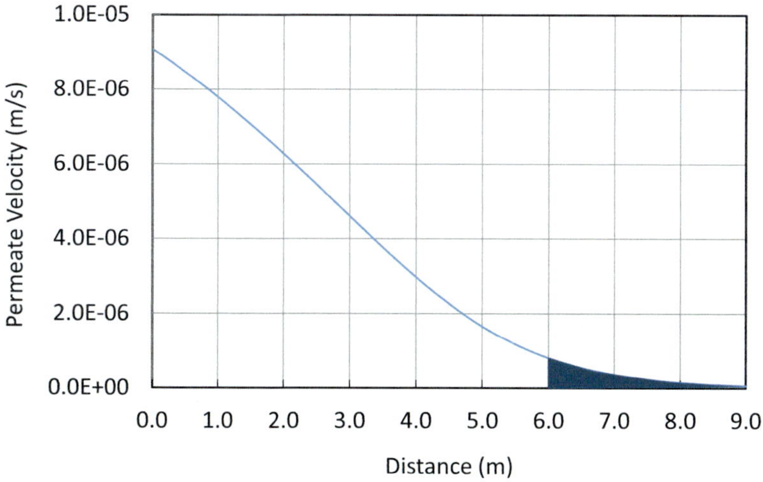

FIGURE 7.6
Permeate velocity along a membrane channel for brackish water desalination calculated with parameters given in Table 7.4. The contribution to permeate production by the second stage is shown by the shadowed area.

It can be seen that wall concentration diverges from the average concentration in most parts of the membrane channel. It indicates that there is concentration polarization occurring in the membrane channel. However, the wall concentration converges to the average concentration by the exit of the channel. It indicates that the process is almost controlled by thermodynamic

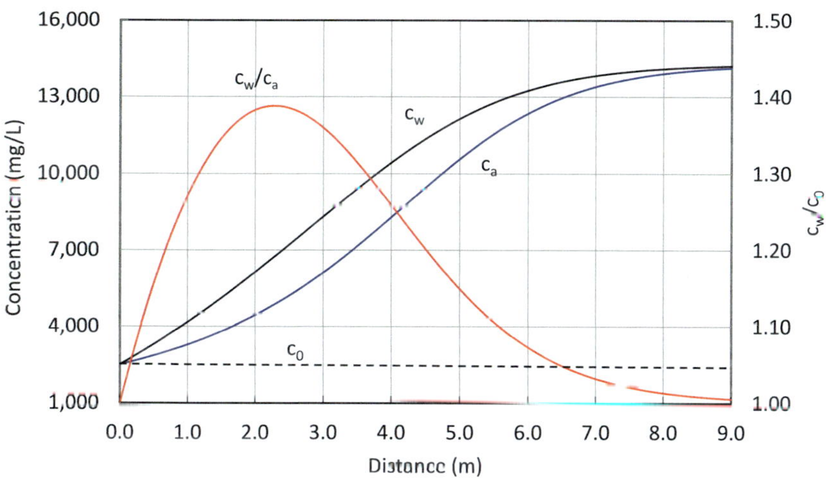

FIGURE 7.7
Average salt concentration and wall salt concentration along a membrane channel for brackish water desalination calculated with parameters given in Table 7.4. The concentration polarization factor c_w/c_0 and feed salt concentration c_0 are indicated in the figure.

restriction. Though concentration polarization develops, it does not have a strong impact on the overall recovery of the membrane channel under the simulation conditions!

By the way, the feed concentration c_0 is also indicated on the graph by the break line. If the feed concentration c_0 was used as the reference line for concentration polarization, the ratio c_w/c_0 can be greater than 5 by the exit end of the membrane channel. However, this big value does not indicate an uneven transverse distribution of salt concentration along the channel height, which is what the concentration polarization truly means. It is more a result of the accumulation of retained salt longitudinally along the membrane channel. Actually, salt concentration is rather evenly distributed along the channel height by the exit end of the membrane channel in this case because the ratio c_w/c_a almost equals 1. It is obvious that the ratio c_w/c_0 cannot be used as an indicator or measurement for concentration polarization in the long narrow membrane channel.

7.4.2 Seawater Desalination

The role of concentration polarization is also investigated in a long narrow channel for seawater desalination. Parameters used in the simulation are presented in Table 7.5. The membrane resistance of 3×10^{11} Pa·s/m for seawater desalination is obviously much higher than the one in brackish water desalination. Usually, a single stage of eight elements in series is used for seawater desalination and, therefore, L = 8 m. The osmotic pressure coefficient of 73.9 Pa/(mg/L) for seawater is used. Other parameters are the typical values in practical seawater desalination.

Distribution of tangential velocity of the simulation results is first presented in Figure 7.8. It can be seen in the figure that the tangential flow velocity declines about 50% in the first 5 m in the membrane channel. The numerical solution shows that the final tangential flow velocity is 0.0488 m/s. The recovery of this membrane channel is then determined to be 51.2%.

TABLE 7.5

Parameters Used in Seawater Desalination

Parameter	Symbol	Value	Unit
Membrane resistance	R_m	3.00×10^{11}	Pa·s/m
Channel height	H	3.50×10^{-4}	m
Channel length	L	8	m
Salt concentration	c_0	35,000	mg/L
Osmotic pressure coefficient	f_{os}	73.5	Pa/(mg/L)
Feed tangential velocity	u_0	0.1	m/s
Pressure	ΔP	55	bar
Diffusion coefficient	D	1.29×10^{-9}	m^2/s

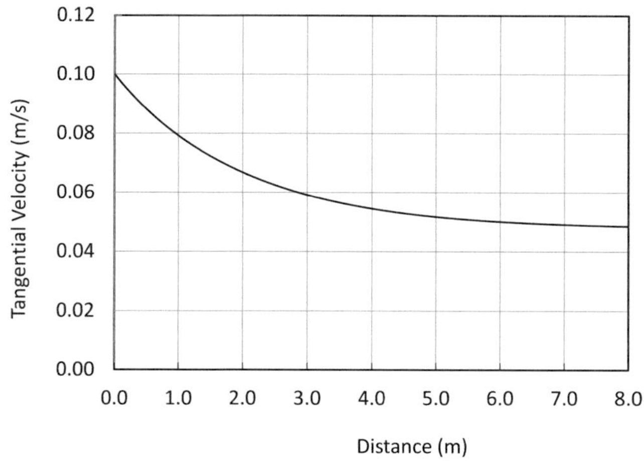

FIGURE 7.8
Tangential velocity along the membrane channel for seawater desalination calculated with parameters given in Table 7.5.

The distribution of permeate velocity in the channel for seawater desalination is presented in Figure 7.9. The permeate velocity is not zero but close to it by the end of the membrane channel, which means that the channel is almost controlled by thermodynamic restriction. There is only one stage for seawater desalination. However, the contribution to permeate production from the last two elements (x > 6) is very limited (~3.0%), as highlighted by the shadowed area in the figure.

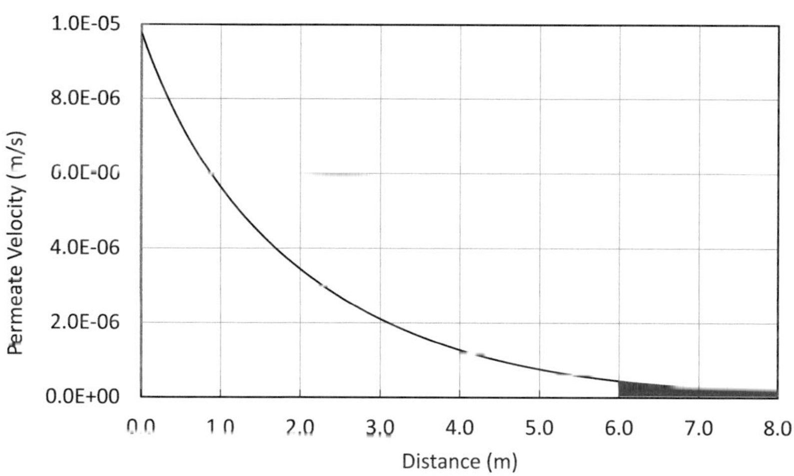

FIGURE 7.9
Permeate velocity in the membrane channel for seawater desalination calculated with parameters given in Table 7.5.

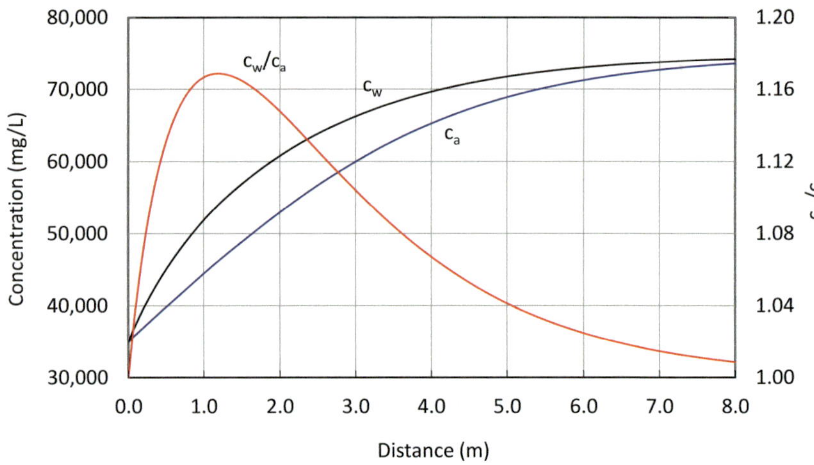

FIGURE 7.10
Salt concentrations and polarization factor along the membrane channel for seawater desalination calculated with parameters given in Table 7.5.

Wall salt concentration c_w and average salt concentration c_a are presented in Figure 7.10 with the polarization factor c_w/c_a. As shown in the figure, the polarization ratio increases rapidly from the entrance to the maximum value of 1.17 at the location about 1.2 m. It then gradually decreases downstream along the membrane channel. Therefore, the impact of concentration polarization on the overall recovery is also almost negligible because of the small value of the polarization factor throughout the membrane channel.

7.4.3 Impact Assessment of Concentration Polarization

It was shown that the impact of concentration polarization on system recovery is small in both brackish water and seawater membrane processes when they are working in or near the regime of thermodynamic restriction. How about concentration polarization under other conditions? Concentration polarization in a brackish desalination membrane channel is investigated with the numerical model for different applied pressures covering both mass transfer and thermodynamic restriction regimes. Parameters except for pressure are the same as in Table 7.4. The pressures under investigation are 3, 4, 5, 6, 8, 11, and 14 bar, respectively. One reason to select these pressures is to compare the results with those obtained previously without considering concentration polarization. In this way, the impact of concentration polarization on performance can be assessed.

The maximum and average polarization factors of the whole membrane channel are listed in Table 7.6. The polarization factor at the exit of the membrane channel is also listed in the table because it signals thermodynamic

TABLE 7.6

Polarization Factors Under Various Applied Pressures

Δp (bar)	3	4	5	6	8	11	14
Max c_w/c_a	1.01	1.04	1.07	1.11	1.21	1.39	1.59
Ave c_w/c_a	1.01	1.03	1.06	1.08	1.13	1.17	1.20
Exit c_w/c_a	1.01	1.03	1.05	1.05	1.03	1.01	1.00
Recovery (%)	20.5	36.8	50.3	60.7	73.6	81.9	85.8

restriction. Recovery is also provided as the indicator of overall performance of the membrane channel at every applied pressure.

Both maximum (peak) and average polarization factors increase with applied pressure. Though the peak polarization factor is as high as 1.59 at a pressure of 14 bars, the average polarization factor is only 1.20. The polarization factor at the exit increases from 1.01 under 3 bars to 1.05 under 5 bars and then decreases with pressure until it reaches 1.00 under a pressure of 14 bars. A polarization factor of 1.00 at the exit means that concentration polarization has no impact at all on system performance.

The recoveries determined numerically with concentration polarization fully considered are listed in Table 7.7 in the first row under the applied pressure. The recoveries determined previously with the state equation are listed in the second row, where concentration polarization is not considered. The relative errors are listed on the bottom of the table. It can be seen that concentration polarization affects the performance most in the medium range of applied pressure. The largest relative error of −5.1% occurs at an applied pressure of 6 bars. The impact of concentration polarization on performance is relatively small under other applied pressures, especially the preferred high pressure commonly used in practical membrane desalination processes. The recoveries with and without considering concentration polarization are graphically presented in Figure 7.11.

From simulations with the numerical method, it is safe to say that the development of concentration polarization in the long narrow reverse osmosis channel is very limited. Even though the peak polarization factor occurs at the applied pressure of 14 bars, the recovery of the membrane channel is not affected by concentration polarization. The reason is that the membrane

TABLE 7.7

Comparisons of Recovery Calculated with State Equation and Numerical Model

Δp (bar)	3	4	5	6	8	11	14
Numerical Model (%)	20.5	36.8	50.3	60.7	73.6	81.9	85.8
State Equation (%)	20.9	38.3	52.9	63.9	75.6	82.5	86.2
Error (%)	−2.1	−3.8	−5.0	−5.1	−2.7	−0.7	−0.4

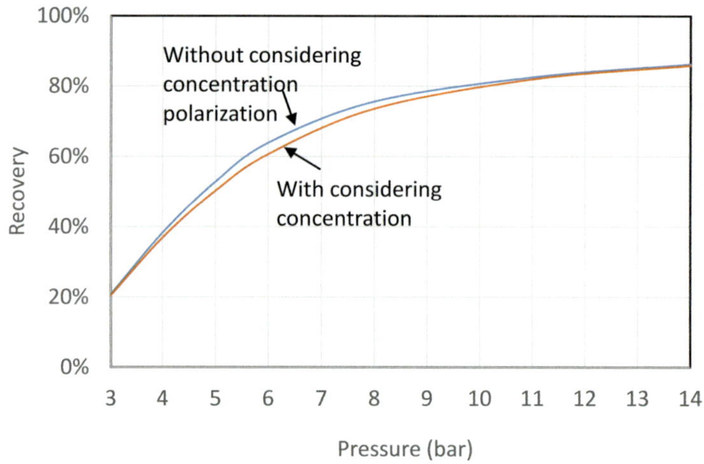

FIGURE 7.11
Recoveries of brackish water desalination calculated with and without considering concentration polarization. Parameters in Table 7.4 are used in the calculation.

process works under thermodynamic restriction under such high pressure, which is indicated by the polarization factor (1.00) on the exit of the membrane channel.

Example 7.3

Wall concentration and average concentration in a seawater desalination membrane channel are provided in the table below. (1) Calculate the average polarization factor c_w/c_a with the data provided. (2) What is the average ratio of c_w/c_0?

x (m)	c_a (mg/L)	c_w (mg/L)
0.0	35,000	35,000
0.8	42,500	49,200
1.6	49,500	57,500
2.4	55,700	62,900
3.2	60,900	66,700
4.0	64,900	69,200
4.8	67,900	71,000
5.6	70,000	72,100
6.4	71,500	72,900
7.2	72,500	73,400
8.0	73,100	73,800

SOLUTION

The salt concentration at x = 0 is the feed salt concentration c_0. The ratios of c_w/c_a and c_w/c_0 at all locations are calculated and listed in the two additional columns on the right side of the table as below. The average values are provided on the bottom line.

x (m)	c_a (mg/L)	c_w (mg/L)	c_w/c_a	c_w/c_0
0.0	35,000	35,000	1.00	1.00
0.8	42,500	49,200	1.16	1.41
1.6	49,500	57,500	1.16	1.64
2.4	55,700	62,900	1.13	1.80
3.2	60,900	66,700	1.10	1.91
4.0	64,900	69,200	1.07	1.98
4.8	67,900	71,000	1.05	2.03
5.6	70,000	72,100	1.03	2.06
6.4	71,500	72,900	1.02	2.08
7.2	72,500	73,400	1.01	2.10
8.0	73,100	73,800	1.01	2.11
	Average value		1.07	1.83

The average polarization factor c_w/c_a of this membrane channel is 1.07. However, the average value of the ratio c_w/c_0 is 1.83, which is significantly higher than the true polarization factor.

7.5 Problems

1. When permeate flux is constant along the membrane channel, derive the formulas for tangential velocity and transverse velocity in a membrane channel.
2. The permeate flux is constant 2.5×10^{-6} m/s in a membrane channel of half height of 0.5 mm. Determine the tangential and transverse velocities along the central line of the channel and at the distance 0.1 mm from the membrane surface at x = 1.0, 2.0, 3.0, 4.0, 5.0, and 6.0 m. The tangential velocity at the entrance of the membrane channel x = 0 is 0.12 m/s.

3. Calculate the changes of driving pressure as pure water flows along a channel of impermeable walls with the parameters given in the table below.

Parameter	Symbol	Value	Unit
Channel height	H	3.50×10^{-4}	m
Channel length	L	6	m
Feed tangential velocity	u_0	0.12	m/s
Applied pressure	Δp	10	bar
Viscosity	μ	10^{-3}	Pa·s

4. In Question 3, if the channel walls are membranes with resistance of 10^{11} Pa·s/m and the variation in driving pressure along the channel can be neglected, calculate
   a. Permeate flux, and
   b. Average tangential velocity along the membrane channel.
5. Derive the expressions for tangential flow and transverse flow in a membrane channel with coordinates as indicated in the figure below.

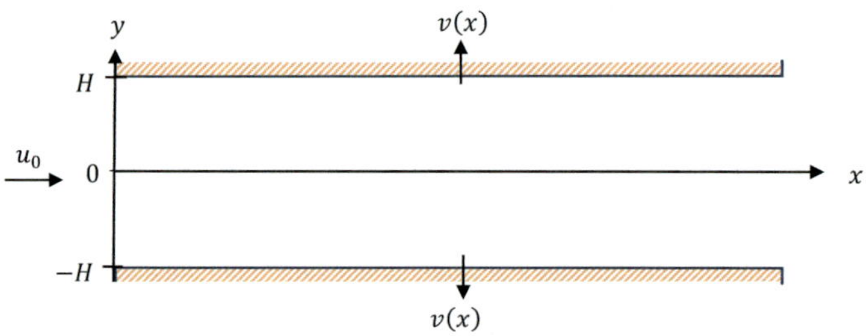

6. Assume constant permeate velocity along the membrane channel, and derive the expression for tangential flow velocity in the channel. The height (2 H) of the membrane channel is 0.80 mm. The incoming tangential flow velocity is 0.1 m/s at the entrance, and the constant permeate velocity is 5×10^{-6} m/s, respectively. Determine the tangential velocities at the points 0.1, 0.2, 0.3, and 0.4 mm from the membrane surface at a distance of 1 m from the entrance of the membrane channel.
7. Concentration polarization refers to the phenomenon that salt concentration distributes unevenly in the direction transverse to the membrane surface. Is the ratio of wall concentration to initial feed concentration c_w/c_0 a meaningful indicator for concentration polarization inside a long narrow membrane channel? Why?
8. A membrane channel is working at 70% recovery and controlled by thermodynamic restriction. What is the ratio of wall concentration

to initial feed concentration c_w/c_0 by the exit end of the membrane channel? How about the ratio of wall concentration to average concentration c_w/c_a at the same point?

9. A membrane channel has a recovery of 75% with a feed salt concentration of 2000 mg/L. What is the average salt concentration by the exit of the membrane channel? If the polarization factor is 1.12 by the exit, what is the wall salt concentration there?
10. If the wall salt concentration of a membrane channel equals the average salt concentration before reaching the exit of the channel, what mechanism controls the performance of this membrane channel?
11. A membrane channel with channel height (2 H) of 0.7 mm has permeate velocities of 8×10^{-6} and 1×10^{-6} m/s at the entrance and exit of the channel, calculate the PN at the entrance and exit. The diffusion coefficient of the salt is 1.29×10^{-9} m²/s.
12. If the feed salt concentration and recovery of the membrane channel as given in Question 10 are 2500 mg/L and 55%, respectively, calculate the salt concentration on the membrane surface by the exit of the channel.
13. When transverse flow velocity at a point in a membrane channel is constant 4×10^{-6} m/s, what is the Peclet number at that point? The diffusion coefficient of the salt is 1.29×10^{-9} m²/s and channel height is 0.35 mm.
14. For permeate velocity of 5×10^{-6} m/s and a diffusion coefficient of the salt of 1.29×10^{-9} m²/s, how big is the channel height at which the PN can be approximated by the Peclet number with error smaller than 0.1%?
15. The feed salt concentration and average salt concentration at a point are 2500 and 7500 mg/L, respectively. What is the amount of retained salt at that point?
16. The permeate flux is 5×10^{-6} m/s at a point in a membrane channel with the variable transverse velocity. The channel height and salt diffusion coefficient are 0.35 mm and 1.29×10^{-9} m²/s, respectively. Determine
    a. the Peclet number at the point, and
    b. the PN at that point.
17. The Peclet numbers are 4.3 and 2.0 at two points in a membrane channel. The corresponding average salt concentrations are $2.5c_0$ and $4.0c_0$, respectively. What are the wall salt concentrations on the membrane surface at the two points?
18. Carrying out a numerical simulation of a long narrow membrane channel with 100 steps in channel length. Show graphically permeate flux, tangential velocity, wall salt concentration, and average

concentration as functions of location along the membrane channel. The parameters of the membrane channel are given in the table below.

Parameter	Symbol	Value	Unit
Membrane resistance	R_m	1.00×10^{11}	Pa·s/m
Channel height	H	3.50×10^{-4}	m
Channel length	L	9	m
Salt concentration	c_0	3000	mg/L
Osmotic pressure coefficient	f_{os}	77	Pa/(mg/L)
Feed tangential velocity	u_0	0.12	m/s
Pressure	Δp	10	bar
Diffusion coefficient	D	1.29×10^{-9}	m^2/s

19. Find out the maximum concentration polarization factor (c_w/c_a) in the membrane channel in Question 18. Also find out the concentration polarization factor by the exit end of the membrane channel.
20. What are the recovery and the average permeate flux of the membrane channel in Question 18?
21. Calculate the recovery and average permeate flux with the state equation for the membrane channel as in Question 18 without considering concentration polarization. Compare the results with these in Question 20.
22. Numerical simulation of a membrane seawater desalination channel with parameters is given in the table below.

Parameter	Symbol	Value	Unit
Membrane resistance	R_m	2.50×10^{11}	Pa·s/m
Channel height	H	3.50×10^{-4}	m
Channel length	L	8	m
Salt concentration	c_0	38,000	mg/L
Osmotic pressure coefficient	f_{os}	73.5	Pa/(mg/L)
Feed tangential velocity	u_0	0.12	m/s
Pressure	Δp	60	bar
Diffusion coefficient	D	1.29×10^{-9}	m^2/s

From the numerical solution, determine (a) the recovery, (b) the average permeate flux, and (c) the maximum concentration polarization factor.

23. Calculate the recovery of the membrane channel as given in Question 22 with the equation under thermodynamic restriction. Compare the result with the one obtained numerically in Question 22. What can you say about the comparison?

24. In the membrane desalination channel, salt concentration increases as water flows along the membrane channel because of the loss of water from the channel as permeate. On the same time, salt concentration increases transversely toward the membrane surface due to concentration polarization induced by permeate flow. Which of the tangential or transverse salt variations has more profound impact on the permeate flux of the membrane channel? Give an example to support your opinion.
25. Under what circumstances does the impact of concentration polarization on permeate flux become insignificant?

References

1. Song, L.F., & Elimelech, M. (1995). Theory of concentration polarization in crossflow filtration. Journal of the Chemical Society, Faraday Transactions, 91, 3389–3398.
2. Song, L.F., & Yu, S.C. (1999). Concentration polarization in crossflow reverse osmosis. AIChE Journal, 45, 921–928.
3. Elimelech, M., & Bhattacharjee, S. (1998). A novel approach for modeling concentration polarization in crossflow membrane filtration based on equivalence of osmotic pressure model and filtration theory. Journal of Membrane Science, 145, 223–241.
4. Bhattacharya, S., & Hwang, S.T. (1997). Concentration polarization, separation factor, and Peclet number in membrane processes. Journal of Membrane Science, 132, 73–90.
5. Gekas, V., & Hallstrom, B. (1987). Mass transfer in the membrane concentration polarization layer under turbulent cross flow: I. Critical literature review and adaptation of existing sherwood correlations to membrane operations. Journal of Membrane Science, 30, 153–170.
6. Murthy, Z.V.P., & Gupta, S.K. (1997). Estimation of mass transfer coefficient using a combined nonlinear membrane transport and film theory model. Desalination. 109, 39–49.
7. Sablani, S.S., Goosen, M.F.A., Al-Belushi, R., & Wilf, M. (2001). Concentration polarization in ultrafiltration and reverse osmosis: A critical review. Desalination, 141, 269–289.
8. Johnson, J.S., & McCutchan, J.W. (1972). Concentration polarization in the reverse osmosis desalination of sea water. Desalination, 10, 147–156.
9. Lyster, E., & Cohen, Y. (2007). Numerical study of concentration polarization in a rectangular reverse osmosis membrane channel: Permeate flux variation and hydrodynamic end effects. Journal of Membrane Science, 303, 140–153.
10. Song, L. (2010). Concentration polarization in a narrow reverse osmosis membrane channel. AIChE Journal, 56, 143–149.
11. Liu, C., Morse, A.N., Rainwater, K.A., & Song, L. (2014). Modeling of concentration polarization in a reverse osmosis channel with parabolic crossflow. Water Environment Research, 86, 56–62.

12. Lin, W., Zhang, Y., Li, D., Wang, X.-M., & Huang, X. (2022). Performance enhancement of spiral-wound reverse osmosis membrane elements with novel diagonal-flow feed channels. Desalination, 524, 115447.
13. Li, M.H., Bui, T., & Chao, S. (2016). Three-dimensional CFD analysis of hydrodynamics and concentration polarization in an industrial RO feed channel. Desalination, 397, 194–204.
14. Li, Y.-L., & Tung, K.-L. (2008). CFD simulation of fluid flow through spacer-filled membrane module: Selecting suitable cell types for periodic boundary conditions. Desalination, 233, 351–358.
15. Cao, Z., Wiley, D.E., & Fane, A.G. (2001). CFD simulations of net-type turbulence promoters in a narrow channel. Journal of Membrane Science, 185, 157–176.
16. Caldwell, D.R. (1973). Thermal and Fickian diffusion of sodium chloride in a solution of oceanic concentration. Deep Sea Research and Oceanographic Abstracts, 20, 1029–1039.
17. Zhoug, W., Song, L., & Tay, K.G. (2006). A numerical study on concentration polarization and system performance of spiral wound RO membrane modules. Journal of Membrane Science, 271, 38–46.
18. Song, L., Hong, S., Hu, J., Ong, S.L., & Ng, W.J. (2002). Simulations of full-scale reverse osmosis membrane process. Journal of Environmental Engineering, 128, 960–966.
19. Bartels, C., Hirose, M., & Fujioka, H. (2008). Performance advancement in the spiral wound RO/NF element design. Desalination, 221, 207–214.

8

Membrane Fouling

Membrane fouling refers to performance deterioration of reverse osmosis processes with time. It often manifests itself as permeate flux decline or pressure increase when other conditions of the membrane processes remain unchanged [1–5]. It is currently one of the major concerns in this field because it can significantly increase the difficulty of operation and the cost of water desalination [6–9].

8.1 Membrane Fouling in Reverse Osmosis Desalination Process

8.1.1 Membrane Fouling and Foulants

Membrane fouling is caused by species in feed water that are not the targets, or at least not the major targets to be removed by the reverse osmosis membrane. Fouling, literally the deterioration of performance of a reverse osmosis desalination process, means something bad or abnormal occurring in the processes. As we discussed in Part I, cake formation in microfiltration or ultrafiltration should not be considered membrane fouling because it is a legitimate or normal behavior of particles in the processes. It is not the case with reverse osmosis desalination. The target species to remove in desalination is the dissolved species, mainly sodium chloride. These species do not deposit to form a cake layer on the membrane surface because their sizes are much smaller than the critical size for cake formation. Feed water that contains only sodium chloride will not cause membrane fouling or any performance deterioration of the reverse osmosis system with time.

Fouling, when it occurs in a reverse osmosis desalination process, must be caused by other components in feed water, such as colloids and suspended substances [10–14]. They are outsiders or intruders into the ideal desalination membrane process that is supposed to deal with the dissolved species only. The sizes of these intruders can be above the critical size for cake formation under the pressure normally employed in reverse osmosis desalination processes. Therefore, these species would deposit on the membrane surface to form a cake layer and accumulate inside the membrane process. Therefore, the performance deterioration caused by these intruders is legitimately named "membrane fouling."

DOI: 10.1201/9781003743569-10

Therefore, two very different separation processes are going on simultaneously in the real-world reverse osmosis systems. One is the intended separation of the dissolved species. The dissolved species are usually high in concentration and can develop a strong concentration polarization layer that, in turn, can substantially reduce permeate flux. However, these species cannot form a cake layer and, therefore, will not accumulate in the membrane process. The other is the unintended separation with the suspended species. These species are much lower in concentration compared to the dissolved species. Their impact on permeate flux is not visible immediately because concentration polarization by these species is so weak that it has virtually no effect on system performance. However, these species can form a cake layer and accumulate inside the membrane processes. After a sufficiently long time, the cake layer grows to an extent that increases the membrane resistance noticeably. At this moment, the permeate flux decreases under the same applied pressure, or the applied pressure has to be raised to maintain constant permeate flux.

The two separations interact with each other in different ways. The separation of dissolved species controls permeate flux, or the permeate flux in a reverse osmosis process is designed for the separation of dissolved species. Permeate flux determines the quantity of suspended species brought to the membrane surface. In other words, fouling rate is controlled by permeate flux that is designed for the separation of dissolved species. On the other hand, the deposition of suspended species on the membrane surface increases membrane resistance. Permeate flux or the capacity to separate dissolved species is affected by the increased membrane resistance caused by the accumulation of the suspended species.

Membrane fouling in this book is restricted to the phenomenon of flux decline or pressure increase in reverse osmosis desalination processes due to the increase in membrane resistance caused by deposition of foulants on the membrane surface. The "foulants" refer to the species that exist in a suspended state in feed water. Precipitation of sparingly soluble minerals (scaling) is excluded. The reproduction and growth of microorganisms are also excluded. The latter two are controlled by totally different mechanisms from these for the suspended species. Usually appropriate acidification and disinfection can handle them satisfactorily. The persistent headache and challenge in reverse osmosis processes is the effective control of membrane fouling by suspended species.

8.1.2 Membrane Fouling Control

Membrane fouling is the single most serious obstacle currently for more economical application of reverse osmosis processes in water desalination. Membrane fouling not only significantly reduces separation efficiency but also increases system complexity and operation difficulty of desalination processes. A membrane process cannot be operated sustainably if membrane fouling cannot be properly and economically managed. Therefore, effective control of membrane fouling is essential to the successful operation of reverse osmosis desalination plants.

Many measures or methods are developed for fouling mitigation and control in reverse osmosis desalination processes [15–20]. Pretreatment of feed water is employed in almost every reverse osmosis desalination plant as a principal measure for fouling mitigation. Pretreatment is the treatment of feed water to remove or reduce the quantity of foulants in it with physical and chemical operations/processes before the reverse osmosis unit. Feed water can be treated with a conventional pretreatment method or with a membrane pretreatment method. The conventional pretreatment method uses granular filter to remove foulants with configurations and media similar to these used for freshwater treatment for potable uses. The membrane pretreatment method uses microfiltration and ultrafiltration membranes to separate foulants from feed water.

Another common measure for fouling control is membrane cleaning. When membrane fouling develops to a certain extent, indicated by flux decline or pressure increase, cleaning is required to move the foulant and to restore membrane permeability. Associated with membrane cleaning, monitoring and detecting facilities and cleaning in place (CIP) facilities need to be installed. An adverse impact of membrane cleaning is that the use of strong detergents or chemicals to remove foulants from the membrane may damage the membranes and reduce their service time. As a result, membranes may have to be replaced more frequently.

Considerable cost is associated with the measures for fouling mitigation and control, including the down time for membrane cleaning and replacement. In the economic analysis and optimization of membrane desalination processes conducted in Chapter 6, it was shown that the cost associated with pretreatment can be a large contributor to the total cost for water desalination.

Currently, a big challenge for the selection of the appropriate pretreatments is the lack of accurate characterization methods to assess fouling strengths of the feed water [21–25]. The existing methods are inadequate to articulate fouling strength in an appropriate form or manner. The currently available fouling indices are not capable of predicting fouling development in full-scale reverse osmosis processes. Consequently, the competence of a pretreatment method cannot be adequately evaluated in advance but can only be found after commencement of the reverse osmosis plants.

8.2 Fouling Characterization of Feed Water

8.2.1 Existing Fouling Characterization Methods

Currently there are two methods available to measure and assess the fouling property of feed water, namely the silt density index (SDI) [26–28] and modified fouling index (MFI) [29–32]. A brief description of each method is provided below.

8.2.1.1 Silt Density Index

The SDI is the most commonly used and widely accepted method to characterize the fouling property of feed water. The test to measure the SDI of feed water is conducted by filtering feed water through a 47 mm diameter disc with a 0.45 μm membrane in a dead-end mode at a constant pressure of 30 psi (2.07 bar). Three time intervals are recorded in the test, which are the time t_i needed to collect 500 ml filtrate initially at the beginning of the test, the time t_f needed to collect the same volume filtrate finally by the end of the test, and the elapse time T between the two collections. Then the SDI is calculated as follows:

$$SDI = \frac{1 - t_i/t_f}{T} \times 100 \tag{8.1}$$

where SDI is the silt density index, t_i is the initial time in seconds required to collect the 500 ml sample, t_f is the final time in seconds required to collect the second sample, and T is the total elapsed test time. The elapsed time T usually is 15 minutes under normal conditions. It can be reduced to 10 or 5 minutes for waters with strong fouling properties.

The SDI test is simple and straightforward. At the same time, it is also the least reliable or meaningful method in assessing fouling property of feed water. It cannot be used to predict fouling development in the practical reverse osmosis desalination systems because there is no intrinsic relationship between them. In addition, the use of a 0.45 μm membrane in the standard test cannot catch foulants that are smaller than the pore size of the membrane. However, because of its simplicity in test and calculation, SDI still remains the primary fouling index for fouling characterization even though it is known to be fundamentally weak.

8.2.1.2 Modified Fouling Index

MFI was developed with a theoretical basis of cake formation on the membrane surface by the foulants in feed water. The test is conducted by filtering feed water through a 0.45 μm membrane in the dead-end mode at a constant pressure of 30 psi (2.07 bar). Different from the test for SDI, volumes of filtrate V_i are measured at a series of times t_i in the MFI test. MFI cannot be calculated with a simple formula as does in the case of SDI. A graph has to be developed by plotting t_i/V_i versus V_i. The graph appears as a curve with a linear middle section. Then the slope of the middle linear section is taken as the MFI. In principle, MFI can be written as

$$MFI = \frac{\mu I}{2\Delta p S^2} \tag{8.2}$$

where *MFI* is the modified fouling index, μ is the water viscosity, I is the cake resistivity, Δp is the test pressure, and S is the effective membrane area. The unit of MFI is s/L^2. It is a wired unit directly derived by curve fitting.

The measurement of filtrate volumes at a series of times and the generation of a graph to determine the MFI make it a much more laborious and time-consuming test procedure. It makes an automatic measuring and recording system necessary in the test. Another obvious drawback of MFI is that the value of MFI is not independent from the test pressure and membrane area, as shown in Eq. (8.2). In principle, the fouling property should be a property of feed water only and would be better characterized by a parameter independent of measuring conditions, such as pressure and membrane area. Besides, it is difficult to interpret MFI in plain language about its connection to membrane fouling. Therefore, MFI is just a parameter identified graphically from fouling tests without a clearly defined physical meaning.

Though MFI seems to have a certain fundamental basis in some sense, it is still not widely accepted and used in the membrane desalination community. The vague physical meaning and inconvenience in measurement and use by engineers in practical desalination plants might be important reasons.

8.2.2 A Simple and Effective Fouling Characterization Method

A more effective characterization method for fouling strength of feed water is introduced hereafter [33–36]. This characterization can be easily carried out with a simple device and straightforward calculation. It also has a clearly defined physical meaning and can be readily used to predict fouling development in full-scale reverse osmosis desalination processes.

A concept of *fouling strength* of feed water is introduced here to facilitate the characterization of membrane fouling in reverse osmosis processes. Fouling strength is defined as the increment in membrane resistance after a unit volume of permeate is filtrated per unit area of membrane, i.e.,

$$S_f = \frac{R_{mf} - R_{m0}}{(V/A)} = \frac{\Delta R_m}{(V/A)} \tag{8.3}$$

where S_f is the fouling strength of feed water, R_{m0} and R_{mf} are the remembrance resistances at the beginning and the end of test, respectively, ΔR_m is the increment in membrane resistance during the test, V is the volume of permeate collected during the test, and A is the membrane area. Fouling strength has a basic unit of Pa·s/m^2 when the basic metric units are used for pressure and permeate flux. A more convenient unit is bar·min/m^2 with the following conversion formula from the basic unit:

$$1\ bar \cdot min/m^2 = 10^5\ Pa \times 60\ s/m^2 = 6 \times 10^6\ Pa \cdot s/m^2 \tag{8.4}$$

The fouling strength so defined is directly measurable with a device similar to that for the SDI test. The only suggestion or recommendation for the test is to use membranes with a smaller pore size than 0.45 μm in the SDI test. From the pressure used in the reverse osmosis process, the critical particle size for cake formation can be calculated. Pore size of the membrane should be small enough to reject all foulants. A desirable property for the candidate membrane is as small as possible membrane resistance. The smaller the membrane resistance, the smaller the volume of permeate needed to generate a notable change in membrane resistance.

It is not difficult to demonstrate that fouling strength is proportional to foulant concentration in feed water. The increment of membrane resistance is the resistance of the fouling (cake) layer. It is well established that the resistance of the cake layer is proportional to the amount of foulant on the membrane surface. When the same volume of filtrate V and membrane area A are used in the tests of two water samples, i.e., we have

$$\frac{\Delta R_{m1}}{\Delta R_{m2}} = \frac{m_{f1}}{m_{f2}} = \frac{Vc_{f1}}{Vc_{f2}} = \frac{c_{f1}}{c_{f2}}$$

where m_f is the mass of foulant per unit area of membrane, c_f are foulant concentrations in feed water, and the subscripts 1 and 2 refer to sample 1 and sample 2, respectively. Then the ratio of fouling strengths of these two feed water samples is

$$\frac{S_{f1}}{S_{f2}} = \frac{\Delta R_{m1}/(V/A)}{\Delta R_{m2}/(V/A)} = \frac{\Delta R_{m1}}{\Delta R_{m2}} = \frac{c_{f1}}{c_{f2}}$$

Therefore, the fouling strength can be written as

$$S_f = kc_f \tag{8.5}$$

where k is a constant coefficient. The proportionality of fouling strength to foulant concentration is an essential requirement of a good parameter for fouling characterization of feed water. With this property, the parameter of fouling strength can be used as an effective parameter to assess the effectiveness of the pretreatment method in the removal of foulants from feed water.

8.2.3 Easy Test Procedure for Fouling Strength

The test for fouling strength can be done with a dead-end membrane device similar to that for the SDI test. The filtration test is conducted at a constant pressure Δp (e.g., 30 psi) on a flat sheet membrane of known area A. Membranes of 0.45 μm pore size can be used, but membranes of smaller pore size would generate better results. Two times are measured in the test. The first time t_1 is spent collecting a volume of filtrate (e.g., 500 ml) with clean water. The second time t_2 is spent collecting the same volume of filtrate with the feed water under test. The fouling strength of the feed water is calculated with the fouling strength model:

$$S_f = 2\Delta p \left(\frac{A}{V}\right)^2 (t_2 - t_1) \tag{8.6}$$

where Δp is the pressure used in the test, V is the volume of filtrate collected, A is the membrane area, and t_1 and t_2 are the times spent for collecting filtrates of the same volume with clean water and feed water, respectively. Though pressure Δp and membrane area A appear in Eq. (8.4), the value of S_f is independent of them. In other words, fouling potential S_f will have the same value when the test is conducted at different pressures with different membrane areas.

One beauty of the parameter of fouling strength is that it is readily usable to predict fouling development in reverse osmosis processes. From the definition of the fouling strength in Eq. (8.3), the membrane resistance at any time can be calculated with

$$R_m = R_{m0} + S_f \frac{V}{A} = R_{m0} + S_f \int_0^t v\, dt' \tag{8.7}$$

where v is the permeate flux, t is the operation time, and t' is the dummy integration variable. Eq. (8.7) complements Eq. (8.6) in applying it to predict fouling development in practical reverse osmosis processes.

Example 8.1

In a test for fouling strength with a membrane area of 17.35 cm^2 at a pressure of 2 bars, 24 and 28 minutes are recorded to collect 500 ml of filtrate with clean water and feed water in the test. (1) Calculate the fouling strength, and (2) calculate the time that a reverse osmosis process operated at constant flux can operate before the pressure increases by 15% of its original value. The average permeate flux is 5×10^{-6} m/s and the initial resistance of the reverse osmosis membrane is 10^{11} Pa·s/m.

SOLUTION

1. The fouling strength is calculated with Eq. (8.6):

$$S_f = 2\Delta P\left(\frac{A}{V}\right)^2 (t_2 - t_1) = 2 \times 2\ bar \times \left(\frac{17.35\ cm^2 \times \dfrac{m^2}{10^4\ cm^2}}{500\ ml \times \dfrac{m^3}{10^6\ ml}}\right)^2 (28\ min - 24\ min)$$

$$= 192.9\ bar \cdot min/m^2$$

2. When the permeate is constant, Eq. (8.7) becomes

$$R_{mt} = R_{m0} + S_f v \Delta t$$

The pressure increase is equivalent to the resistance increase. Then the time for operation is calculated as

$$\Delta t = \frac{R_{mt} - R_{m0}}{S_f v} = \frac{0.15 R_{m0}}{S_f v} = \frac{0.15 \times 10^{11}\, Pa \cdot s/m \times \dfrac{min}{60\, s} \times \dfrac{bar}{10^5\, Pa}}{192.9\ bar \cdot min/m^2 \times 5 \times 10^{-6}\, m/s \times \dfrac{60\, s}{min}}$$

$$= 43{,}200\, min = 30\, d$$

The fouling strength of the feed water is 192.9 $bar \cdot min/m^2$. The membrane system can operate 30 days before the pressure rises 15%.

8.2.4 Features of Fouling Strength

Different from SDI and MFI, the parameter of fouling strength has a clearly defined physical meaning. It is the increment in membrane resistance when a unit volume of water filtrates through a unit membrane area. Though the test is conducted with a specific membrane area and at a specific pressure, the value of fouling strength is independent of both. This independence provides flexibility to carry out the test with different pressures and membranes. For example, high pressure can be used to shorten the test time. Membranes with different pore sizes can be used to see if all foulants are caught by membranes.

A stunning beauty of this characterization method is the extreme simplicity of its test procedure. It is even simpler than the test for SDI. Both the new and SDI tests can be obtained with the same simple device and straightforward calculation. There is no need for the measurements of filtrate volume at multiple times as in the MFI test, which probably requires an automatic sampling system and a computer to complete properly. There is no need to convert the data to t/V or whatever intermediate parameters and to identify the fouling strength graphically as done to obtain the parameter MFI.

SDI measurement needs to record three time intervals for each test. Two time intervals are required for fouling strength when there is only a single sample in the test. Fouling strength has a big advantage when multiple samples are measured. The time for clean water filtration only needs to be measured once. If there are n samples in the test, the total measurements needed in fouling strength tests are $n + 1$, while the SDI test requires $3n$ measurements for the same number of water samples.

Another marked advantage of the fouling strength is its readiness to be used for predicting fouling development in practical reverse osmosis processes. There is no need to transform it to other forms or make any modifications. It can be directly used in a simple equation to calculate the development of membrane fouling in the full-scale reverse osmosis processes. Though MFI also contains the needed information for fouling prediction, it also contains information specific to the test conditions, such as pressure and membrane characteristics (area and resistance). This test-specific information has to be filtered out before it can be used for fouling prediction in the full-scale reverse osmosis processes. The equations and calculations with MFI for fouling development in full-scale reverse osmosis processes are much more complex and demanding than those with the fouling strength.

In conclusion, not only the test procedure of the fouling strength is simpler than SDI and MFI, but also it carries the needed information in a very concise and appropriate form for fouling development in full-scale reverse osmosis processes. It will become clear with the discussion of fouling development in full-scale membrane channels later.

8.2.5 Derivation of the Fouling Strength Model

The simpleness of the test procedure comes from the fouling strength model Eq. (8.6). Is this model fundamentally sound or just like the model for SDI without a solid theoretical basis? The answer is positive. It is derived rigorously from the first principles.

When there is fouling in a dead-end microfiltration or ultrafiltration membrane process that is operated under the constant pressure mode, permeate flux declines as described by Eq. (4.9):

$$v = \frac{\Delta p}{\sqrt{R_{m0}^{\ 2} + 2S_f \Delta p t}}$$

where v is the permeate flux, and R_{m0} is the resistance of the clean membrane. The original term $r_{cl}\dfrac{c_0}{C_g}$ in Eq. (4.9) is replaced by fouling strength S_f in the equation. The original term is for resistance increase strictly by cake formation of uniform spherical particles. In the current context, this requirement is relaxed to resistance increase by the accumulation of foulants in any form. Eq. (4.9) shows that flux decline is controlled by the fouling strength of feed water. A model to determine fouling strength from experimental tests can be developed rigorously from this relationship.

First, a reference with clean water (e.g., reverse osmosis filtrate), i.e., $S_f = 0$, is established. From Eq. (4.9), now the flux is

$$v = \frac{\Delta p}{R_{m0}}$$

The volume of filtrate collected with clean water filtration is calculated with

$$V = \frac{A \Delta p t_1}{R_{m0}} \tag{8.8}$$

where V is the volume of filtrate, A is the membrane area used in the test, and t_1 is the time to collect the filtrate in the clean water filtration.

The volume of filtrate collected with feed water of fouling strength S_f can be calculated by integration of Eq. (4.9):

$$V = A\int_0^{t_2} v\,dt = A\frac{R_{m0}}{S_f}\left(\sqrt{1+\frac{2S_f\Delta p t_2}{R_{m0}^2}}-1\right) \tag{8.9}$$

where t_2 is the time to collect filtrate with feed water filtration. When the same volume is collected for both clean water and feed water, the two equations can be combined to eliminate volume, i.e.,

$$\frac{A\Delta p t_1}{R_{m0}} = A\frac{R_{m0}}{S_f}\left(\sqrt{1+\frac{2S_f\Delta p t_2}{R_{m0}^2}}-1\right) \tag{8.10}$$

Rearranging

$$1+\frac{S_f\Delta p t_1}{R_{m0}^2} = \sqrt{1+\frac{2S_f\Delta p t_2}{R_{m0}^2}}$$

Squaring both sides

$$\left(1+\frac{S_f\Delta p t_1}{R_{m0}^2}\right)^2 = 1+\frac{2S_f\Delta p t_2}{R_{m0}^2}$$

Rearranging

$$\left(\frac{\Delta p t_1}{R_{m0}^2}\right)^2 S_f^2 - \frac{2\Delta p}{R_{m0}^2}(t_2-t_1)S_f = 0$$

This is a quadratic equation of S_f. There are two roots of the equation. The trivial solution

$$S_f = 0$$

is obviously not the logical answer. Then we have the other root as the solution of this problem:

$$S_f = 2\frac{R_{m0}^2}{\Delta p t_1^2}(t_2-t_1) \tag{8.11}$$

From Eq. (8.8), we have

$$R_{m0} = \frac{A\Delta p t_1}{V}$$

Substituting the expression of R_{m0} into Eq. (8.11) results in

$$S_f = 2\Delta p\left(\frac{A}{V}\right)^2 (t_2 - t_1)$$

This is the fouling strength model Eq. (8.6).

Example 8.2

In order to catch small foulants, a membrane with a pore size of 0.1 μm is used instead of the original membrane with a pore size of 0.45 μm. The resistance of the new membrane is three times as high as the original one. If the other parameters remain the same, how much time is needed to complete the test (assume the original membrane resistance to be 10^9 Pa·s/m)?

SOLUTION

The time needed to collect a certain volume of filtrate with clean water can be derived from Eq. (8.8):

$$V = \frac{A\Delta p t_1}{R_{m0}}$$

$$t_1 = \frac{VR_{m0}}{A\Delta p}$$

The time needed to collect the same volume of filtrate with feed water can be derived from Eq. (8.9):

$$V = A\frac{R_{m0}}{S_f}\left(\sqrt{1 + \frac{2S_f \Delta p t_2}{R_{m0}^2}} - 1\right)$$

Rearranging

$$\frac{VS_f}{AR_{m0}} + 1 = \sqrt{1 + \frac{2S_f \Delta p t_2}{R_{m0}^2}}$$

Squaring on both sides

$$\left(\frac{VS_f}{AR_{m0}} + 1\right)^2 = 1 + \frac{2S_f \Delta p t_2}{R_{m0}^2}$$

$$t_2 = \frac{R_{m0}^2}{2S_f \Delta p}\left[\left(\frac{VS_f}{AR_{m0}} + 1\right)^2 - 1\right] = \frac{1}{2S_f \Delta p}\left[\left(\frac{VS_f}{A} + R_{m0}\right)^2 - R_{m0}^2\right]$$

$$t_2 = \frac{1}{2S_f \Delta p}\left(\frac{VS_f}{A}\right)^2\left[1 + 2\frac{A}{VS_f}R_{m0}\right]$$

The test conditions are $\Delta p = 2\ bar = 2\times10^{5}\,Pa$, $V = 500\ ml = 5\times10^{-4}\,m^{3}$, $A = 1.735\times10^{-3}\,m^{2}$, $R_{m0} = 3\times10^{9}\,Pa\cdot s/m$.

Calculate the times for two feed water samples with $S_f = 100, 200\ bar.min/m^{2}$, respectively.

Time with clean water is the same:

$$t_1 = \frac{5\times10^{-4}\,m^{3}\times3\times10^{9}\,Pa\cdot s/m}{1.735\times10^{-3}\,m^{2}\times2\times10^{5}\,Pa} = 4323\,s = 72.0\ min$$

For $S_f = 100\ bar.min/m^{2} = 6.0\times10^{8}\ Pa\cdot s/m^{2}$

$$t_2 = \frac{1}{2\times6.0\times10^{8}\,Pa\cdot s/m^{2}\times2\times10^{5}\,Pa}\left(\frac{5\times10^{-4}\,m^{3}\times6.0\times10^{8}\,Pa\cdot s/m^{2}}{1.735\times10^{-3}\,m^{2}}\right)^{2}$$

$$\left[1+2\frac{1.735\times10^{-3}\,m^{2}}{5\times10^{-4}\,m^{3}\times6.0\times10^{8}\,Pa\cdot s/m^{2}}\times3\times10^{9}\,Pa\cdot\frac{s}{m}\right]$$

$$= 4447\,s = 74.1\ min$$

For $S_f = 200\ bar.min/m^{2} = 1.2\times10^{9}\ Pa\cdot s/m^{2}$

$$t_2 = \frac{1}{2\times1.2\times10^{9}\,Pa\cdot s/m^{2}\times2\times10^{5}\,Pa}\left(\frac{5\times10^{-4}\,m^{3}\times1.2\times10^{9}\,Pa\cdot s/m^{2}}{1.735\times10^{-3}\,m^{2}}\right)^{2}$$

$$\left[1+2\frac{1.735\times10^{-3}\,m^{2}}{5\times10^{-4}\,m^{3}\times1.2\times10^{9}\,Pa\cdot s/m^{2}}\times3\times10^{9}\,Pa\cdot\frac{s}{m}\right]$$

$$= 4572\,s = 76.2\ min$$

The time needed in the test is about linearly related to the membrane resistance used in the test. A smaller volume of filtrate can be used to complete the test in a shorter time with membranes of higher resistance.

8.3 Fouling Modeling in a Long Narrow Channel

The prediction of fouling development in Example 8.1 is an oversimplified case where the membrane process is treated as a homogeneous system. Permeate flux was assumed uniform throughout the membrane channel so that fouling progresses with time evenly on the entire membrane channel. Fouling development in a practical reverse osmosis membrane channel is actually much more complicated. Fouling first occurs in the front section of a membrane channel, and the fouling rate decreases along the membrane channel with permeate flux. There might be a section of membrane channel that is unused at the beginning of operation when the membrane process is initially controlled by thermodynamic restriction. Fouling gradually develops downstream as permeate flux declines in the front section due to resistance increase. Using the fouling strength of feed water, this dynamic process of membrane fouling can be described in detail with a properly formulated model for fouling development along the membrane channel.

8.3.1 Dynamics of Membrane Fouling

Though membrane fouling is commonly described as flux decline or pressure increase, the essence is actually the increment in membrane resistance due to accumulation of foulant on the membrane surface. Therefore, membrane fouling can be modeled by tracking the change of membrane resistance with time.

The fouling rate, which is the rate of membrane resistance increase, at any point x in a membrane channel is described by

$$\frac{dR_m(x,t)}{dt} = S_f v(x,t) \tag{8.12}$$

Such a concise model for fouling development becomes possible largely due to the introduction of the parameter of fouling strength. Because membrane fouling is a dynamic process, the variables in a membrane channel are not only functions of x but also dependent on time t. That is the reason that the argument t is added in functions.

Eq. (8.12) is obtained by taking the derivative of Eq. (8.7) with respect to time t at any point x. It can be seen that the increasing rate of membrane resistance is proportional to permeate flux and fouling strength of feed water. The initial condition for this problem is usually the clean membrane with a constant resistance through the whole channel length:

$$R_m(x,0) = R_{m0} \tag{8.13}$$

If the membrane at the beginning is not clean, the initial membrane resistance at $t = 0$ can be a function of x.

The fouling strength of a feed water S_f can be determined by the test procedure described previously. Permeate flux $v(x,t)$ is needed in Eq. (8.12) to determine membrane resistance $R_m(x,t)$. The numerical model initially presented in Chapter 5 can be used to calculate accurately permeate flux in the membrane channel. To maintain the main focus on membrane fouling, concentration polarization is not considered here. But it is not difficult mathematically to include concentration polarization in the calculations because the numerical solution process is virtually the same.

There are originally three Euler's difference equations in the numerical model for tangential velocity, permeate flux, and salt concentration as follows:

$$\Delta p_{i+1,j} = \Delta p_{i,j} - \frac{3K\mu}{H^2} u_{i,j} \Delta x$$

$$c_{i+1,j} = c_{i,j} + r \frac{c_{i,j} v_{i,j}}{H u_{i,j}} \Delta x \tag{8.14}$$

$$u_{i+1,j} = u_{i,j} - \frac{v_{i,j}}{H} \Delta x$$

The osmotic pressure and permeate flux are calculated at the same spatial step:

$$\Delta \pi_{i+1,j} = f_{os} c_{i+1,j}$$

$$v_{i+1,j} = \frac{\Delta p_{i+1,j} - \Delta \pi_{i+1,j}}{R_{mi+1,j}} \tag{8.15}$$

Because all variables are also functions of time, another subscript is added to the variables to indicate the time step with j meaning the current time step.

Euler's difference form of Eq. (8.12) is added to the numerical model to evolve to the next time step j + 1.

$$R_{mi,j+1} = R_{mi,j} + S_f v_{i,j} \Delta t \tag{8.16}$$

where Δt is the time step.

Now the numerical model is constructed for two-dimensional problems. The grid points used in the calculation are schematically described in Figure 8.1. The spatial grid points are in the horizontal direction indexed by *"i"* while the time grid points are in the vertical direction indexed by *"j."*

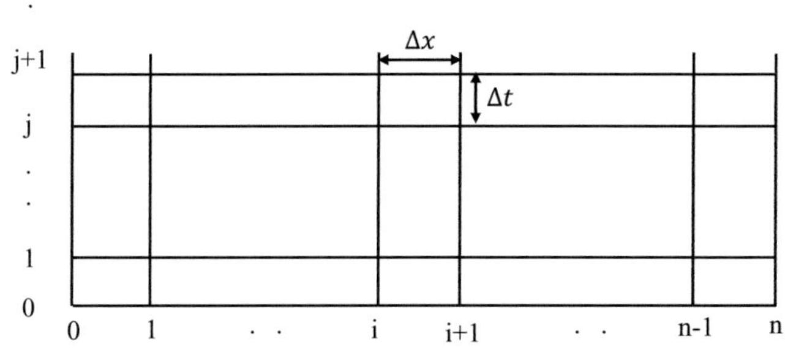

FIGURE 8.1
Grid points used in the numerical model for membrane fouling.

Boundary conditions for the spatial difference equations (Eq. 8.14) are

$$\Delta p_{0,j} = \Delta p_0 \ (j = 0, 1, 2, \ldots)$$
$$c_{0,j} = c_0 \ (j = 0, 1, 2, \ldots)$$
$$u_{0,j} = u_0 \ (j = 0, 1, 2, \ldots)$$

The osmotic pressure and permeate flux at entrance $i = 0$ are calculated by

$$\Delta \pi_{o,j} = f_{os} c_{0,j} \ (j = 0, 1, 2, \ldots)$$
$$v_{0,j} = \frac{\Delta p_{0,j} - \Delta \pi_{0,j}}{R_{m0}} \ (j = 0, 1, 2, \ldots)$$

The initial condition required for the only time difference equation (Eq. 8.16) is the discretized form of Eq. (8.13):

$$R_{mi,0} = R_{m0} \ (i = 0, 1, 2, \ldots, n)$$

Up to now, the numerical model for fouling development has been completely set up.

The procedure for numerical simulation of membrane fouling is described below:

1. At time step j = 0,
   a. Calculate all five variables $\Delta p_{i,0}$, $c_{i,0}$, $u_{i,0}$, $\Delta \pi_{i,0}$, and $v_{i,0}$ using Eqs. (8.14) and (8.15) with initial membrane resistance R_{m0} for all spatial grid points i = 1, 2, …, n.
2. For time step j > 0,
   a. Update the membrane resistance with Eq. (8.16) from $R_{mi,j-1}$ to $R_{mi,j}$ (i = 1, 2, …, n) using $v_{i,j-1}$ obtained in the previous time step j – 1.
   b. Calculate all five variables $\Delta p_{i,j}$, $c_{i,j}$, $u_{i,j}$, $\Delta \pi_{i,j}$, and $v_{i,j}$ using Eqs. (8.14) and (8.15) with the updated membrane resistance $R_{mi,j}$ for all spatial grid points i = 1, 2, …, n.
3. Repeat 2 until the time period of concern is completed.

By the end, the numerical solution is the spatial distributions of five variables plus membrane resistance at every time step. By inspecting the evolutions of profiles of the variables and membrane resistance, valuable information about membrane fouling can be obtained.

8.3.2 Fouling Development in a Long Narrow Channel

A case of fouling development in a long narrow membrane channel is studied with the parameters shown in Table 8.1. Particularly, the new parameters that are needed for membrane fouling simulations are the feed water fouling

TABLE 8.1

Parameters Used in the Simulation of Fouling Development

Parameter	Symbol	Value	Unit
Membrane resistance	R_m	1.0×10^{11}	Pa·s/m
Channel height	H	3.5×10^{-4}	m
Salt concentration	c_0	2500	mg/L
Fouling strength	S_f	2.40×10^{9}	Pa·s/m²
Osmotic pressure coefficient	f_{os}	77	Pa/(mg/L)
Feed tangential velocity	u_0	0.1	m/s
Pressure	Δp	9	bar
Channel length	L	9	m
Spatial step size	Δx	0.09	m
Time period	T	400	day
Time step size	Δt	20	day

strength of 400 bar·min/m² (2.40×10^9 Pa·s/m²) and the simulation period of 400 days. The rest of the parameters remain the same as in Example 5.2 in Chapter 5. A time step size of 20 days is used here for demonstration purposes. Smaller time steps should be used for higher accuracy with real-world problems. The numerical simulations are carried out in the Microsoft Excel spreadsheet.

The distribution of membrane resistance along the channel length at different times is presented in Figure 8.2. It can be seen that membrane fouling starts from a front section of the channel and develops downstream with time. The membrane fouling reaches $x = 6$ m in slightly less than 40 days and develops to the exit of the membrane channel in about 100 days. After fouling is started at a location, the membrane resistance there grows with time as fouling develops. Membrane resistance at any time has its maximum value at the entrance of the membrane channel and decreases downstream along the membrane channel.

8.3.3 A Grace Period of Permeate Flux

As a result of fouling development with time, distribution of permeate flux along the channel changes accordingly. As shown in Figure 8.3, the most uneven distribution of permeate flux occurs at the very beginning of operation with a clean membrane. Permeate is produced mostly in the first half of the channel. The membrane channel beyond x = 6 (the second stage) does not contribute much to permeate production at this moment. As time passes, permeate flux declines in the leading section of the membrane channel due to membrane fouling. The rear section of the membrane channel is observed to increase in permeate flux, and a larger membrane area contributes to

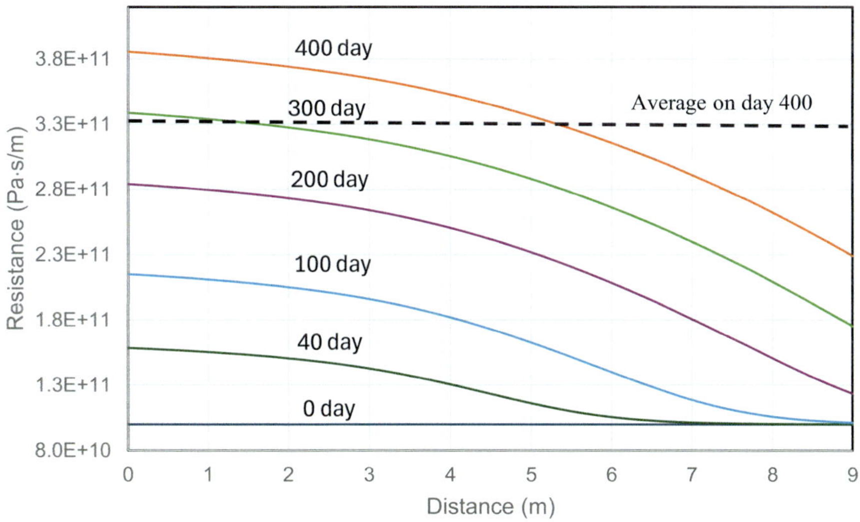

FIGURE 8.2
Resistance profile along channel length at different times. Parameters in Table 8.1 are used in simulation.

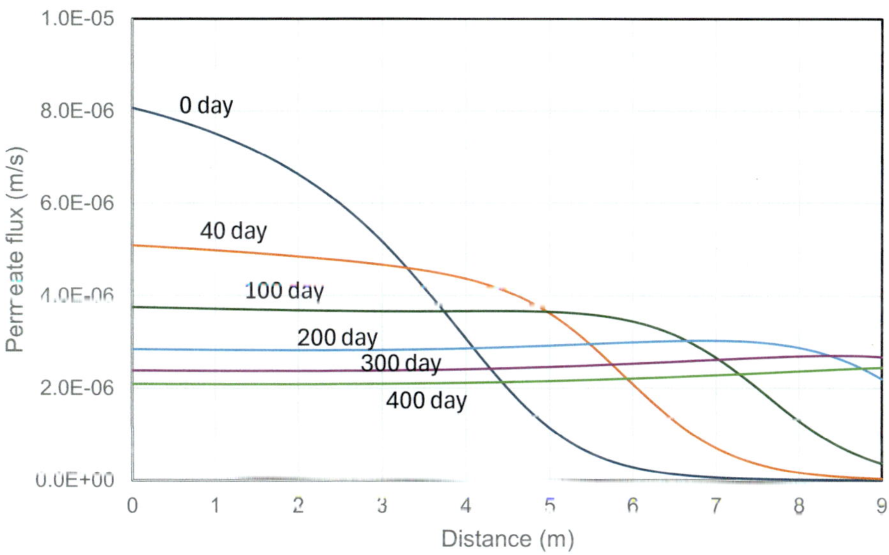

FIGURE 8.3
Distribution of permeate flux along channel length at different times. Parameters in Table 8.1 are used in simulation.

permeate production. By day 100, all membrane in the membrane channel contributes to permeate production.

It is interesting to find that permeate flux distributes more evenly over the channel length as fouling develops. On day 200, the permeate flux is almost constant in the leading section of 0–3 m. Because of the even permeate flux, membrane resistance increases at the same rate according to Eq. (8.16). Therefore, the permeate flux will remain even in this section from that time on. Longer channel sections of constant flux can be seen at days 300 and 400. This observation reveals that fouling in the membrane channel tends to even permeate flux over the whole channel length.

Average permeate flux of the membrane channel as a function of time is presented in Figure 8.4. A striking feature in Figure 8.4 is that there is a "grace period," in which the average permeate flux remains unchanged even though membrane fouling is ongoing from the very beginning. This phenomenon of a grace period is a particular characteristic of membrane fouling in the long narrow membrane channel. It does not occur in the lab-scale membrane device with a small piece of membrane, where a decline in permeate flux will be immediately observed if there is membrane fouling.

The grace period occurs in the long narrow membrane channel when it works initially in the thermodynamic restriction regime. In this case, a certain membrane area in the channel is not being used at the beginning of operation. For example, Figure 8.3 shows that the membrane in the channel with distance $x > 7$ m from the entrance is not in use at day 0 (because permeate

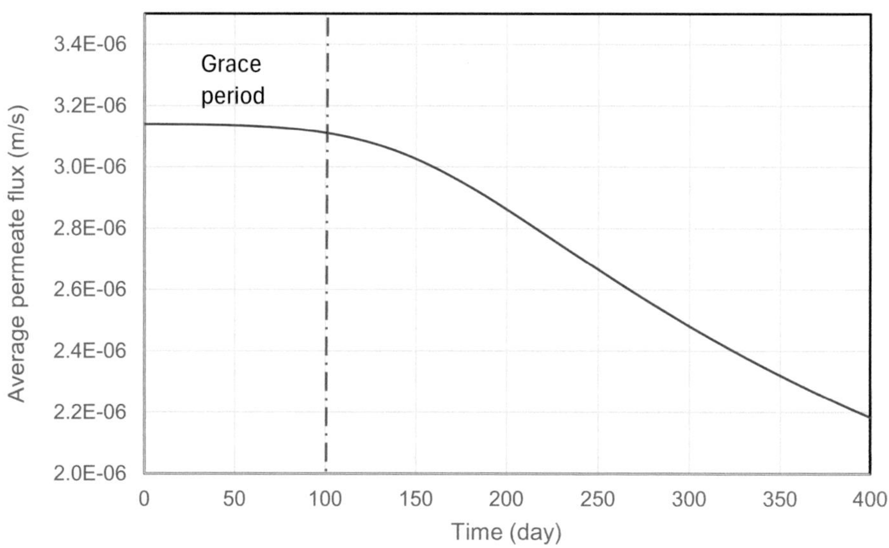

FIGURE 8.4
Average permeate flux over 400 days. There is a grace period of about 100 days featured with constant average permeate flux indicated on the graph. Parameters in Table 8.1 are used in simulation.

TABLE 8.2

Fouling Development in a Membrane Channel

Time (day)	0	40	100	200	300	400
Recovery (%)	80.7	80.7	80.0	73.6	63.8	56.1
Average v ($\times 10^{-6}$ m/s)	3.14	3.14	3.11	2.86	2.48	2.18
Flux change (%)	0.0%	−0.1	−0.9	−8.9	−21.0	−30.5

flux is zero). This portion of membrane area is gradually being used as membrane fouling develops in the membrane channel. However, the progress of the fouling is not shown on the overall performance parameter, the recovery or average permeate flux, as long as there is still unused membrane area in the channel. This is the reason for the appearance of the grace period. Figure 8.3 also shows that all membrane area in the channel is being used for permeate production by day 100. Then further fouling will cause a decline in the average permeate flux from that time on. Correspondingly, a grace period of about 100 days is observed in Figure 8.4 as indicated by the vertical broken line.

The decline of average permeate flux is exaggerated visually in Figure 8.4 because only a small flux range of 2.0×10^{-6}–3.5×10^{-6} m/s is plotted in the figure. A better way to show the change in average flux is to present the values at different times, as in Table 8.2. It can be seen that the flux decline is less than 1% from the initial values to the end of the grace period at day 100 of operation. This variation might be too small to detect in the practical reverse osmosis processes, considering variations of environmental parameters, such as salt concentration and temperature of feed water. The decline of average permeate flux becomes obvious at day 200 and at later days. The average permeate flux drops by about 30.5% at day 400 of the operation. However, checking the membrane resistance profile at day 400 back in Figure 8.2, the average membrane resistance, which is indicated with a horizontal broken line, is 330% as high as that of the clean membrane. It is astonishingly striking to see that the increment in membrane resistance is so much more out of proportion than the decline in average permeate flux.

Example 8.3

The reverse osmosis desalination process in the above case study has a grace time of 100 days. (1) Determine the average resistance of the membrane by the end of the grace period. (2) What is the characteristic pressure at the end of the grace period?

SOLUTION

The calculation of the average membrane resistance at the end of the grace period starts with Eq. (8.7):

$$R_{mt} = R_{m0} + S_f v \Delta t$$

From Table 8.1, the fouling strength of the feed water in this case study is

$$S_f = 2.4 \times 10^0 \, Pa \cdot s/m^2$$

Because average flux in the grace period is constant and is controlled by thermodynamic restriction,

$$\begin{aligned} v &= R\frac{u_0 H}{L} = \left(1 - \frac{\Delta\pi_0}{\Delta p}\right)\frac{u_0 H}{L} \\ &= \left(1 - \frac{77 \, Pa/(mg/L) \times 2500 \, mg/L}{9 \times 10^5 \, Pa}\right) \times \frac{0.1 \, m/s \times 0.35 \times 10^{-2} \, m}{9 \, m} \\ &= 3.14 \times 10^{-6} \, m/s \end{aligned}$$

The operating time before the end of the grace period is

$$\Delta t = 100 \, d \times \frac{24 \, h}{1 \, d} \times \frac{60 \, min}{1 \, h} \times \frac{60 \, s}{1 \, min} = 8.64 \times 10^6 \, s$$

Then

$$\begin{aligned} R_{mt} &= R_{m0} + S_f v \Delta t \\ &= 10^{11} \, Pa \cdot s/m + 2.4 \times 10^9 \, Pa \cdot s/m^2 \times 3.14 \times 10^{-6} \, m/s \times 8.64 \times 10^6 \, s \\ &= 1.65 \times 10^{11} \, Pa \cdot s/m \end{aligned}$$

The characteristic pressure at the end of the grace period is calculated with Eq. (5.26):

$$\begin{aligned} \Delta p^* &= \frac{u_0 H R_{mt}}{L} = \frac{0.1 \, m/s \times 0.35 \times 10^{-2} \, m \times 1.65 \times 10^{11} \, Pa \cdot s/m}{9 \, m} \\ &= 6.42 \times 10^5 \, Pa = 6.42 \, bar \end{aligned}$$

The average membrane resistance and characteristic pressure are $1.65 \times 10^{11} \, Pa \cdot s/m$ and 6.42 *bars* at the end of the grace period, respectively.

By the way, the initial characteristic pressure is 3.89 bars and the ratio $\Delta p/\Delta p^* = 2.57$. It was definitely in the thermodynamic restriction regime. At the end of the grace period, the ratio $\Delta p/\Delta p^* = 1.56$. It means that the process is not completely controlled by thermodynamic restriction.

8.4 Regime Transition

A long narrow membrane channel can work in two possible regimes. The recovery or average permeate flux of the membrane channel in the thermodynamic regime is affected only by feed water salt concentration and applied pressure. Characteristics of the membrane channel do not have any impact at all. A membrane channel under thermodynamic restriction means that the maximum amount of permeate under the given pressure has been produced before the exit of the membrane channel. In this case, there is some extra membrane area unused in the membrane channel. Membrane fouling is unseeable from the overall performance indicator of the membrane channel, either recovery or average permeate flux. When permeate production is reduced by membrane fouling in the front section of the channel, some of the unused membrane will be summoned to join the service. Anyway, the maximum amount of permeate will be produced, and the recovery of the membrane channel remains unchanged.

On the other hand, the recovery or average permeate flux of a membrane channel in the mass transfer regime is affected not only by feed salt concentration and applied pressure but also by characteristics of the membrane channel, including membrane resistance, crossflow velocity, and channel height and length. The capacity of the membrane channel under the mass transfer regime is insufficient to produce the amount of permeate to raise the osmotic pressure up to the applied pressure. When membrane fouling occurs, the capacity is further reduced and less permeate would be produced. Therefore, fouling can be immediately reflected by the overall performance indicator.

However, there is a limit to covering the ongoing membrane fouling in a membrane channel working in the thermodynamic restriction regime. Fouling will become visible when there is no more unused membrane area in the channel. At that moment, the membrane channel is said to transit from the thermodynamic restriction regime to the mass transfer regime. The regime transition is illustrated and discussed with the help of the filtration coefficient introduced below.

8.4.1 Filtration Coefficient

When a membrane channel is working in the *mass transfer regime*, recovery of the membrane channel can be determined by

$$R = \frac{L}{u_0 H}\bar{v} = \frac{L}{u_0 H} \times \frac{1}{L}\int_0^L \frac{\Delta P - \Delta\pi}{R_m(x)}dx = F(\Delta P - \Delta\pi) \tag{8.17}$$

where F is the filtration coefficient, which is defined by

$$F = \frac{1}{u_0 H}\int_0^L \frac{1}{R_m(x)}dx \tag{8.18}$$

In the mass transfer regime, the net driving force $\Delta P - \Delta\pi$ can be approximately treated as constant, and a linear relationship can be assumed between recovery and net driving force. When recovery is plotted versus pressure, the filtration coefficient is the slope of the straight line for recovery.

Development of membrane fouling in a membrane channel can be indicated by declining *F*, the filtration coefficient, with time or reducing slope of the recovery line. It should be emphasized that the constant net driving force in Eq. (8.17) is employed for qualitative analysis and graphical illustration of regime transition only. If high accuracy is pursued, the numerical model of the membrane channel introduced previously should be used because the strict linear relationship does not exist between recovery and pressure even in the mass transfer-controlled regime.

8.4.2 Graphical Illustration of Regime Transition

Recoveries of a membrane channel under thermodynamic restriction and mass transfer are schematically presented in Figure 8.5. The curved line represents the recovery under thermodynamic restriction, while the straight line is the recovery controlled by mass transfer. There is also a vertical straight line to indicate the pressure at which the membrane channel is working. Fouling development and regime transition in a membrane process will be illustrated with these three lines.

Intersection points of recovery lines with the vertical pressure line provide the two candidates for recovery of the membrane channel. The smaller one is the recovery because it is the limiting factor for the channel. This is obviously true for membrane processes. When pressure is low, the recovery line of mass transfer is smaller than that of thermodynamic restriction. Because the membrane channel cannot produce enough permeate, there is no chance to reach the recovery as indicated by the thermodynamic restriction line. Under sufficiently high pressure, the recovery line of thermodynamic restriction is smaller and becomes the controlling mechanism. The higher recovery as indicated by the mass transfer recovery cannot be reached simply because there is no more water left in the channel for permeate production.

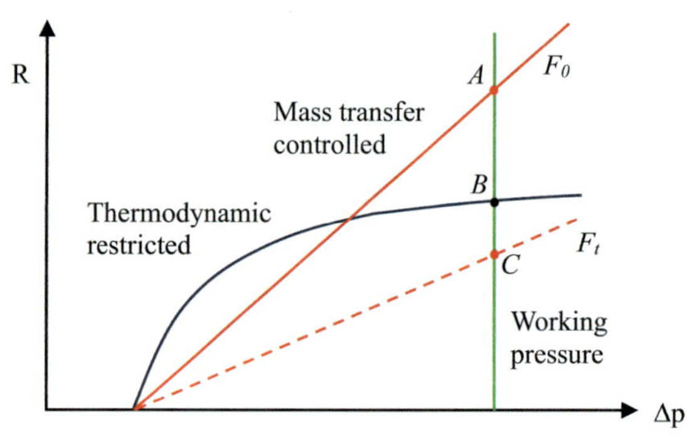

FIGURE 8.5
Schematic for regime change from thermodynamic restriction to mass transfer due to membrane fouling.

As shown in Figure 8.5, the mass transfer straight line labeled F_0 intersects with the vertical pressure line at point "A." The line labeled F_0 represents the initial filtration coefficient for the clean membrane. The thermodynamic restriction curve line intersects with the working pressure line at point "B." Because "B" is lower than "A," the recovery of this membrane process at this moment is "B," which means that the membrane channel works under thermodynamic restriction initially.

The filtration coefficient reduces with the progress of membrane fouling, and the slope of the mass transfer line declines with time. However, as long as the intersection point of the mass transfer recovery line with the pressure line is above point "B," the process continues to work under thermodynamic restriction with a constant recovery at point "B." When membrane fouling passes a critical point, the mass transfer recovery line, as indicated by the broken straight line labeled F_t, intersects with the working pressure at point "C." Now thermodynamic restriction is no longer the controlling mechanism of the membrane channel because point "C" is lower than point "B." At this moment, point "C" represents recovery of this membrane channel, which signals a regime transition, in which mass transfer becomes the controlling mechanism of the membrane channel.

As an illustration of the graphical method to show the regime transition of a membrane channel as a result of membrane fouling, the filtration coefficients are calculated from membrane resistances on different days in the case study presented in Figure 8.2. The integrations in Eq. (8.18) are done numerically, and results are presented in Table 8.3.

The recovery lines for filtration coefficients in Table 8.3 are presented in Figure 8.6 together with the thermodynamic restriction recovery curve for the same case. The graph shows that the membrane channel is controlled by thermodynamic restriction on days 0, 40, 100, and 200. In this period, the recovery of the membrane channel remains constant. However, on days 300 and 400, the membrane channel is controlled by mass transfer, and the recovery decreases with time. Figure 8.6 shows that the regime transition occurs sometime between days 200 and 300. Actually, this observation is not accurate. By observing Figures 8.3 and 8.4, it is clear that the regime transition occurs before day 200 because there is no unused membrane area in the channel (from Figure 8.3) and the average permeate flux declines obviously with time (from Figure 8.4). Therefore, though the filtration coefficient can be conveniently used to explain qualitatively regime transition due to membrane fouling, it is not very accurate quantitatively.

TABLE 8.3

The Filtration Coefficients at Different Times of Operation

Day	0	40	100	200	300	400
F (1/bar)	0.257	0.211	0.168	0.120	0.094	0.079

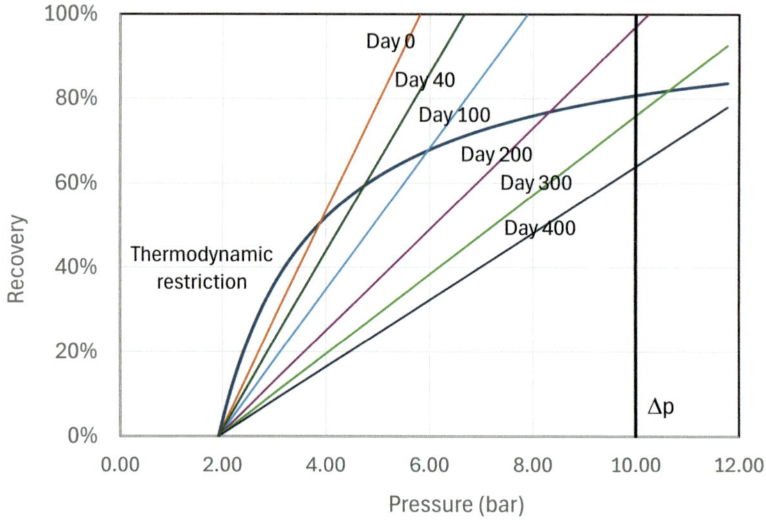

FIGURE 8.6
Recoveries and controlling mechanisms at different times.

8.4.3 Graphical Explanation for the Effectiveness of Membrane Cleaning

Membrane cleaning is a common practice to remove foulants from the fouled membranes when the permeate flux is reduced to a certain level under constant pressure operation mode or pressure increases to a certain level with constant flux operation mode. The extent to which permeate flux is restored is commonly used as an indicator for the effectiveness of membrane cleaning. Full restoration of permeate flux is usually regarded as complete removal of foulants. It stuns people when fouling comes back sooner than before after membrane cleaning. It does not make much sense that membrane fouling "accelerates" under the same operating conditions after membrane cleaning.

This phenomenon can be explained graphically with filtration coefficient lines coupled with flux profiles with time, as shown in Figure 8.7(a) and (b), respectively. Average permeate flux in Figure 8.7 is equivalent to recovery as previously used in Figure 8.6. The use of average permeate flux here is only because it is more commonly used as an indicator for cleaning efficiency. A membrane at the beginning with a clean membrane has a relatively large filtration coefficient, indicated by the mass transfer filtration line F_0. The intersection point of this filtration line with the working pressure line (does not show in the graph) is higher than the intersection point of the thermodynamic restriction curve with the working pressure line. Therefore, the membrane channel is initially working under thermodynamic restriction. Point "A" indicates the average permeate flux controlled by thermodynamic restriction. The average permeate flux will remain constant, though membrane fouling starts from the very beginning of operation.

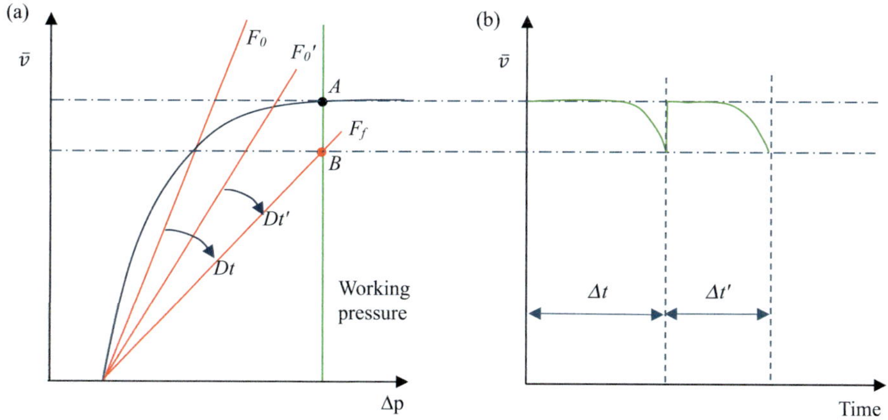

FIGURE 8.7
Schematic for restoration of (a) filtration coefficient and (b) average flux with membrane cleansing.

Fouling development can be visualized in Figure 8.7(a) as the slope of the filtration line for mass transfer is decreasing with time. But it cannot be seen in the average permeate flux as long as the filtration coefficient line does not pass point "A." When the filtration coefficient line for mass transfer passes point "A" and intersects with the vertical pressure line at a point lower than point "A," average permeate flux starts to decline with time, and membrane fouling becomes "visible" in average permeate flux.

Membrane cleaning starts when the filtration coefficient line F_t reaches point "B." If the accumulated foulants could be fully removed, the filtration coefficient line would return to the original position indicated by F_0. However, in case the foulants cannot be fully removed, the filtration coefficient line cannot return back to the original location F_0. However, the average permeate flux can still be fully stored to its initial value as long as the filtration line returns back to the thermodynamically restricted regime, as indicated by F_0'. Because of the remaining foulants from the previous round of operation, the time $\Delta t'$ needed to reach point "B" again is shorter than the initial time Δt. The change of slope of the filtration coefficient line and the corresponding profile of permeate flux with time are illustrated in the graphs in Figure 8.7(a) and (b), respectively.

Example 8.1

The efficiency of membrane cleaning can be assessed with clean water filtration at a reduced pressure to produce a similar permeate flux as it works at normal operation conditions with feed water (containing salt). It was found that membrane resistance cannot be returned to the original value by membrane cleaning. The membrane resistance after cleaning is about 1.5 times as high as its original value. By careful calculation, it is found that only 75% of the foulant was removed by membrane cleaning. Is this method better for assessing cleaning effectiveness than the use of the average permeate flux?

SOLUTION

Yes. Definitely. Because there is only one mass transfer mechanism to control permeate flux with clean water filtration. Resistances of the initially clean membrane and the membrane after cleaning can be accurate determined. The applied pressure has to be reduced so that the permeate flux is about the same as in normal operation. Otherwise, the cake layer on the membrane surface may be under different hydraulic conditions in which the resistance may change. From the data, it can be deduced that the 50% increment in membrane resistance is caused by the remaining 25% foulants stuck in the membrane process.

When the test is done under normal operation conditions, there is a chance that the process is controlled by thermodynamic restriction, in which membrane resistance is irrelevant to the permeate flux. The permeate flux can be fully restored as long as a partial removal of foulant is sufficient to return the process to the thermodynamic restriction regime. It cannot be used to accurately assess the effectiveness of membrane cleaning.

8.5 Problems

1. What is membrane fouling in reverse osmosis processes? What are the foulants? Are foulants the major targets to be separated by the membrane?
2. Sodium ions and chloride ions are the major species to be separated by reverse osmosis membranes in water desalination. Can these two ions be foulants in any circumstances?
3. Cake layer formed with particulates is the major form of membrane fouling. The applied pressure as low as 6 bars can be used in brackish water desalination. What is the critical size of particulates to become foulant under this pressure?
4. Sodium chloride salt can form strong concentration polarization; is it a form of membrane fouling? Provide reasons to support your opinion.
5. How does membrane fouling adversely affect the performance of a membrane desalination process?
6. What are the common methods employed in reverse osmosis processes to control membrane fouling?
7. Why there is a need for accurate assessment of fouling propensity or tendency of feed water in membrane desalination processes?

8. What are desired properties of the characterization method for fouling propensity or tendency of feed water in membrane desalination processes?
9. Compare advantages and disadvantages of the two existing methods for assessment of the fouling strength of feed water, SDI and MFI.
10. A piece of microfiltration membrane of 0.45 μm pore size is used in the standard test methods for both SDI and MFI. There are suggestions to use membranes of much smaller pore sizes in the tests. Do you support the suggestions? Why?
11. What is the physical meaning of the new characterization parameter "fouling strength (S_f)"?
12. The fouling strength of feed water can be measured with a dead-end membrane device similar to that used for the SDI test. Are the fouling strength measurements affected by the pressure and membrane area used in the test?
13. The fouling strength of feed water is determined with a dead-end membrane device. The total volume of filtrate collected is 200 ml over a membrane of area 17.35 cm^2. Membrane resistance changes from 1.0×10^9 to 1.1×10^9 Pa·s/m. Calculate the fouling strength of the feed water.
14. A reverse osmosis desalination process is operated in the constant flux mode of 3×10^{-6} m/s. The fouling strength of the feed water is 300 bar·min/m^2. The clean membrane resistance is 1.0×10^{11} Pa·s/m. Determine the operation time for driving pressure to increase by 10%.
15. Both the MFI and fouling strength (S_f) are based on the theory of cake filtration. Compare the physical soundness and practical execution of the two methods.
16. A microfiltration membrane with an area of 10^{-3} m^2 is operated in a dead-end device under pressure of 2 bars. The membrane resistance is 10^9 Pa·s/m, and the fouling strength of feed water is 250 bar·min/m^2. (1) Determine the time needed to collect 250 ml of filtrate. (2) What is the difference of times for collecting the same volume of filtrate with feed water and clean water ($S_f = 0$).
17. A microfiltration membrane with an area of 10^{-3} m^2 is operated in a dead-end device under pressure of 3 bars. The membrane resistance is 10^9 Pa·s/m, and the fouling strength of feed water is 250 bar·min/m^2. (1) Determine the time needed to collect 250 ml of filtrate. (2) What is the difference of times for collecting the same volume of filtrate with feed water and clean water ($S_f = 0$).
18. A microfiltration membrane with an area of 10^{-3} m^2 is operated in a dead-end device under pressure of 2 bars. The membrane resistance is 10^9 Pa·s/m, and the fouling strength of feed water is 500 bar·min/m^2. (1) Determine the time needed to collect 250 ml of filtrate. (2) What is

the difference of times for collecting the same volume of filtrate with feed water and foulant-free clean water ($S_f = 0$).

19. Compare the times for collecting the required volume of filtrate with feed water and the time differences in Questions 16, 17, and 18.
20. In a dead-end test device with a membrane area of 17.35 cm^2 and a pressure of 2 bars, 20 and 24 minutes are recorded to collect 250 ml of filtrate with clean water and feed water in the test. Calculate the fouling strength of the feed water.
21. In the test for fouling strength, the times to collect 250 ml filtrate are 20 and 24 minutes with clean water and feed water, respectively. If 500 ml filtrate is collected under the same operation condition, what will be the times for clean water and feed water?
22. From the fouling strength model, what linear relationship exists between the volume of filtrate with feed water and the time needed for collection?
23. The pressure increase is 1 bar in 30 days due to membrane fouling under constant flux operation mode. The permeate flux is 3×10^{-6} m/s, and the initial membrane resistance is 1.2×10^{11} Pa·s/m.
24. The reverse osmosis membrane is working at a constant pressure with a feed water of fouling strength S_f. Derive the expression of permeate velocity as a function of time. Calculate the permeate flux at 10 days of operation when $S_f = 5.0 \times 10^9\, Pa \cdot s/m^2$, $\Delta P = 10\, bar,\ \Delta\pi_0 = 2\, bar,\ and\ R_m = 1.1 \times 10^{11}\, Pa \cdot s/m$.
25. A reverse osmosis membrane channel of 9 m long and 0.35 mm high is operated at a pressure of 10 bars. The feed tangential velocity is 0.1 m/s, and feed concentration is 2500 mg/L. The membrane resistance is 1011 Pa·s/m, and the fouling strength of feed water is 100 bar·min/m^2. Assume foulant deposits evenly over the whole membrane channel, and find the time of operation when the characteristic pressure of the membrane channel becomes equal to the applied pressure.

References

1. Jiang, S., Li, Y., & Ladewig, P.B. (2017). A review of reverse osmosis membrane fouling and control strategies. Science of the Total Environment, 595, 567–583.
2. Chong, T.H., Wong, F., & Fane, A.G. (2008). Implications of critical flux and cake enhanced osmotic pressure (CEOP) on colloidal fouling in reverse osmosis: Experimental observations. Journal of Membrane Science, 314, 101–111.

3. Ho, J.S., Sim, L.N., Gu, J., Webster, R.D., Fane, A.G., & Coster, H.G. (2016). A threshold flux phenomenon for colloidal fouling in reverse osmosis characterized by transmembrane pressure and electrical impedance spectroscopy. Journal of Membrane Science, 500, 55–65.
4. Li, H., Yu, P., & Luo, Y. (2015). Fouling mechanisms and primary foulant constituents in reverse osmosis membrane reclamation of a petrochemical secondary effluent. Desalination and Water Treatment, 54, 3200–3210.
5. Niriella, D., & Carnahan, R.P. (2006). Influence of operating parameters on clay fouling of a RO membrane. Separation Science and Technology, 41, 25–41.
6. Tayeh, Y.A. (2024). A comprehensive review of reverse osmosis desalination: Technology, water sources, membrane processes, fouling, and cleaning. Desalination and Water Treatment, 320, 100882.
7. Liu, H., Gu, J., Wang, S., Zhang, M., & Liu, Y. (2020). Performance, membrane fouling control and cost analysis of an integrated anaerobic fixed-film MBR and reverse osmosis process for municipal wastewater reclamation to NEWater-like product water. Journal of Membrane Science, 593, 117442.
8. Shahonya, I., Nangolo, F., Erinosho, M., & Angula, E. (2021). Scaling and Fouling of Reverse Osmosis (RO) Membrane: Technical Review. In: Awang, M., Emamian, S.S. (eds) Advances in Material Science and Engineering: Lecture Notes in Mechanical Engineering. Springer, Singapore.
9. Tanuwidjaja, D., Jin, X., Huang, X., Marambio-Jones, C., Jawor, A., Zhang, M., Jiang, S.C., Cheng, R.C., & Hoek, E.M. (2020). Comparison of membrane fouling and cleaning in one-pass reverse osmosis and two-pass nanofiltration approaches to seawater desalination. Desalination and Water Treatment, 193, 235–250.
10. Chen, G., Wu, Y., Fang, P., Bai, Y., Chen, Z., Xu, Y., Wang, Y., Tong, X., Luo, L., Wang, H., Zhang, Z., Ikuno, N., & Hu, H. (2022). Performance of different pretreatment methods on alleviating reverse osmosis membrane fouling caused by soluble microbial products. Journal of Membrane Science, 641, 119850.
11. Li, Y., Li, M., Xiao, K., & Huang, X. (2020). Reverse osmosis membrane autopsy in coal chemical wastewater treatment: Evidences of spatially heterogeneous fouling and organic-inorganic synergistic effect. Journal of Cleaner Production, 246, 118964.
12. Zhou, F., Wang, W., Li, K., Yang, W., Lee, J., Xie, B., Wu, B., Ren, H., Hong, S., & Zhan, M. (2024). Controlling of irreversible fouling and mechanism in a hybrid ceramic membrane bioreactor (CMBR)-reverse osmosis (RO) process for textile wastewater reclamation. Desalination, 586, 117914.
13. Pandey, S.R., Jegatheesan, V., Baskaran, K., & Shu, L. (2012). Fouling in reverse osmosis (RO) membrane in water recovery from secondary effluent: A review. Reviews in Environmental Science and Bio/Technology, 11, 125–145.
14. Park, M.J., Pathak, N.B., Wang, C., Tran, V.H., Han, D.S., Hong, S., Phuntsho, S., & Shon, H.K. (2023). Fouling of reverse osmosis membrane: Autopsy results from a wastewater treatment facility at central park, Sydney. Desalination, 565, 116848.
15. Freire-Gormaly, M., & Bilton, A.M. (2019). Impact of intermittent operation on reverse osmosis membrane fouling for brackish groundwater desalination systems. Journal of Membrane Science, 583, 220–230.
16. Hoek, E.M.V., Kim, AS., & Elimelech, M. (2002). Influence of crossflow membrane filter geometry and shear rate on colloidal fouling in reverse osmosis and nanofiltration separations. Environmental Engineering Science, 19, 357–372.

17. Li, Z., Ni, G.W., Wang, J., Zhou, L., Yang, Q., Li, P., Gao, A., Yu, T., Cheng, L., & Bi, X. (2023). Fouling characteristic of reverse osmosis membrane for reclaimed water treatment operating under cold winter condition. Desalination, 549, 116309.
18. Peters, C.D., Li, D., Mo, Z., Hankins, N.P., & She, Q. (2022). Exploring the limitations of osmotically assisted reverse osmosis: Membrane fouling and the limiting flux. Environmental Science & Technology, 56(10), 6678–6688.
19. Shang, C., Pranantyo, D., & Zhang, S. (2020). Understanding roughness – Fouling relationship in reverse osmosis: Mechanism and implications. Environmental Science & Technology, 54(8), 5288–5296.
20. Massons-Gassol, G., Gilabert-Oriol, G., Johnson, J.E., & Arrowood, T.L. (2017). Comparing biofouling development in membrane fouling simulators and spiral-wound reverse osmosis elements using river water and municipal wastewater. Industrial & Engineering Chemistry Research, 56, 11628–11633.
21. Teng, Y., & Yong Ng, H. (2024). Prediction of reverse osmosis membrane fouling in water reuse by integrated adsorption and data-driven models. Desalination, 576, 117353.
22. Weinrich, L., Haas, C.N., & LeChevallier, M.W. (2013). Recent advances in measuring and modeling reverse osmosis membrane fouling in seawater desalination: A review. Journal of Water Reuse and Desalination, 3, 85–101.
23. Ling, B., Xie, P., Ladner, D., & Battiato, I. (2021). Dynamic modeling of fouling in reverse osmosis membranes. Membranes, 11, 349.
24. Rezaei, L., Dehghani, M., hessam Hassani, A., & Alipour, V. (2020). Seawater reverse osmosis membrane fouling causes in a full scale desalination plant; through the analysis of environmental issues: Raw water quality. Environmental Health Engineering and Management, 7(2), 119–126.
25. Probstein, R.F., Chan, K.K., & Cohen, R. (1981). Model and Preliminary Experiments on Membrane Fouling in Reverse Osmosis. In: Turbak, A.F. (ed.) Synthetic Membranes and Their Applications (2nd ed.). American Chemical Society, Washington, DC, pp. 131–145.
26. Alhadidi, A., Blankert, B., Kemperman, A.J.B., Schippers, J.C., Wessling, M., & van der Meer, W.G.J. (2011) Effect of testing conditions and filtration mechanisms on SDI. Journal of Membrane Science, 381, 142–151.
27. Nahrstedt, A., & Camargo-Schmale, J. (2008). New insights into silt density index and modified fouling index measurements. Water Supply, 8, 401–411.
28. ASTM (2007). ASTM Standard (D 4189 – 07): Standard Test Method for Silt Density Index (SDI) of Water, D19.08 on Membranes and Ion Exchange Materials.
29. Schippers, J.C., & Verdouw, J. (1980). The modified fouling index, a method of determining the fouling characteristics of water. Desalination, 32, 137–148.
30. Schippers, J.C., Hanemaayer, J.H., Smolders, C.A., & Kostense, A. (1981). Predicting flux decline of reverse osmosis membranes. Desalination, 38, 339.
31. Boerlage, S.F.E, Kennedy, M.D., Aniye, M.P., Abogrean, E.M., Galjaard, G., & Schippers, J.C. (1998). Monitoring particulate fouling in membrane systems. Desalination, 118, 131–142.
32. Boerlage, S.F.E., Kennedy, M., Aniye, M.P., & Schippers, J.C. (2003). Applications of the MFI-UF to measure and predict particulate fouling in RO systems. Journal of Membrane Science, 220, 97–116.
33. Tay, K.G., & Song, L. (2005). A more effective method for fouling characterization in a full-scale reverse osmosis process. Desalination, 77, 95–107.

34. Chen, K.L., Song, L., Ong, S.L., & Ng, W.J. (2004). The development of membrane fouling in full-scale RO processes. Journal of Membrane Science, 232, 63–72.
35. Song, L., & Tay, K.G. (2011). Advanced Membrane Fouling Characterization in Full-Scale Reverse Osmosis Processes. In: Wang, L.K., Chen, J.P., Hung, YT., Shammas, N.K. (eds) Membrane and Desalination Technologies: Handbook of Environmental Engineering, vol 13. Humana Press, Totowa, NJ.
36. Song, L., Chen, K.L., Ong, S.L., & Ng, W.J. (2004). A new normalization method for determination of colloidal fouling potential in membrane processes. Journal of Colloid and Interface Science, 271, 426–433.

Part III

Membrane Transport Phenomena

This part of the book deals with water and solute transport across membranes, which are essential to understand membrane separation processes. While water can transport across porous membranes by convection and diffusion under hydraulic pressure gradient and concentration gradient, it transports through the osmotic (semipermeable) membranes purely through convection because a concentration gradient does not exist inside the membrane. Concentration gradient manifests itself as negative pressure that sucks water through the membrane. The effectiveness of negative pressure on water transport usually is lower than that of hydraulic pressure possibly due to the breakup of water continuity inside the membrane.

Salinity energy is a type of energy that exists between water bodies of different salinities that can be harvested by means of controlled mixing. Salinity energy has not been harvested because of the low efficiency of the available technologies. Conventional pressure retarded osmosis, which is the most promising technology for salinity energy harvest, has membrane power densities that are too small to be economically feasible mainly due to the very low water flux under osmotic pressure. There is a chance for the salinity energy to be harvested profitably if the chemical potential (osmotic pressure) in draw solution can be used more effectively on something else than the driving force for water transport across membrane. After all, it is well-known that osmotic pressure is ineffective for water transport across membrane. Some interesting ideas are discussed on more effective uses of osmotic pressure in salinity energy harvest.

DOI: 10.1201/9781003743569-11

Though it is well established that ion transports across membrane are governed by Nernst–Planck equation coupled with Poisson equation, it is very difficult to obtain the true (analytical or numerical) solutions for the real-world cases. One reason is that the unrealistic assumptions of electroneutrality and constant electric field are commonly used to reduce the complexity and difficulty of mathematical models. However, the simplified mathematic models are no longer the genuine reflection or description of the real-world ions transport problems. A challenge in rigorous formulation of ion transport across membrane is the specification of boundary conditions of the governing equations. Irrelevant or wrong solutions for ion transport may be obtained if the boundary conditions are inappropriately imposed. A boundary updating scheme is introduced to ensure that the solutions of the governing equations are obtained with the appropriate boundary conditions.

The following topics will be covered in this part of the book:

Chapter 9. Water Transport Through Membranes

Chapter 10. Salinity Energy Harvest with Osmotic Membranes

Chapter 11. Solute Transport Through Membrane

9

Water Transport Through Membranes

Transport of water, or in general liquids, through a membrane is essential for membrane separation processes [1–4]. Mechanisms that control water transport are discussed in this chapter. Hydraulic pressure is usually the primary driving force to drive water across membrane in the pressure-driven membrane processes. However, some complications occur when the membranes reject solutes in water. In addition to hydraulic pressure, the chemical potential difference of water can also induce water flow through membranes. The manner and magnitude of this "chemical driving force" are of great interest to understand water transport across membranes [5–9].

9.1 Osmotic Membrane and Osmotic Pressure

9.1.1 Osmotic Membrane

Osmotic membranes are a special type of membranes that reject solutes in water, such as uncharged molecules and charged ions. Osmotic membranes are also named semipermeable membranes. Rejection of solutes is usually done by size exclusion because solutes are usually bigger in size than water molecules. For example, the hydrated sodium ion and chloride ion are 0.36 and 0.33 nm in size [10], respectively, while the water molecule has a size of 0.27 nm [11,12]. Membranes with pore sizes in between will allow water to pass but reject most ions. There might be other mechanisms to reject the solutes, such as electrical interactions.

Due to the semipermeability of osmotic membranes, another driving force named osmotic pressure arises for water transport across the membranes in addition to hydraulic pressure. Because of the osmotic pressure, water transport through osmotic membranes behaves differently from the porous membranes that do not reject the solutes [13–19]. Besides the different nature of osmotic pressure from hydraulic pressure as a driving force for water transport, water transport is also inevitably influenced by concentration polarization, which occurs inherently with osmotic membranes. Therefore, special attentions are needed for water transport across osmotic membranes.

To facilitate discussions of mechanisms for water transport, the concept of a rigid ideally semipermeable osmotic membrane is defined and introduced here. The term "rigid" means the membrane does not swell and the porosity

DOI: 10.1201/9781003743569-12

and pore sizes remain constant. The term "ideally semipermeable" means that solutes are completely rejected at the membrane surface and cannot enter the membrane matrix. In other words, there are no solutes inside the membrane. Though actual membrane matrix is not relevant to the concept and discussion of osmotic membranes, there is no harm in imagining an osmotic membrane as a thin film with pores interconnected from one surface to another of the membrane. The pore sizes are big enough for water to pass but small enough to prevent ions and solutes from entering and passing through.

9.1.2 Osmotic Pressure as a Driving Force

When an osmotic membrane separates two solutions with different solute or ion concentrations, water spontaneously flows across the membrane from low-concentration side to high-concentration side. This phenomenon is called osmosis. The driving force for osmotic flow is commonly named "osmotic pressure." Osmotic pressure is a chemical property of solutions that is different in nature from hydraulic pressure, which is a well-defined mechanical force vertical on a surface. The difference between osmotic pressure and hydraulic pressure is shown in Figure 9.1. The membrane that is fixed in the channel is not deformed when there is an osmotic flow of water through the membrane under an osmotic pressure difference. It means that the membrane does not experience a force on it. On the contrary, the membrane is deformed when the water flow is driven by a hydraulic pressure difference.

The kinetic nature of osmotic pressure to drive water across a membrane is still a mystery. However, the magnitude of osmotic pressure of a solution is well defined. The osmotic flow can be reduced by adding hydraulic pressure on the high-concentration side of the membrane. Water flow induced by concentration difference can be completely stopped when the hydraulic pressure on the high-concentration side is increased to a certain value, as shown in Figure 9.2. The osmotic pressure of the solution, which is a thermodynamic property of a solution in contact with an osmotic membrane, is defined as the value of a hydraulic pressure difference that ceases the osmotic flow across the membrane when the other side of the membrane is pure water.

FIGURE 9.1
Membrane does not deform under osmotic flow (left) but deforms under hydraulic flow (right).

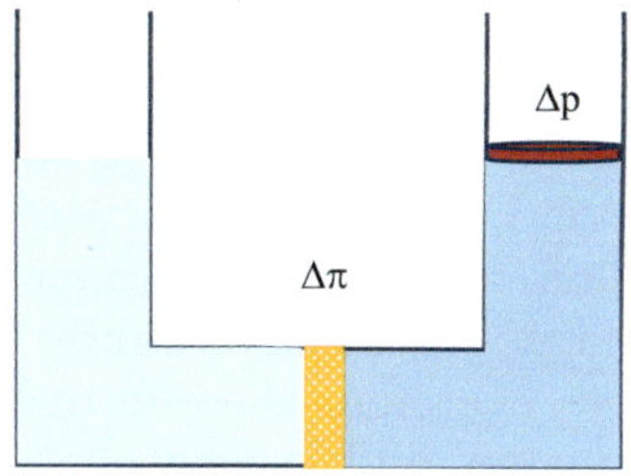

FIGURE 9.2
The hydraulic pressure added on solution that stops water flow is the osmotic pressure of a solution when the other side of membrane is pure water.

Calculation of the osmotic pressure of a solution was introduced in Chapter 1. Briefly, osmotic pressure of a solution can be calculated in two ways. When the chemical composition of solution is known, its osmotic pressure can be calculated with van't Hoff equation Eq. (1.5):

$$\Delta\pi = RT\sum_{i=1}^{n} m_i$$

where $\Delta\pi$ is the osmotic pressure (with pure water as the reference), R is the gas constant, T is the absolute temperature, and m_i is the molar concentration of species i. When the units of Pa and mol/L are used for osmotic pressure and molar concentration, the value of gas constant is $R = 8.31 \times 10^3$ Pa·L/K·mol.

Alternatively, osmotic pressure of a solution can be simply calculated with a linear correlation to the total concentration of all total solutes in it with the empirical Eq. (1.6):

$$\Delta\pi = f_{os}c$$

where f_{os} is the osmotic coefficient and c is the total dissolved solids (TDS) concentration. Eq. (1.6) is more convenient to use because it does not require the detailed composition of the solution.

The low effectiveness of osmotic pressure as driving force for water transport across a membrane remains a big concern or puzzle to resolve. It has been experimentally demonstrated that permeate flux through an osmotic membrane under osmotic pressure is much smaller than that under hydraulic pressure of the same magnitude. In addition, the nominal permeability (the ratio of net driving pressure to permeate flux) of a membrane drops significantly with increasing osmotic pressure while it remains relatively constant with varying hydraulic pressure.

Internal concentration polarization is commonly blamed for the low permeate flux under osmotic pressure. An osmotic membrane usually has a thin skin layer and a relatively thick supporting layer. The skin layer is the function layer that rejects ions while the supporting layer is a much more porous layer that provides physical support to the skin layer. According to the concept and speculation of internal concentration polarization, the lower permeate flux under osmotic pressure is because a profoundly strong concentration polarization would develop in the porous supporting layer under osmotic pressure. Salt concentration difference (equivalently osmotic pressure difference) across the skin layer can be significantly reduced by this internal concentration polarization. Therefore, the effective driving force for water to pass through the membrane can be much smaller than the total osmotic pressure of the solution.

This explanation is questionable because concentration polarization occurs regardless of the nature of driving force. As long as there is separation, there is concentration polarization. It is internal concentration polarization if it occurs in the porous supporting layer of the membrane or external concentration polarization if it occurs in the water outside the skin layer. There is no solid fundamental support that internal concentration polarization is particularly favorable to develop under osmotic pressure. In fact, it is experimentally demonstrated that low water flux under osmotic pressure remains when the supporting porous layer is significantly reduced or even completely eliminated (no internal concentration polarization at all). Therefore, there is still a need for better understanding of kinetic nature of osmotic pressure on water transport.

9.2 Water Transport in Porous Membrane

Let us first study water transport through a porous membrane. A porous membrane is permeable to both water and solutes without preference. The primary movement of water in this case is convection along the pressure gradient. However, if there is a concentration gradient of water along the membrane thickness, a diffusive movement of water under the concentration gradient will superpose on the convective movement.

9.2.1 Average Diffusive Velocity

In general, the chemical potential gradient of a species is a driving force for the species to move in the mixture of multicomponent. Chemical potential of a species in solution is defined as follows:

$$\mu_i = \mu_i^0 + RT\ln\left(c_i/c_i^0\right) \tag{9.1}$$

where μ_i is the chemical potential per mole of species i, μ_i^0 is the chemical potential at baseline concentration, c_i is the concentration (activity coefficient is assumed unity) of species i, R is the gas constant, T is the absolute temperature, and c_i^0 is the concentration of species i at the standard condition.

Because a porous membrane is permeable to both water and solute, gradients of both chemical potentials can exist inside the membrane.

Diffusive force for a species to move through a membrane is the negative gradient of chemical potential, i.e.,

$$F_d = -\frac{RT}{c_i}\frac{dc_i}{dx} \tag{9.2}$$

where F_d is the diffusive force on one mole of the species.

When the impact of porous membrane on transport of the species is negligible (very loose), and the species can be treated as spherical particles in a dilute solution, the frictional force on a mole of the species moving through relatively still water can be calculated with Stokes' law:

$$F_f = -6N_a\pi\eta a_i\langle v_i\rangle = -k_{di}\langle v_i\rangle \tag{9.3}$$

where F_f is the frictional drag force asserted on one mole of species i, N_a (= 6.022×10^{23}) is the Avogadro's number, η is the viscosity, a_i is the radius of the species i, k_{di} is the friction coefficient with the unit of kg/(s·mole), and $\langle v_i\rangle$ is the average velocity of species i. The diffusive force and the frictional force are equal in value to each other but opposite in direction when the species are moving at constant velocity. Therefore, the average velocity of the species can be found from Eqs. (9.2) and (9.3) as

$$\langle v_i\rangle = -\frac{RT}{k_{di}}\frac{1}{c_i}\frac{dc_i}{dx} \tag{9.4}$$

From the Einstein–Stokes equation, the leading coefficient in Eq. (9.4) is the diffusion coefficient, i.e.,

$$D_i = \frac{RT}{k_{di}} \tag{9.5}$$

Substituting Eq. (9.5) into Eq. (9.4) results in

$$\langle v_i\rangle = -D_i\frac{1}{c_i}\frac{dc_i}{dx} \tag{9.6}$$

Diffusion coefficient is a parameter defined rigorously by Eq. (9.5) and Eq (9.3). It is fundamentally independent of concentration.

9.2.2 Conjugation of Chemical Potential and Diffusion

Chemical potential is essentially the Gibbs free energy per mole of materials (molar free energy). An intrinsic property of chemical potential is its dependence on concentration, as expressed by Eq. (9.1). When the species are placed in fields of external forces, such as pressure field, gravity field, and

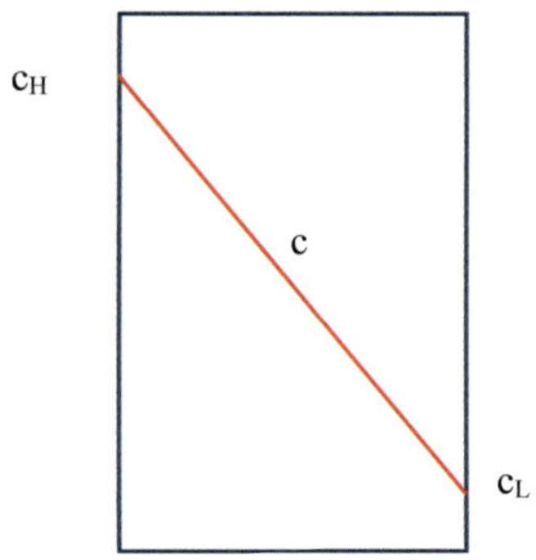

FIGURE 9.3
Concentration is a straight line across membrane when the gradient of chemical potential is the only driving force for mass transport.

electrical field (for charged species only), the molar free energy is superposed by the external fields. It is common to consider the external fields in addition to chemical potential when the direction of reaction or movement is judged by the free energy minimization principle of a system. However, different potential components may change the states of the system in different ways.

The chemical potential expressed as Eq. (9.1) is the only potential or potential component that is responsible for diffusive transport. It can be delineated with a simple diffusion process presented in Figure 9.3. A solute transports by diffusion through a layer of water. Solute concentrations on two sides of the water layer are maintained constant. The water layer is maintained stagnant in which solute transports by diffusion only.

At steady state, concentration profile is a straight line in the water layer. According to Fick's law, solute flux is constant across whole thickness of the water layer. In a kinetic view of this process, however, because the solute concentration decreases from the left side to the right side of water layer, the average solute velocity must increase with reducing concentration to maintain the same solute flux across whole thickness of the water layer. Such a dependence of velocity on concentration can only be possible when the chemical potential difference is the driving force, as indicated by Eq. (9.2).

In general, solute flux is the product of concentration and velocity no matter what the force is. Therefore, the diffusive flux is given as

$$J_i = \langle v_i \rangle c_i = -D_i \frac{1}{c_i} \frac{dc_i}{dx} \times c_i = -D_i \frac{dc_i}{dx} \tag{9.7}$$

Eq. (9.7) is Fick's law. Fick's law and chemical potential as defined in the form of Eq. (9.1) are uniquely conjugated to each other. Any other potentials would not lead to Fick's law. In other words, there are no other driving forces or potential components that can induce diffusive transport if diffusion is defined by Fick's law.

For convenience, the concept of chemical potential is often extended to include other potential components in some applications. However, other potentials in the extended potential are irrelevant to diffusive transport. Mechanical potential and electric potential are two examples that are often combined with chemical potential. The mechanical–chemical potential can be written as

$$\mu_i = \mu_i^0 + RT \ln c_i + V_i p \tag{9.8}$$

where V_i is the molecular volume of species *i* and *p* is the pressure (mechanical force per unit area). The last term in Eq. (9.8) is the mechanical potential component related to pressure.

The electrochemical potential, which is a frequently used term in electrochemistry, is written as

$$\mu_i = \mu_i^0 + RT \ln c_i + z_i F\varphi \tag{9.9}$$

where z_i is the valence of species (ion) *i*, *F* is the Faraday constant, and φ is the electrostatic potential. It can be shown that none of the two extended potentials can lead to Fick's law Eq. (9.7).

Example 9.1

Calculate the diffusive friction coefficient of sodium chloride through a layer of stagnant water at temperature is 20°C.

SOLUTION

The diffusive coefficient of sodium chloride at 20°C is calculated with Eq. (7.21):

$$D = (0.44 + 0.0423 \times 20) \times 10^{-9}\, m^2/s = 1.29 \times 10^{-9}\, m^2/s$$

Assume the friction force of sodium chloride in water follows Stokes's law. By definition of the friction coefficient,

$$k_d = \frac{F_{fric}}{\langle v \rangle} = \frac{6N_a \pi\eta a \langle v \rangle}{\langle v \rangle} = 6N_a \pi\eta a$$

According to Einstein Stokes' law

$$6N_a \pi\eta a = \frac{RT}{D}$$

Then

$$k_d = \frac{RT}{D} = \frac{8.31\, m^3 \cdot Pa/K.mole \times 293\ K}{1.29 \times 10^{-9}\, m^2/s} = 1.89 \times 10^{12}\, m \cdot Pa \cdot s/mole$$

9.2.3 Water Flux

The total potentials for water transfer in membrane are the chemical potential and mechanical potential, as given by Eq. (9.8). The diffusive water flux can be determined by applying Eq. (9.7) to water transport:

$$J_d = -D_w \frac{dc}{dx} = -\frac{RT}{k_{dw}} \frac{dc}{dx}$$

where J_d is the diffusive water flux, k_{dw} is the diffusive friction coefficient of water, and D_w is the diffusion coefficient of water, respectively. When molar fraction is used as concentration unit, the basic unit for water flux is m/s or $m^3/m^2 \cdot s$.

Hydraulic pressure, as the principal driving force for water transport through membrane, produces the convective water flux. The driving force for convective water transport is therefore obtained by taking derivative of the mechanical potential, which is the last term in Eq. (9.8):

$$F_c = -V_w \frac{dp}{dx} \tag{9.10}$$

where F_c is the convective (mechanical) force that does not have the term "$1/c$" as the diffusive force from chemical potential in Eq. (9.2). There is no doubt that mechanical potential is not a driving force for diffusive water transport. It is actually the driving force for convective water transport. Balancing the mechanical force with the frictional force, the convective transport velocity is

$$v = -\frac{V_w}{k_f} \frac{dp}{dx}$$

where v is the convection velocity and k_f is the friction coefficient of viscous flow with the membrane matrix. The convective water flux driven by mechanical potential is

$$J_c = vc_w = -\frac{V_w c_w}{k_f} \frac{dp}{dx} \tag{9.11}$$

where J_c is the convective water flux. Therefore, the total water flux under both chemical potential (concentration) and mechanical potential (hydraulic pressure) is

$$J = J_c + J_d = -\frac{V_w c_w}{k_f} \frac{dp}{dx} - \frac{RT}{k_d} \frac{dc_w}{dx}$$

When the pressure difference and concentration difference are ΔP and Δc across a membrane of thickness l are used, the difference form of the above equation is

$$J = -\frac{V_w}{k_f\, l}\left(c_w \Delta p + \frac{k_f}{k_d}\frac{RT}{V_w}\Delta c_w\right) \tag{9.12}$$

A new parameter of the membrane is introduced as

$$A = \frac{V_w}{k_f\, l} \tag{9.13}$$

where A is the permeability of membrane with the basic metric unit of m/(s·Pa). Membrane resistance R_m is the reciprocal of membrane permeability A.

The first item in the bracket in Eq. (9.12) is the partial pressure on water. Because the molar fraction of water is roughly unity for most water filtration problems, we have

$$c_w \Delta p \cong \Delta p$$

With van't Hoff equation, the second term in the bracket of Eq. (9.12) can be rewritten as

$$\frac{k_f}{k_d}\left(\frac{RT}{V_m}\Delta c_w\right) = \frac{k_f}{k_d}\left(-\frac{RT}{V_m}\Delta c_s\right) = -\frac{k_f}{k_d}\Delta\pi \tag{9.14}$$

where $c_s\,(=1-c_w)$ is the solute concentration in molar fraction. Eq. (9.12) can be rewritten as

$$J = A\left(\Delta p - \frac{k_f}{k_d}\Delta\pi\right) = \frac{1}{R_m}\left(\Delta p - \frac{k_f}{k_d}\Delta\pi\right) \tag{9.15}$$

Eq. (9.15) is the water flux through the porous membrane under both hydraulic pressure and osmotic pressure. Usually the minus sign does not show in the final expression, but the flow direction is understood from the high pressure to the low pressure.

Example 9.2

A membrane separates a solution of 3 bar osmotic pressure and pure water. The membrane thickness is 100 μm and resistance of 10^{11} Pa·s/m. The diffusion coefficient of water is 2×10^{-9} m²/s and a hydraulic pressure of 5 bar is applied on pure water side of the membrane. There are strong tangential flows on the membrane surfaces to keep solute concentrations constant on the membrane surface. Calculate (1) the diffusive friction coefficient of water. (2) convective friction coefficient of water. (3) Determine the convective, diffusive, and total water fluxes.

SOLUTION

1. The diffusive friction coefficient for water to transport through the porous membrane can be calculated from Einstein–Stokes' law:

$$k_d = \frac{RT}{D} = \frac{8.31\, m^3 \cdot Pa/K \cdot mole \times 293\, K}{2 \times 10^9\, m^2/s} = 1.22 \times 10^{12}\, m \cdot Pa \cdot s/mole$$

2. It is known that the molar volume of water $V_w = 18 \times 10^{-6}\, m^3/mole$ The convective friction coefficient of water is calculated from Eq. (9.13):

$$k_f = \frac{V_w}{A\, l} = \frac{R_m V_w}{1}$$

$$= \frac{10^{11}\, Pa \cdot s/m \times 18 \times 10^{-6}\, m^3/mole}{10^{-4}\, m}$$

$$= 1.80 \times 10^{10}\, m \cdot Pa \cdot s/mole$$

3. Using Eq. (9.15),
The convective flux by the first term in Eq. (9.15)

$$J_p = \frac{\Delta p}{R_m} = \frac{5 \times 10^5\, Pa}{10^{11}\, Pa \cdot s/m} = 5 \times 10^{-6}\, m/s$$

The diffusive flux by the second term in Eq. (9.15) is as follows:

$$J_d = -\frac{1}{R_m}\frac{k_f}{k_d}\Delta\pi$$

$$= -\frac{1}{10^{11}\, Pa \cdot s/m} \times \frac{1.80 \times 10^{10}\, m \cdot Pa \cdot s/mole}{1.22 \times 10^{12}\, m \cdot Pa \cdot s/mole}\left(-3 \times 10^5\, Pa\right)$$

$$= 4.44 \times 10^{-8}\, m/s$$

The total water flux is as follows:

$$J = J_p + J_d = 5 \times 10^{-6}\, m/s + 4.44 \times 10^{-8}\, m/s = 5.04 \times 10^{-6}\, m/s$$

Even though there are both diffusive and convective water transports in a membrane, the contribution from diffusion is less than 1% under the given conditions.

9.3 Concentration Polarization of Solutes with Osmotic Membranes

Because an osmotic membrane prevents solutes from entering and passing through it, the rejected solutes accumulate and form a layer of higher

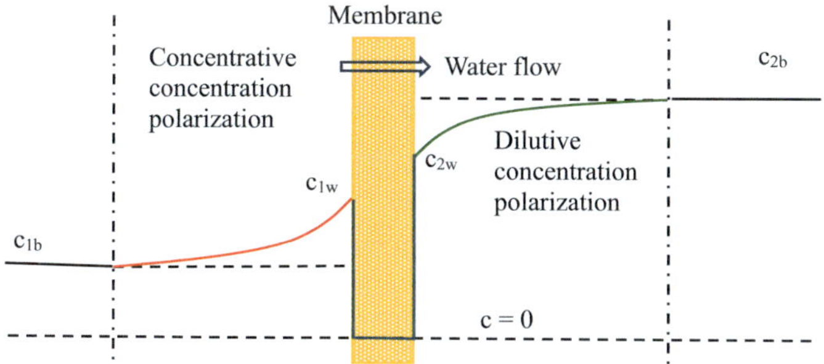

FIGURE 9.4
Schematic for concentrative and dilutive concentration polarizations.

concentration than in the bulk solution on the feed side of membrane. On the other side of the membrane, concentration of solutes in a layer adjacent to membrane surface is lower than the bulk solution due to the dilution effect of water through the membrane.

An even more complicated situation occurs with the osmotic membranes of composite structure. Composite membranes have a thin function layer that rejects solutes and a thick supporting layer that does not. Concentration polarization can develop in porous supporting layer of composite membranes. Because the supporting layer is a part of the membranes, this type of concentration polarization is named "internal concentration polarization" [20–23]. Alternatively, concentration polarization that develops in solution outside the membrane surfaces is named "external concentration polarization."

As schematically depicted in Figure 9.4, water flows into the left surface and out from the right surface of an osmotic membrane. The concentrative concentration polarization develops on the left hand side of the membrane with a higher wall concentration c_{1w} than the bulk concentration c_{1b} because of the accumulation of solutes brought to membrane surface by water flow. On the other hand, the dilutive concentration polarization develops on the right hand side of the membrane with a wall concentration c_{2w} lower than the bulk concentration c_{2b} due to the loss of solutes from membrane surface carried away by water flow. Concentration of solutes inside the membrane is zero because solutes cannot enter the osmotic membrane.

Because the phenomena of concentration polarization of solutes with osmotic membranes do not occur with porous membranes, water transport through osmotic membranes differs significantly from that through porous membranes.

9.3.1 External Dilutive Concentration Polarization

There is only concentrative concentration polarization on the feed side in reverse osmosis desalination processes. Dilutive concentration polarization

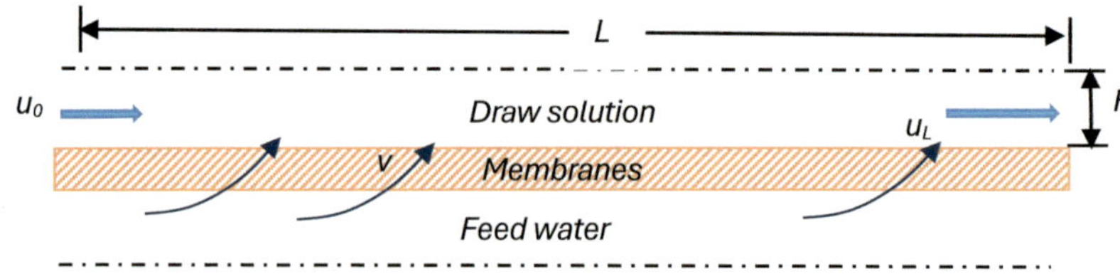

FIGURE 9.5
Water transport into draw side in a forward osmosis membrane channel.

does not exist because on the other side of membrane is permeate, not another solution of different concentration. Dilutive concentration polarization often occurs in forward osmosis processes where feed water passes through the membrane and mixes with the draw solution of high concentration of solutes. The process is schematically shown in Figure 9.5. In this figure, a half of forward membrane channel is presented with draw solution flowing inside the channel. The upper broken line indicates the central line of the membrane channel. The draw flow enters the channel at tangential velocity u_0 and leaves the channel at tangential velocity of u_L. Feed water that is drawn into the channel through membrane is flowing outside of the membrane channel. The length and height of the membrane channel are L and H as indicated in the figure.

Dilutive concentration polarization in a forward osmosis process can be calculated with the similar numerical model previously used for concentrative concentration polarization in the reverse osmosis membrane channels. However, one question remaining for forward osmosis is the calculation of permeate flux. Unlike the basic filtration equation Eq. (1.11) that is experimentally proven and widely accepted for reverse osmosis processes, there is no such equation for forward osmosis processes. The most common treatment of this problem is to assume that the permeate flux is proportional to the net osmotic pressure difference, i.e.,

$$v = \frac{\Delta\pi - \Delta p}{R_m} \tag{9.16}$$

It will be shown later that Eq. (9.16) cannot correctly predict the permeate flux in general. It is used here only to demonstrate the simulation procedure for dilutive concentration polarization. Eq. (9.16) can be replaced by a more accurate equation when it becomes available. Now we change the symbol of permeate flux from J to v to be consistent with the usage in previous chapters. In the forward osmosis process, the direction of permeate flow is from the feed water to the draw solution.

Because there is no forward osmosis process in practical operation, the configuration similar to seawater reverse osmosis desalination process is assumed. It is assumed that membrane elements of a nominal length of 1 m are used in series to form an 8-m-long forward osmosis membrane channel. Draw solutions are provided in one end and the diluted draw solution comes out of the channel from the other end. More than sufficient feed water is provided on the feed side of the membrane.

Similar to concentrative concentration polarization in reverse osmosis membrane channels, a numerical model can be developed for dilutive concentration polarization in a forward membrane channel. The three difference equations for pressure, concentration, and tangential velocity are

$$\Delta p_{i+1} = \Delta p_i - \frac{3K\mu}{H^2} u_i \Delta x$$

$$c_{a,\,i+1} = c_{a,\,i} - r\frac{c_{a,\,i} v_i}{Hu_i}\Delta x$$

$$u_{i+1} = u_i + \frac{v_i}{H}\Delta x$$

The only difference of these equations from those for concentrative concentration polarization is the opposite sign of v_i related terms in the second and third equations.

The other variables on step i + 1 are calculated by

$$\Delta\pi_{i+1} = f_{os} c_{w,i}$$

$$v_{i+1} = \frac{\Delta\pi_i - \Delta p_i}{R_m}$$

$$Pe_{i+1} = \frac{v_i H}{D}$$

$$PN_{i+1} = f\left(Pe_i\right)$$

$$c_{w,i+1} = c_0 + PN_i\left(c_{a,\,i} - c_0\right)$$

All are the same as for concentrative concentration polarization except the calculation of permeate flux, which is done using the discretized form of Eq. (9.16).

Numerical simulation of dilutive concentration polarization can be obtained when the initial conditions at entrance of the forward osmosis channel, which are the applied pressure Δp_0, salt concentration c_0, and tangential velocity u_0, are specified. A sample calculation is carried out with the parameters of a forward osmosis channel as provided in Table 9.1.

Wall concentration c_w and average salt concentration c_a from the numerical simulation, together with polarization factor c_w/c_a, are plotted along the channel length in Figure 9.6. Comparing Figure 9.6 with Figure 7.10, it can be seen that dilutive concentration polarization in a forward osmosis channel behaves totally different from concentrative concentration polarization in a reverse osmosis channel. The polarization factor c_w/c_a of dilutive concentration polarization in Figure 9.6 decreases monotonically along the membrane

TABLE 9.1

Parameters of a Forward Osmosis Process

Parameter	Symbol	Value	Unit
Membrane resistance	R_m	3.00×10^{11}	Pa·s/m
Channel height	H	3.50×10^{-4}	m
Channel length	L	8	m
Draw solution salt concentration	c_d	35,000	mg/L
Feed water salt concentration	c_f	0	mg/L
Osmotic pressure coefficient	f_{os}	73.9	Pa/(mg/L)
Feed crossflow velocity	u_0	0.1	m/s
Pressure	Δp	0	bar
Diffusion coefficient	D	1.29×10^{-9}	m²/s

channel while there is a peak value in the polarization factor of concentrative concentration polarization in Figure 7.10. Furthermore, the dilutive concentration polarization is much stronger with more than 30% lower in the wall concentration than the average concentration by exit end of the membrane channel. On the contrary, the peak polarization factor is only 18% for the concentrative concentration polarization.

Simulation results show clearly that dilutive concentration polarization is not a simple mirror symmetry of concentrative concentration polarization.

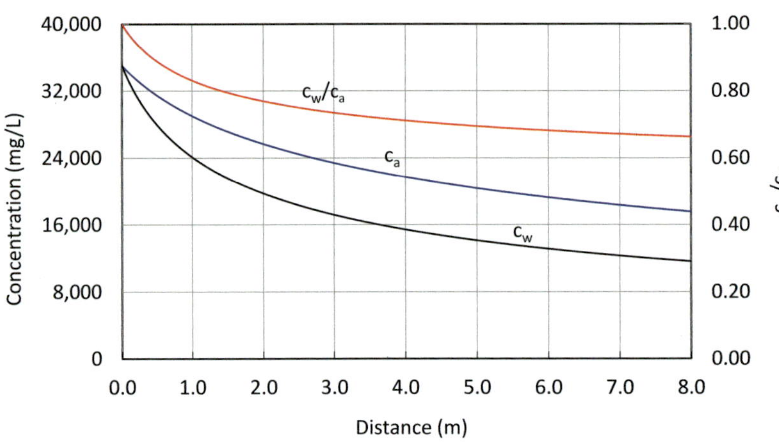

FIGURE 9.6
Average concentration, wall concentration, concentration factor along a forward osmosis membrane channel, indicating dilutive concentration polarization. The parameters used in the calculation are given in Table 9.1.

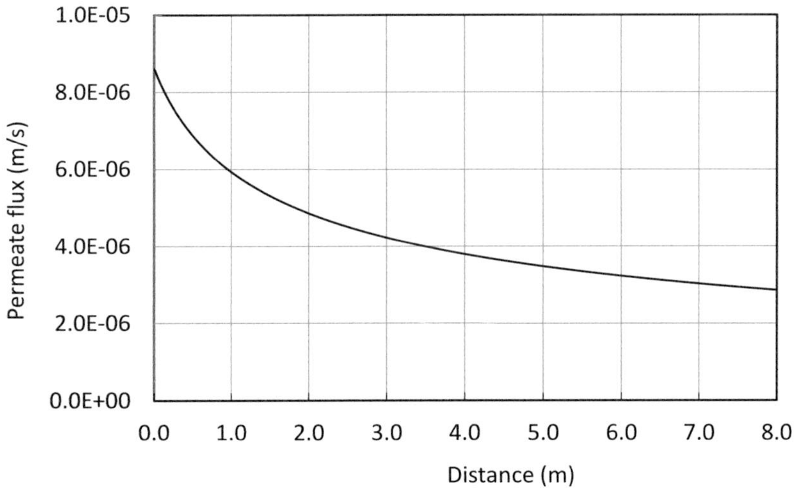

FIGURE 9.7
Permeate flux along a forward osmosis membrane channel. The parameters used in the calculation are given in Table 9.1.

Permeate flux in the forward osmosis membrane channel is shown in Figure 9.7. It also behaves quite differently compared to that in the reverse osmosis membrane channel as shown in Figure 7.9. The permeate flux approaches zero in the reverse osmosis membrane channel due to thermodynamic restriction. But there is substantial permeate flux by exit end of the forward osmosis membrane channel. It reflects that the permeate flux is only controlled by osmotic pressure. There are no other mechanisms, such as thermodynamic restriction in reverse osmosis processes, to jointly control the permeate flux in forward osmosis processes.

9.3.2 Internal Concentration Polarization

Internal concentration polarization develops inside the porous supporting layer of a membrane. The influence of hydraulic conditions over membrane surface, such as tangential flow or shear rate, on internal concentration polarization is minimized. Therefore, it behaves quite differently from external concentration polarization that is substantially affected by the hydraulic conditions over the membrane surface.

Dilutive and concentrative internal concentration polarizations are schematically presented in Figure 9.8. Water flow velocity is v with flow direction indicated in the figure. Solute concentrations in bulk solution and at the interface of skin layer with supporting porous layer are indicated with c_0 and c_i, respectively. Thickness of supporting porous layer of the membrane l is also indicated in the figure.

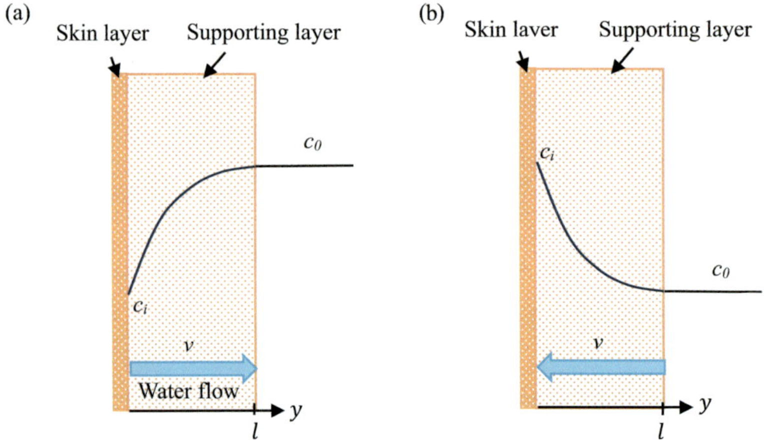

FIGURE 9.8
(a) Dilutive and (b) concentrative internal concentration polarizations developed in the porous supporting layer of a membrane.

Unlike the external concentration polarization layer, the thickness of internal concentration layer is restricted between the two interfaces of porous supporting layer with skin layer and solution, respectively. Because there is no tangential flow to take away solutes in the supporting layer, the only transport direction of solutes is vertical to the membrane surface. Concentration distribution in the supporting layer can be confidently described with one-dimensional mass balance equation:

$$vc - D\frac{dc}{dy} = 0 \tag{9.17}$$

When the origin of y coordinate is chosen at the interface of porous layer with skin layer as shown in Figure 9.8, the boundary condition at the interface with the solution is

$$c = c_0 \;\; at \;\; y = l \tag{9.18}$$

The solute concentration at the interface of supporting layer with skin layer is

$$c_i = c_0 e^{-\frac{vl}{D}} \tag{9.19}$$

Flow velocity leaving the skin layer is taken as positive in Eq. (9.19).

By the way, tortuosity of the porous supporting layer does not appear in Eq. (9.19) because of the high porosity and large pore sizes of the porous supporting layer [24]. The supporting layer of composite polyamide membranes is about 50 μm thick with pore size between 20 and 100 nm [25]. The hydrated inorganic ions, which are about 0.4–0.6 nm in sizes [10] can move freely in the porous supporting layer. In this case, there is no need to consider tortuosity because it is virtually about unity.

Example 9.3

A polysulfone supporting layer of an osmosis membrane has a thickness of 100 μm. Calculate the concentration polarization factor for dilutive and concentrative internal concentration polarizations of sodium chloride, respectively, for the range of permeate flow velocity from 10^{-6} to 10^{-5} m/s at temperature 20°C.

SOLUTION

It is known from the question that the thickness of porous supporting layer is

$$l = 100 \times 10^{-6}\, m$$

The diffusion coefficient of sodium chloride at 20°C is calculated with Eq. (7.71):

$$D = (0.44 + 0.0423 \times 20) \times 10^{-9}\, m^2/s = 1.29 \times 10^{-9}\, m^2/s$$

Sample calculation for $v = 10^{-6}\, m/s$,

Polarization factor for dilutive internal concentration polarization is

$$\frac{c_i}{c_0} = e^{-\frac{vl}{D}} = e^{-\frac{10^{-6}\, m/s \times 100 \times 10^{-6}\, m}{1.29 \times 10^{-9}\, m^2/s}} = 0.925 = 92.5\%$$

Polarization factor for concentrative internal concentration polarization is

$$\frac{c_i}{c_0} = e^{\frac{vl}{D}} = e^{\frac{10^{-6}\, m/s \times 100 \times 10^{-6}\, m}{1.29 \times 10^{-9}\, m^2/s}} = 1.081 = 108.1\%$$

The results for all permeate velocities are listed in the table and presented in the figure below.

v (μm/s)	0	1	2	4	6	8	10
Dilutive c_i/c_0 (%)	100	92.5	85.6	73.3	62.8	53.8	46.1
Concentrative c_i/c_0 (%)	100	108	116	136	159	185	217

In conclusion, the polarization factor is unlikely smaller than 50% or greater than 200% for dilutive or concentrative internal concentration polarization, respectively, because the permeate velocity is usually smaller than 10^{-5} m/s through osmotic membranes.

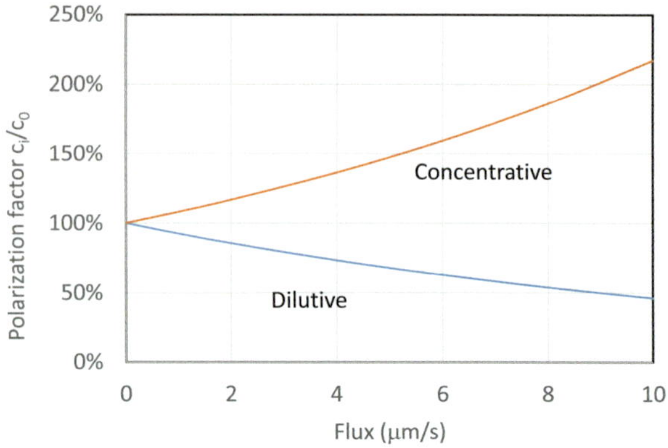

9.4 Water Transport in Osmotic Membrane

9.4.1 Role of Chemical Potential for Water Transport

Hydraulic pressure is the only driving force for water transport through osmotic membranes because a concentration gradient does not exist inside the membrane [26–30]. However, the chemical potential (osmotic pressure) difference of solutions on two sides of the membrane remains a major driving force for water transport because it can convert itself to a (negative) hydraulic pressure at the membrane–solution interfaces. Let us elaborate on this fascinating conversion below.

At the interface of an osmotic membrane with a solution, an equilibrium of water can be always assumed. It means that the total (chemical plus mechanical) potential of water on two sides of the interface is equal to each other [31–34], i.e.,

$$\mu_0^{(m)} + V_w p^{(m)} = \mu_0 + RT\ln\left(c_w/c_{\mathrm{w}}^0\right) + V_w p$$

where $\mu_0^{(m)}$ and μ_0 are the chemical potentials of pure water at the interface in the membrane side and solution side, respectively, $p^{(m)}$ and p are the hydraulic pressure inside the membrane and in the solution at the interface, respectively, and c_w and c_w^0 are the water concentrations in solution and pure water, respectively. The logarithm term does not show in left side of the equation because it is zero due to the absence of solute inside the membrane. Because the chemical potential of pure water is the same in membrane and solution, the equation reduces to

$$p^{(m)} = \frac{RT}{V_w} ln\left(c_w/c_w^0\right) + p = p - \pi \qquad (9.20)$$

where π $(= -RTln(c_w/c_w^0)/V_w)$ is the osmotic pressure of the solution according to Eq. (9.14). Note that the pure water concentration c_w^0 is unity when the molar fraction is used as concentration unit.

Eq. (9.20) states that, at the interface, the hydraulic pressure inside the membrane is smaller than the hydraulic pressure in the solution by an osmotic pressure of the solution. Eq. (9.20) is essential to understand the role of chemical potential in water transport across osmotic membranes.

9.4.2 Driven by Net Hydraulic Pressure

Water transport through osmotic membranes can be readily handled when it is driven by net hydraulic pressure even with the participation of osmotic pressure. Consider an osmotic membrane in a reverse osmosis process that separates two solutions with osmotic pressures π_1 and π_2, respectively. The hydraulic pressures applied on the two solutions are p_1 and p_2, respectively.

With Eq. (9.20), the hydraulic pressures inside the membrane at the interfaces with solutions are determined by, respectively,

$$\begin{aligned} p_1^{(m)} &= p_1 - \pi_1 \\ p_2^{(m)} &= p_2 - \pi_2 \end{aligned} \tag{9.21}$$

As shown in Figure 9.9, when hydraulic pressures added on both sides of the membrane are greater than osmotic pressures in solutions, the pressures inside of the membrane at the interfaces $p_1^{(m)}$ and $p_2^{(m)}$ are both positive. It means that water transport through the membrane is driven by net hydraulic pressure.

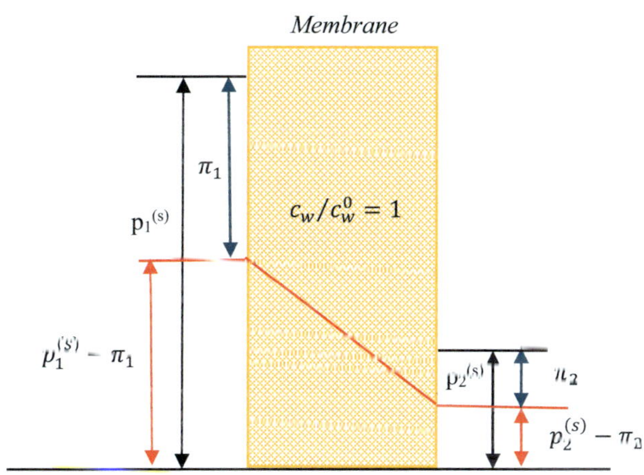

FIGURE 9.9
Net pressure for water transport in reverse osmosis processes is positive across the entire membrane thickness.

As indicated in the figure, the water activity (c_w/c_w^0) is unity throughout membrane thickness. There is no driving force for diffusive water transport. Water can only be transported by convection driven by the hydraulic pressure difference across the membrane. Therefore, the water flux is

$$v = A\Delta p^{(m)} = A\left(p_1^{(m)} - p_2^{(m)}\right) \tag{9.22}$$

Substituting Eq. (9.21) into Eq. (9.22) results in

$$v = A\left(\Delta p - \Delta\pi\right) \tag{9.23}$$

where Δp and $\Delta\pi$ are the hydraulic pressure difference and osmotic pressure difference across the membrane, respectively, i.e.,

$$\Delta p = p_1 - p_2$$

and

$$\Delta\pi = \pi_1 - \pi_2$$

Comparing Eq. (9.23) with Eq. (9.15), it can be seen that osmotic pressure plays a much bigger role for water transport in osmotic membranes than in porous membranes because $k_f/k_d \ll 1$.

Eq. (9.23) is a well-established phenomenological equation for water flux through osmotic membranes operated under reverse osmosis mode, in which the driving force is net hydraulic pressure. The equation has been validated for various types of membranes, solutes, and concentrations. Permeability A remains relatively constant in a wide range of hydraulic pressure.

9.4.3 Driven by Net Osmotic Pressure

On the contrary, the calculation of permeate flux through an osmosis membrane under net osmotic pressure is an extremely challenging task. An equation parallel to Eq. (9.23) is commonly written for this case:

$$v = A_{os}\left(\Delta\pi - \Delta p\right) \tag{9.24}$$

where A_{os} is the permeability of a membrane under net osmotic pressure. The use of different symbols for permeability of the same membrane under net hydraulic pressure and net osmotic pressure is to emphasize that they are different under two driving forces. When an expression is given in the form of Eq. (9.23) or Eq. (9.24), it usually means that the permeate flux v is proportional to the net driving pressure, implying a constant leading coefficient. However, the linear relationship between flux and net osmotic pressure does not actually exist. The osmotic permeability itself is a strong function of net osmotic pressure. Therefore, Eq. (9.24) cannot be practically used to calculate the permeate flux under net osmotic pressure.

A mediation to Eq. (9.24) is to assume that the membrane permeability under net osmotic pressure is the same as that under net hydraulic pressure. The nonlinearity between the permeate flux and net osmotic pressure is attributed to the internal concentration polarization that can reduce effective osmotic pressure significantly. Furthermore, because the internal concentration polarization in the porous layer on one side of the membrane has more profound impact than the external concentration polarization in solution on the other side of the membrane, osmotic pressures on two sides of the membrane become two independent variables [35].

The solute concentration at the interface of skin layer with porous layer is reduced by a factor $e^{-\frac{vl}{D}}$ from the draw solution by dilutive internal concentration polarization as indicated by Eq. (9.19), so does the effective osmotic pressure. Therefore, permeate flux is then calculated as

$$v = A\left(\pi_d e^{-\frac{vl}{D}} - \pi_f - \Delta p\right) \tag{9.25}$$

where A is the hydraulic permeability of membrane, and π_d and π_f are osmotic pressures of draw solution and feed water, respectively. The impact of external concentration polarization on the feed side is assumed negligibly small. Eq. (9.25) shows that the role of osmotic pressure in the osmotically driven membrane processes cannot be represented by a single osmotic pressure difference between two membrane surfaces. The hydraulic permeability A can be used in Eq. (9.25) because the internal concentration polarization is explicitly considered. Eq. (9.25) is an implicit equation for permeate flux because it appears on both sides of the equation. Explicit permeate flux can be determined from Eq. (9.25) with the successive substitution method introduced previously in Chapter 7.

Example 9.4

A pressure-retarded osmosis process uses seawater and pure water as draw solution and feed water, respectively. Parameters of the process are membrane permeability $A = 10^{-12}$ m/(s·Pa), thickness of porous supporting layer of the membrane l = 100 nm, diffusion coefficient $D = 1.29 \times 10^{-9}$ m^2/s, and hydraulic retarded pressure (which is the pressure added on draw solution) Δp = 10 bars. Calculated the permeate flux in the process.

SOLUTION

The osmotic pressure of draw solution is

$$\pi_d = f_{os} c_d = 73.9\ Pa/(mg/L) \times 35{,}000\ mg/L = 2.59 \times 10^6\ Pa$$

The osmotic pressure of feed water assuming negligible salinity is

$$\pi_f = f_{os}c_d = 73.9\,Pa/(mg/L) \times 0\,mg/L = 0\,Pa$$

Permeate flux in this pressure retarded osmosis process can be determined with the successive substitution method. Let the initial guess $v^{(0)} = 1 \times 10^{-5}\,m/s$, then the first substitution is

$$v^{(1)} = A\left[\pi_d \exp\left(-\frac{v^{(0)}l}{D}\right) - \pi_f - \Delta p\right]$$

$$= 10^{-12}\,m/(s \cdot Pa)\left[2.59 \times 10^6\,Pa \times \exp\left(-\frac{1 \times 10^{-5}\,m/s \times 10^{-4}\,m}{1.29 \times 10^{-9}\,m^2/s}\right) - 0 - 10^6\,Pa\right]$$

$$= 1.91 \times 10^{-7}\,m/s$$

The subsequent substitutions are given in the table below.

Substitution	Permeate flux (m/s)
0	1.00×10^{-5}
1	1.91×10^{-7}
2	1.55×10^{-6}
3	1.29×10^{-6}
4	1.34×10^{-6}
5	1.33×10^{-6}
6	1.33×10^{-6}

Permeate flux converges at the sixth substitution. The final result is

$$v = 1.33 \times 10^{-6}\,m/s$$

Unfortunately, the permeate fluxes predicted by Eq. (9.25) as illustrated in this example are usually orders of magnitude higher than the experimentally observed fluxes in forward osmosis and pressure-retarded osmosis processes. There exist many modifications or variations of Eq. (9.25) with the intentions to account more accurately for the impact of internal concentration polarization on permeate flux. None of them are commonly accepted as a reliable method for permeate flux through a membrane driven by net osmotic pressure.

9.5 Transmutation

A filtration process with osmotic membranes under net osmotic pressure is a transmutation or an alienation of hydraulic pressure-driven membrane

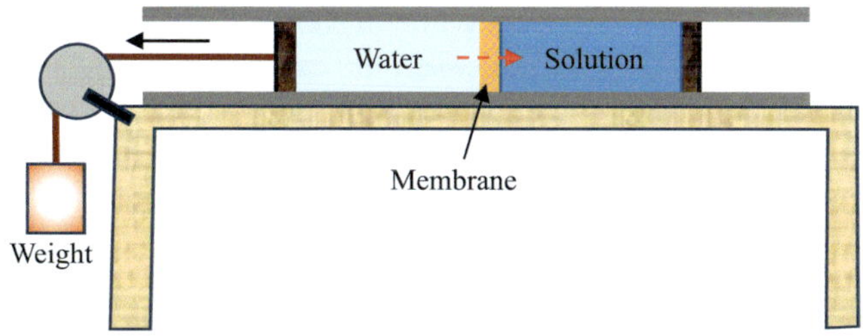

FIGURE 9.10
Water in the membrane system driven by osmotic pressure is under negative hydraulic pressure (tension).

processes. The behavior and performance of such a process cannot be described with the concepts or models that are commonly developed for membrane processes driven by net hydraulic pressure.

9.5.1 Negative Hydraulic Pressure

When the osmotic pressure of a solution is greater than the hydraulic pressure on one or both sides of an osmotic membrane, such as in forward osmosis and pressure-retarded osmosis, negative hydraulic pressure develops inside the membrane. The negative pressure at the interface of the membrane with a solution is not a pure theoretical speculation. It can be demonstrated experimentally with a device as shown in Figure 9.10.

As shown in the figure, an osmotic membrane is fixed in the middle of a channel to separate a solution on the right side from pure water on the left side of the membrane. A weight is attached to the piston on the pure water side through a pulley. It can be shown that when the pressure (force per area) on the piston added by the weight is equal to the osmotic pressure of the solution, there is no water flow through the membrane. At this moment, pure water on the left side of membrane is stretched (under tension) or under a negative pressure. This negative pressure on pure water is needed to balance the negative pressure at the interface of the membrane with the solution.

9.5.2 Breakup of Water Continuity

Water under negative pressure is in an unstable state and tends to break up (cavitation). Pure water under particular circumstances may keep its continuity under negative hydraulic pressure (in metastable state) due to the lack of the starting points to break up. However, it usually is not the case in membrane processes. Feed water in membrane processes commonly contains a certain amount of dissolved air that is equilibrated with air at the pressure of 1 atm, say 17 L of air per cubic meter water at 20°C [36]. The dissolved air will be released when the pressure on water reduces to zero or turns negative. The tiny bubbles formed by the released air can serve as the starting points for the breakup of water continuity. Unless there are other mechanisms to inhibit, the breakup of water continuity or cavitation is inevitable under negative hydraulic pressure inside the membrane [32,33]. The phenomena of "dehydration" or cavitation were experimentally observed in membrane processes driven by net osmotic pressure [37,38].

Water will evaporate to fill up the voids or gaps in water. This mixed liquid and vapor state of water inside the membrane is unstable, too. The negative hydraulic pressure on water disappears with the breakup of water continuity. The vapor pressure of water is too small, which is about 0.023 atm at 20°C [39], to support or sustain the bubbles to resist external pressure. The applied hydraulic pressure, or the atmospheric pressure in case of no applied hydraulic pressure, would squash out the bubbles to restore water continuity. When water continuity is restored, the negative hydraulic pressure resurges, which then breaks up water continuity again. This process of breakup and restoration of water continuity will repeat and go on unceasingly inside the membrane. The pressure change inside a membrane is schematically described in Figure 9.11. The pressure in water jumps randomly among three levels of negative pressure $-\Delta\pi$ (which equals the osmotic pressure in value), vapor pressure Δp_v, and the atmospheric pressure of 1 atm or the applied pressure.

Water transport is only possible on those parts of the membrane that are under negative pressure at that moment. The other parts with pressures equal to vapor pressure or atmospheric pressure are no longer effective for water transport due to the breakup of water continuity. Therefore, the average permeate flux under negative hydraulic pressure heavily depends on membrane area fraction and the time duration in which water continuity is maintained. During this dynamic breakup and restoration process, water continuity may not be fully restored in certain regions of the membrane due

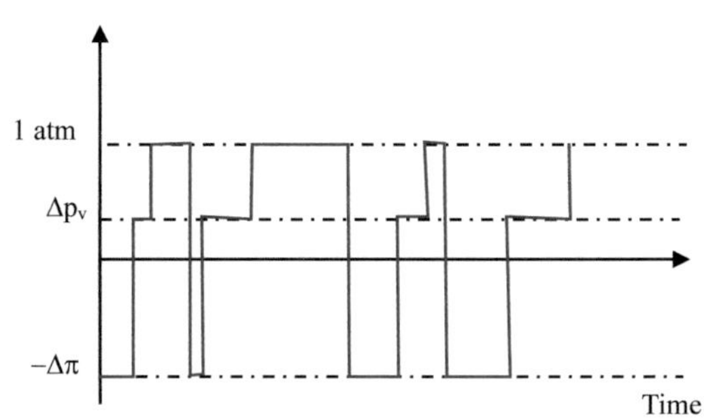

FIGURE 9.11
In a membrane process driven by osmotic pressure, pressure inside the membrane jumps randomly between three pressure levels: the negative pressure induced by chemical potential, vapor pressure, and external hydraulic pressure.

to the accumulation of extra amount of air or vapor or other reasons [37,38]. These regions will be excluded permanently for further water transport.

9.5.3 Failure of Models Based on Continuous Homogeneous Liquid Phase

From the above analysis, an osmotic membrane under negative pressure is actually a heterogeneous system with gaseous blocks and liquid blocks intermixed together. Water flux in this case is impossible to be calculated or predicted with models developed based on continuous homogeneous liquid phase, in which the driving force is uniformly applied on the entire membrane area. The state of heterogeneous liquid and gas phase changes with time as water moves through the membrane. In some sense but more complicated, the breakup of water continuity may analogize the emergence of turbulence from a laminar flow. Similar to that the turbulent flow pattern cannot be described by Navier–Stokes equation coupled with continuity equation, water transport under net osmotic pressure is no longer describable with the models based on a continuous homogeneous liquid phase. There are some stochastic natures to the occurrence and scale of cavitation. By "stochastic," it means that the regions in the membrane and moments for cavitation to occur are unpredictable. As a result, permeate flux cannot be described even using empirical models with fitting parameters. Although one set of experimental flux data can be described by a model with fitting parameters, the model and parameters cannot be used for other sets of experimental data of the same membrane under similar conditions.

9.6 Problems

1. What are osmotic membranes?
2. What is the nature of the osmotic pressure of a solution?
3. An osmotic membrane separates pure water from a solution of salt concentration of 2000 mg/L. The salt concentration at the membrane surface with pure water is virtually zero. According to the principle of chemical potential, the diffusive velocity of salt is infinitely large as salt concentration approaches to zero. Does it mean that the salt flux across the membrane surface is undefined?
4. Calculate the difference of total potentials of an ion across a membrane. Ion concentrations on the two sides of the membrane are 0.1 and 0.01 mol/L, respectively. The pressure difference and potential difference across the membrane are 3 bars and 35 mV, respectively.

5. Why diffusion as described by Fick's law is driven by the gradient of (the strict) chemical potential but not the extended chemical potential that includes the mechanical potential (pressure-related) term?
6. The solute concentrations on the two surfaces of a porous membrane are 1000 and 100 mg/L. The membrane thickness is 100 μm and the concentration is linearly distributed along the membrane thickness. The solute size is assumed to be spherical in shape with radius of 1 nm. Calculate
   a. the diffusive force on the solute along the membrane thickness,
   b. the friction coefficient assuming Stokes' law,
   c. the diffusion coefficient,
   d. the average diffusive velocity along the membrane thickness, and
   e. the solute diffusive flux across the membrane.
7. Salt concentration is 2000 mg/L on one side and is negligibly small on the other side of the membrane, which has a thickness of 20 nm. The diffusion coefficient of water in membrane is assumed to be the largest possible as the unrestricted self-diffusion coefficient is 2.0×10^{-9} m²/s at 20°C. Determine
   a. the diffusive water flux through a membrane
   b. the friction coefficient due to diffusion, and
   c. the hydrodynamic radius of a water molecule.
8. The pressure across a membrane of 20 nm thickness is 5 bars. Pure water flux across the membrane is measured to be 2×10^{-5} m/s. Calculate (a) the mechanical force on water, (b) the friction coefficient of water bulk flow, and (c) the permeability of the membrane.
9. The flux of pure water across a membrane is 3×10^{-5} m/s at a pressure of 3 bars. What is the permeability of the membrane?
10. With a porous membrane that allows both water and solute to transport freely through it, there exist both pressure and solute concentration gradients across the membrane thickness. The volume fraction of solute in solution is much smaller than that of water. What is the major driving force for water transport across the membrane?
11. A supporting layer of an osmosis membrane has a thickness of 50 μm. Calculate the concentration polarization factor for concentrative internal concentration polarizations of sodium chloride, respectively, for permeate flow velocities of 1×10^{-6}, 5×10^{-6}, and 1×10^{-5} m/s at temperature 20°C.
12. Calculate the concentration polarization factor for dilutive internal concentration polarizations with the same parameters as in Question 11. Compare the results with those from Question 11.

13. Calculate numerically the dilutive concentration in a forward osmosis membrane channel with the parameters given in the table below.

Parameter	Symbol	Value	Unit
Membrane resistance	R_m	2.50×10^{11}	Pa·s/m
Channel height	H	3.50×10^{-4}	m
Channel length	L	5	m
Draw solution salt concentration	c_d	40,000	mg/L
Feed water salt concentration	c_f	0	mg/L
Osmotic pressure coefficient	f_{os}	73.9	Pa/(mg/L)
Feed crossflow velocity	u_0	0.1	m/s
Pressure	Δp	0	bar
Diffusion coefficient	D	1.29×10^{-9}	m²/s

14. Why internal concentration polarization is unlikely the principal cause for low water flux in the osmotically driven membrane processes?
15. The only possible driving force for water movement inside an osmotic membrane is the gradient of hydraulic pressure. Do you agree? Give the reasons to support your opinion.
16. In the device below, a membrane separates a solution from pure water. The chamber on the pure water is completely filled up and sealed. A pressure is installed in the pure water chamber. The chamber on the solution side is open to the atmosphere with a salt concentration of 10,000 mg/L. What would be the pressure measure in the pure water chamber?

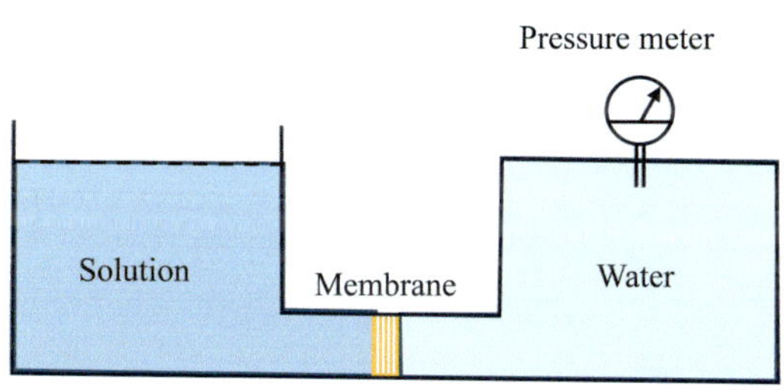

17. Calculate the hydraulic pressure inside an osmotic membrane at the interface with a sodium chloride solution of 5850 mg/L at 20°C.
18. Applied pressure of 5 bars is added to the feed water of 2000 mg/L salt concentration in a desalination process with an osmosis membrane of 10^{11} Pa·s/m resistance. The salt concentration and pressure on the permeate side of the membrane are negligibly small. Calculate the permeate flux across the membrane.

19. In Question 18, if a solution with nonzero salt concentration flushes over the membrane on the permeate flux, does permeate increase or decrease? Why?
20. A pressure-retarded osmosis process uses seawater and pure water as draw solution and feed water, respectively. Other parameters are given in the table below. Calculate the permeate flux in the process.

Parameter	Symbol	Value	Unit
Membrane permeability	A	50×10^{-12}	m/(Pa·s)
Porous layer thickness	l	10	μm
Diffusion coefficient	D	1.29×10^{-9}	m^2/s
Pressure	Δp	0	bar

21. Water transport driven by net osmotic pressure is fundamentally different from that driven by hydraulic pressure. Elaborate on the differences between these two cases.
22. The amount of dissolved air in water is governed by Henry's law, which is

$$c = k_H p$$

Henry's constant for air is 0.025 g/L·atm at 20°C. Calculate the air solubility in unit of g/L in water at 1, 3, and 5 bars.
23. Convert the results in Question 20 to the unit of L air/L water.
24. A cubic meter of water contains 17 L of air at 1 atm and 20°C. This amount of air will be released to form tiny bubbles in water when pressure on water becomes negative. Calculate the number of bubbles formed when the radius of bubbles is 1 μm.
25. Why the water continuity inside a membrane is essential for water transport driven by osmotic pressure?

References

1. Foglia, F., Frick, B., Nania, M., Livingston, A.G., & Cabral, J.T. (2022). Multimodal confined water dynamics in reverse osmosis polyamide membranes. Nature Communications, 13, 2809.
2. Gao, W., She, F., Zhang, J., Dumée, L.F., He, L., Hodgson, P.D., & Kong, L. (2015). Understanding water and ion transport behaviour and permeability through poly(amide) thin film composite membrane. Journal of Membrane Science, 487, 32–39.
3. Hegde, V.H., Doherty, M.F., & Squires, T.M. (2022). A two-phase model that unifies and extends the classical models of membrane transport. Science, 377, 186–191.

4. Heiranian, M., Fan, H., Wang, L., Lu, X., & Elimelech, M. (2023). Mechanisms and models for water transport in reverse osmosis membranes: history, critical assessment, and recent developments. Chemical Society Reviews, 52, 8455–8480.
5. Paul, D.R., & Ebra-Lima, O.M. (1971). The mechanism of liquid transport through swollen polymer membranes. Journal of Applied Polymer Science, 15, 2199–2210.
6. Shen, M., Keten, S., & Lueptow, R.M. (2016). Dynamics of water and solute transport in polymeric reverse osmosis membranes via molecular dynamics simulations. Journal of Membrane Science, 506, 95–108.
7. Vickers, R., Weigand, T.M., Coronell, O., & Miller, C.T. (2025). Water transport mechanisms during pressure-driven transport through polyamide nanogaps. Physics of Fluids, 37(1), 012020.
8. Wang, L., Cao, T., Dykstra, J.E., Porada, S., Biesheuvel, P.M., & Elimelech, M. (2021). Salt and water transport in reverse osmosis membranes: Beyond the solution-diffusion model. Environmental Science & Technology, 55, 16665–16675.
9. Wijmans, J.G., & Baker, R.W. (1995). The solution-diffusion model: A review. Journal of Membrane Science, 107, 1–21.
10. Boyd, C.E. (2019, Jul. 8). Sizes of substances dissolved or suspended in water. Global Seafood Alliance, Portsmouth, NH.
11. Graziano, G. (2003). On the cavity size distribution in water and *n*-hexane. Biophysical Chemistry, 104, 393–405.
12. Graziano, G. (2004). Water: Cavity size distribution and hydrogen bonds. Chemical Physics Letters, 396, 226–231.
13. Kezia, K., Lee, J., Hill, A.J., & Kentish, S.E. (2013). Convective transport of boron through a brackish water reverse osmosis membrane. Journal of Membrane Science, 445, 160–169.
14. Kim, Y.C., Lee, S., & Park, S. (2016). Effects of membrane transport properties and structure parameter on pressure-retarded osmosis performance. Desalination and Water Treatment, 57, 10111–10120.
15. Kotelyanskii, M., Wagner, N.J., & Paulaitis, M.E. (1998). Atomistic simulation of water and salt transport in the reverse osmosis membrane FT-30. Journal of Membrane Science, 139, 1–16.
16. Oren, Y., & Biesheuvel, P.M. (2017). Theory of ion and water transport in reverse-osmosis membranes. Physical Review Applied, 9, 024034.
17. Xu, J., Li, P., Jiao, M., Shan, B., & Gao, C. (2016). Effect of molecular configuration of additives on the membrane structure and water transport performance for forward osmosis. ACS Sustainable Chemistry & Engineering, 4, 4433–4441.
18. Xu, J., Zhu, C., Wang, Y., Li, H., Huang, Y., Shen, Y., Francisco, J.S., Zeng, X.C., & Meng, S. (2018). Water transport through subnanopores in the ultimate size limit: Mechanism from molecular dynamics. Nano Research, 12, 587–592.
19. Yang, L., Zhang, Q., Tian, Y., Zhang, L., & Zhang, H. (2023). Naturally osmotic water transport across nanopores in relation to pore diameters of forward osmosis membrane. Journal of Nanoparticle Research, 25, 1–8.
20. Johnson, D.J., Suwaileh, W.A., Mohammed, A.W., & Hilal, N. (2018). Osmotic's potential: An overview of draw solutes for forward osmosis. Desalination, 434, 100–120.
21. Shaffer, D.L., Werber, J.R., Jaramillo, H., Lin, S., & Elimelech, M. (2015). Forward osmosis: Where are we now? Desalination, 356, 271–284.

22. Zhou, Z., Wang, Q., Qin, Y., & Hu, Y. (2023, Apr. 11). Internal concentration polarization in the polyamide active layer of thin-film composite membranes. Environmental Science & Technology, 57(14), 5999–6007.
23. Gu, Y., Wang, Y., Zhang, M., Cao, Y., & Liang, P. (2014). Characterization of internal and external concentration polarizations during forward osmosis processes. Desalination, 338, 65–73.
24. Cho, Y.H., Han, J., Han, S., Guiver, M.D., & Park, H.B. (2013). Polyamide thin-film composite membranes based on carboxylated polysulfone microporous support membranes for forward osmosis. Journal of Membrane Science, 445, 220–227.
25. Liu, F., Wang, L., Li, D., Liu, Q., & Deng, B. (2019). A review: The effect of the microporous support during interfacial polymerization on the morphology and performances of a thin film composite membrane for liquid purification. RSC Advances, 9, 35417–35428.
26. Lee, S., Kim, Y.C., Park, S., Lee, S., & Choi, H. (2016). Experiment and modeling for performance of a spiral-wound pressure-retarded osmosis membrane module. Desalination and Water Treatment, 57, 10101–10110.
27. Lonsdale, H.K., Merten, U., & Riley, R.L. (1965). Transport properties of cellulose acetate osmotic membranes. Journal of Applied Polymer Science, 9, 1341–1362.
28. Matsuura, T., Guiver, M.D., Lau, W.J., & Ismail, A.F. (2025). Transport in reverse osmosis membranes: Observations and comments on the pore flow model versus the solution–diffusion model. Frontiers in Membrane Science and Technology, 4, 1513591.
29. Song, Y., Xu, F., Wei, M., & Wang, Y. (2017). Water flow inside polamide reverse osmosis membranes: A non-equilibrium molecular dynamics study. The Journal of Physical Chemistry B, 121(7), 1715–1722.
30. Song, L. (2000). Thermodynamic modeling of solute transport through reverse osmosis membrane. Chemical Engineering Communications, 180, 145–167.
31. Gao, N., Wang, J., & Song, L. (2019). Independence of hydraulic pressures on the feed and draw solutions in the osmotically driven membrane processes. Journal of Membrane Science, 586, 1–6.
32. Zhang, H., Wang, J., Rainwater, K., & Song, L. (2019). Metastable state of water and performance of osmotically driven membrane processes. Membranes, 9, 43.
33. Song, L., Heiranian, M., & Elimelech, M. (2021). True driving force and characteristics of water transport in osmotic membranes. Desalination, 520, 115360.
34. Wang, L., He, J., Heiranian, M., Fan, H., Song, L., Li, Y., & Elimelech, M. (2023). Water transport in reverse osmosis membranes is governed by pore flow, not a solution-diffusion mechanism. Science Advances, 9, eadf8488.
35. Mehta, G.D. (1982). Further results on the performance of present-day osmotic membranes in various osmotic regions. Science, 10, 3–19.
36. Battino, R., Rettich, T., & Tominaga, T. (1984). The solubility of nitrogen and air in liquids. Journal of Physical and Chemical Reference Data, 13, 563–600.
37. Tow, E.W., Rencken, M.M., & Lienhard, J.H. (2016). In situ visualization of organic fouling and cleaning mechanisms in reverse osmosis and forward osmosis. Desalination, 399, 138–147.
38. Mehta, G.D., & Loeb, S. (1979). Performance of permasep B-9 and B-10 membranes in various osmotic regions and at high osmotic pressures. Journal of Membrane Science, 4, 335–349.
39. Lide, D.R. (2004). CRC Handbook of Chemistry and Physics (85th ed.). CRC Press, New York, pp. 6–8.

10

Salinity Energy Harvest with Osmotic Membranes

10.1 Salinity Energy

Salinity energy is the energy that exists between two water bodies of different salinities [1–4]. In the literature, the salinity energy is commonly called the "salinity gradient energy" inaccurately. However, the potential salinity relevant energy between two water bodies is not related to the salinity gradient but to the difference. It is similar to the hydraulic potential energy of water that is related to the difference in hydraulic heads, not the head gradient. The origin of salinity energy is the osmotic pressure (chemical potential) difference that drives water from low salinity to high salinity. Salinity energy can be harvested with controlled mixing of waters of different salinities. One of the feasible technologies for controlled mixing is the osmotic membrane. The efficiency of salinity energy harvest is strongly dependent on water transport through the membrane under different driving forces. This is one reason to place this topic just after the chapter on water transport.

The biggest reservoir of saline water on the Earth is the oceans, where seawater virtually maintains its constant salinity. Seawater has an average salinity of 35,000 mg/L with an osmotic pressure of 25.7 bars [5]. This value of osmotic pressure is usually used to estimate salinity energy for the pair of seawater and freshwater because the osmotic pressure of freshwater is negligibly small compared to seawater. The ultimate amount of energy is thus calculated 0.71 *kWh* when a cubic meter of freshwater is mixed with a sufficiently large volume of seawater.

Similar to widely utilized wind and solar energy, salinity energy is also a clean, renewable energy source. Furthermore, salinity energy has a big advantage over wind and solar energies in that it does not need storage facilities to provide a constant, stable power supply because the river flow rate is not affected much by day and night, and clear and cloudy.

Salinity energy is huge in quantity, considering the amount of freshwater running into the seas every day. The global discharge rate of freshwater into the seas is collected from five sources and summarized in Table 10.1.

The total amount of salinity energy is more than the total electricity consumed globally. Table 10.1 shows that the global salinity energy is estimated

DOI: 10.1201/9781003743569 13

TABLE 10.1

Global Discharge of Freshwater into the Seas and Energy Carried

Source	Dai [6]	Broadley [7]	Trenberth [8]	Emami [9]	Wick [10]
Discharge (m^3/year)	3.73×10^{13}	3.60×10^{13}	4.00×10^{13}	4.10×10^{13}	3.47×10^{13}
Power, TW	3.04	2.94	3.26	3.34	2.83
Yearly energy (TWh)	2.66×10^4	2.57×10^4	2.86×10^4	2.93×10^4	2.48×10^4
Ratio to electricity (%)[a]	116.6	112.5	125.1	128.2	108.4

[a] Ratio of salinity energy to global electricity consumption in 2019.

to be in the range of 2.48×10^4–2.93×10^4 TWh. According to the International Energy Agency, the global electricity consumption in 2019 was 2.28×10^4 TWh [11].

Electricity consumption in the United States was about 3900 TWh in 2021 [12]. The average flow rate of the Mississippi river at New Orleans is about 17,000 m^3/s [13]. The salinity energy of the Mississippi river is calculated to be 383 TWh/year, which is about 10% of the electricity consumption of the whole United States.

In addition to the pair of seawater and freshwater from the rivers, there are other pairs of water bodies with significantly different salinities, such as lakes with high salinity, as listed in Table 10.2 [14]. A considerable amount of salinity energy can be potentially extracted if there are sources of freshwater or water with low salinity available nearby. For instance, the Great Salt Lake in Utah has a salinity of about 30% in the northern arm (Gunnison Bay) [15], and the annual discharge from the rivers to the lake is 3.58×10^9 m^3/year [14]. If the river water is mixed with lake water in the northern arm, the salinity energy involved is 21.9 TWh/year. Considering annual electricity consumption in Utah state was 31.2 TWh [16], the salinity energy of the Great Salt Lake is about 70% of the annual electricity consumed in the whole state of Utah.

TABLE 10.2

Lakes with High Salinity

Water Body	Salinity (%)	Area (Square Miles)
Caspian Sea (Russia, Iran)	1.1	170,000
Aral Sea (Russia)	1.0	25,000
Great Salt Lake (Utah)	14.0	1700
Salton Sea (California)	4.4	380
Dead Sea (Israel)	22.0	390

10.2 Specific Salinity Energy

Like other sources of energy, the potential salinity energy of freshwater cannot be 100 percent extractable in any practical processes because it is impossible to mix it with a finite volume of seawater. The fraction of salinity energy that can be harvested depends on the technology and the operating strategies.

10.2.1 Maximum Salinity Energy

The maximum extractable salinity energy can be calculated with a reversible membrane process [17] as shown in Figure 10.1. An osmotic membrane separates a solution and freshwater of negligible salinity in the cylinder. The other ends of the solution and freshwater in the cylinder are sealed with pistons. A certain amount of hydraulic pressure is applied to the piston on solution, while the piston on freshwater is left pressure-free. The pistons move accordingly as freshwater enters the solution chamber. This process is reversible if the applied hydraulic pressure on the solution is always equal to its osmotic pressure.

With the membrane device as shown in Figure 10.1, salinity energy is realized as the work done by the piston on the left side as freshwater is sucked into the solution chamber. The amount of work can be calculated as a function of the volume of freshwater entering the solution chamber.

With the known initial concentration and volume of solution, concentration in the solution chamber at any time can be determined with,

$$c = \frac{c_0 V_0}{V_0 + V_p} \tag{10.1}$$

where c_0 and V_0 are the initial concentration and volume of solution, c is the salt concentration, and V_p is the volume of freshwater passing through

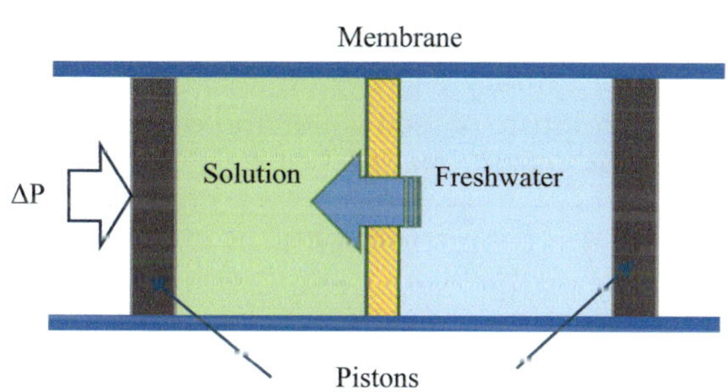

FIGURE 10.1
Schematic of a reversible membrane process for salinity energy harvest.

the membrane. The osmotic pressure of the solution is then calculated with

$$\Delta\pi = f_{os}c = \Delta\pi_0 \frac{V_0}{V_0+V_p} \tag{10.2}$$

where $\Delta\pi$ is the osmotic pressure difference, f_{os} is the osmotic coefficient, and $\Delta\pi_0$ $(= f_{os}c_0)$ is the initial osmotic pressure of the solution.

In a reversible membrane filtration process, the hydraulic pressure added is always equal to the osmotic pressure of the solution. The work done on the piston against the applied hydraulic pressure by expansion of the solution volume is,

$$E = \int_0^{V_p} \Delta p\, dV_p' = \Delta\pi_0 V_0 \int_0^{V_p} \frac{dV_p'}{V_0+V_p'} = \Delta\pi_0 V_0 \ln\left(\frac{V_0+V_p}{V_0}\right)$$

where E is the total energy done by volume V_p of freshwater passing through the membrane, and V_p' is a dummy integration variable. The energy done per unit volume of freshwater, which is named as the *specific energy based on feed water*, i.e.,

$$\hat{E} = \frac{E}{V_p} = \Delta\pi_0 \frac{V_0}{V_p} \ln\left(\frac{V_0+V_p}{V_0}\right) \tag{10.3}$$

Let us define a new parameter of feed fraction as the ratio of freshwater volume to the total volume of the mixed solution

$$f = \frac{V_p}{V_0+V_p} \tag{10.4}$$

Eq. (10.3) can be rewritten as

$$\hat{E} = -\Delta\pi_0 \left(\frac{1-f}{f}\right) \ln(1-f) \tag{10.5}$$

The specific salinity energy depends on the initial osmotic pressure (or more directly the salinity) of the solution and feed fraction. In most cases, about salinity energy, fresh water is a limited resource because there is practically unlimited seawater available. Eq. (10.5) is the maximum possible salinity energy for a unit volume of feed water.

However, the specific salinity energy can also be defined with respect to the draw solution or a mixture of feed water and draw solution. With respect to the draw solution, the specific energy is

$$\hat{E} = \frac{E}{V_0} = -\Delta\pi_0 \ln(1-\mathrm{f}) \tag{10.6}$$

With respect to the mixture of feed water and draw solution, the specific energy is

$$\hat{E} = \frac{E}{V_0+V_p} = -\Delta\pi_0 (1-f) \ln(1-f) \tag{10.7}$$

The specific salinity energy defined with Eq. (10.7) is commonly used to assess the efficiency of engineering processes in the harvest of salinity energy.

Example 10.1

The average seawater salinity is 35,000 mg/L and the river freshwater salinity can be assumed negligible. Calculate the three specific salinity energies defined with respect to the feed, draw solution, and mixture at the feed water fractions of 10 and 50%.

SOLUTION

The seawater osmotic pressure coefficient is $f_{os} = 73.5\, Pa/(mg/L)$. The initial osmotic pressure of seawater is

$$\Delta\pi_0 = f_{os}c_0 = 73.5\, Pa/(mg/L) \times 35{,}000\, mg/L = 2.57 \times 10^6\, Pa$$

For $f = 10\%$, the specific energies with respect to the feed water is,

$$\hat{E} = -\Delta\pi_0 \left(\frac{1-f}{f}\right) \ln(1-f) = -2.57 \times 10^6\, Pa \times \left(\frac{1-0.1}{0.1}\right) \times \ln(1-0.1)$$
$$= 2.44 \times 10^6\, J/m^3 = 0.678\, kWh/m^3$$

The specific energies with respect to the draw solution, and the mixture, are, respectively

$$\hat{E} = -\Delta\pi_0 \ln(1-f) = -2.57 \times 10^6\, Pa \times \ln(1-0.1)$$
$$= 2.71 \times 10^5\, J/m^3 = 0.075\, kWh/m^3$$

The specific energies with respect to the mixture are,

$$\hat{E} = -\Delta\pi_0 (1-f)\ln(1-f) = -2.57 \times 10^6\, Pa \times (1-0.1) \times \ln(1-0.1)$$
$$= 2.44 \times 10^5\, J/m^3 = 0.068\, kWh/m^3$$

For $f = 50\%$, the specific energies with respect to the feed, draw solution, and mixture are, respectively

$$\hat{E} = -\Delta\pi_0 \left(\frac{1-f}{f}\right) \ln(1-f) = -2.57 \times 10^6\, Pa \left(\frac{1-0.5}{0.5}\right) \times \ln(1-0.5)$$
$$= 1.78 \times 10^6\, J/m^3 = 0.495\, kWh/m^3$$

$$\hat{E} = -\Delta\pi_0 \ln(1-f) = -2.57 \times 10^6\, Pa \times \ln(1-0.5)$$
$$= 1.78 \times 10^6\, J/m^3 = 0.495\, kWh/m^3$$

$$\hat{E} = -\Delta\pi_0 (1-f)\ln(1-f) = -2.57 \times 10^6\, Pa \times (1-0.5) \times \ln(1-0.5)$$
$$= 8.92 \times 10^5\, J/m^3 = 0.248\, kWh/m^3$$

Note: the pressure unit Pa is equivalent to the specific energy unit J/m^3. The conversion between J and kWh is 1 kWh = 3.6×10^6 J.

The three specific salinity energies between seawater and river freshwater are calculated in the whole range of feed water fraction. The maximum

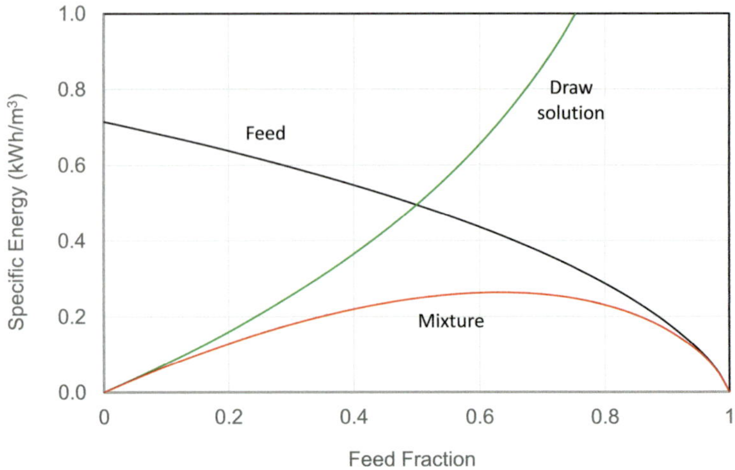

FIGURE 10.2
Three specific energies defined with respect to the feed, draw solution, and mixture as functions of feed fraction.

specific salinity energy with respect to the feed water is obtained as the feed fraction approaches zero, i.e., $f \to 0$, in Eq. (10.5), which is

$$\hat{E} = \lim_{f \to 0}\left[-\Delta\pi_0\left(\frac{1-f}{f}\right)\ln(1-f)\right] = \Delta\pi_{d0}$$
$$= 73.5\,Pa/(mg/L) \times 35{,}000\,mg/L$$
$$= 2.57 \times 10^6\,J/m^3 \times \frac{1\,kWh}{3.6 \times 10^6\,J}$$
$$= 0.71\,kWh/m^3$$

The results are plotted in Figure 10.2. It can be seen that the specific energy with respect to the feed water decreases monotonically with feed fraction and equals zero as the feed water fraction reaches unity. It is obvious that the specific energy is the highest when the feed fraction approaches zero. That means the unit volume of feed water is mixed with an infinitely large volume of draw solution. However, it is practically impossible. In the real world, the cost associated with the draw solution can become forbiddingly high as its volume increases, so that the marginal gain in specific energy is not worth the cost increase.

On the other hand, the specific salinity energy with respect to the draw solution increases monotonically with feed water fraction and approaches infinity as feed water fraction approaches unity. The behavior of specific salinity energy with respect to the mixture is more complicated. It starts from zero at a feed water fraction of zero, increases with feed water fraction to a peak value of 0.263 kWh/m^3 at a feed water fraction of about 0.65, and reduces with further increasing feed water fraction to zero as the feed water fraction approaches unity.

10.2.2 Cost-Corrected Specific Salinity Energy

An important issue in salinity energy study is the economic feasibility or profitability of its harvest. One commonly used criterion is the specific salinity energy defined with respect to a mixture of feed water and draw solution. The underlying reasoning is that salinity energy harvest would be profitable if the specific energy was worth more than the cost to process a unit volume of mixture. As mentioned above, with this criterion, the prospect of beneficial salinity energy harvest is pretty gloomy because the maximum possible specific energy as defined with a mixture is so low. Indeed, the energy ceiling of 0.263 kWh obtained for processing 1 m^3 seems very low to be worth any serious further pursuit. However, comparing the specific energies defined with respect to the feed water and that with respect to the mixture, the whole picture does not look that bad. Thinking about an extreme case in which seawater is freely available, then the specific energy with respect to the mixture would be exactly the same as that with respect to the feed water only!

It is obvious that the use of the specific energy curve for mixture in Figure 10.2 to assess economic feasibility in salinity energy harvest is overly simplified. Actually, the area under the specific energy curve with respect to the feed water in Figure 10.2 is the possible room to play with, which can provide much higher energy for a unit volume of water processed. A concept of cost-corrected specific energy is introduced below to handle this situation.

It is not difficult to understand that the costs to involve freshwater and seawater in a harvest process are usually different. A thorough pretreatment is mandatory for freshwater because it may contain higher concentrations of foulants, and it moves toward the membrane surface in the process. Much less demanding pretreatment can be used for seawater because it may be relatively cleaner and, more importantly, it does not move to the membrane surface but away from the membrane surface in the process. A cost correction index is introduced here to account for the cost difference between freshwater and seawater. For one unit volume of freshwater to be sucked through the membrane, the volume of seawater needed is $(1-f)/f$. The total cost for pretreatment is enlarged from that for freshwater only by the cost correction index, defined as

$$I_{CC} = 1 + \frac{1-f}{f}\frac{c_d}{c_1} \tag{10.8}$$

where I_{CC} is the cost correction index for one unit volume of feed water passing through the membrane, c_1 and c_d are the pretreatment prices of freshwater and seawater, respectively. Eq. (10.8) indicates that the pretreatment cost increases as the feed fraction decreases and approaches infinity as the feed fraction approaches zero unless the pretreatment price for seawater is zero.

The cost-corrected specific energy is obtained by dividing Eq. (10.7) by Eq. (10.8) [17,18]

$$\hat{E} = -\Delta\pi_0 \frac{(1-f)\ln(1-f)}{f + (1-f)\dfrac{c_d}{c_1}} \tag{10.9}$$

Much more relevant economic analysis for salinity energy harvest can be obtained from the cost-corrected specific energy. This is described below with a case study for salinity power generation with seawater and freshwater.

This case study is to investigate the impact of the pretreatment cost ratio of seawater to freshwater c_d/c_1 on maximum specific energy. The pretreatment cost ratios of 0.05, 0.1, 0.25, 0.5, and 1.0 are selected for this case study. The cost-corrected specific energy for each ratio is calculated in the whole range of feed fraction. First, the osmotic pressure of seawater is

$$\Delta\pi_0 = 73.5\,Pa/(mg/L) \times 35{,}000\,mg/L = 0.71\,kWh/m^3$$

Calculate the specific energy for feed fractions from 0.0 with an equal increment of 0.05 in the whole domain.

A sample calculation for $c_d/c_1 = 0.05$ and $f = 0.05$, Using Eq. (10.9),

$$\hat{E} = -0.71\,kWh/m^3 \times \frac{(1-0.05)\times \ln(1-0.05)}{0.05+(1-0.05)\times 0.05} = 0.36\,kWh/m^3$$

The calculations are carried out for all combinations of the pretreatment ratios and the feed fractions. The results are graphically presented in Figure 10.3. The pretreatment cost ratios of seawater to freshwater are indicated on the curves in the figure.

The dark thick line in the figure represents the commonly used specific energy with respect to a mixture of freshwater and seawater, which is obtained with an implied assumption that pretreatment costs are equal for seawater and freshwater. The maximum specific energy in this case is 0.262 kWh/m^3, obtained at a feed fraction of about 65%. Such low specific energy is indeed a serious restriction for the practical utilization of salinity energy.

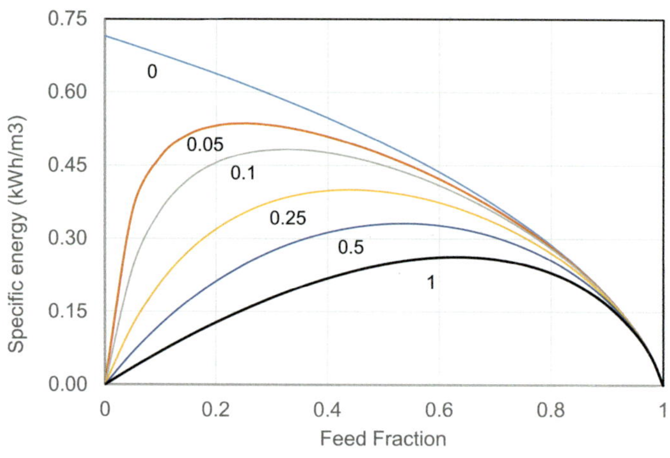

FIGURE 10.3
The cost-corrected specific energies as functions of feed fraction for various pretreatment price ratios of c_d/c_1, which are indicated on the curves.

TABLE 10.3

Maximum Specific Energy at Various c_d/c_1 Ratios

c_d/c_1	0.00	0.05	0.10	0.25	0.50	1.00
$\hat{E}_{Max}$ (kWh/m^3)	0.71	0.54	0.48	0.40	0.33	0.26
Feed fraction	0	0.25	0.35	0.45	0.55	0.65

However, it can be seen in the figure that the curves of cost-corrected specific energies fill up in the area below the specific energy with respect to the feed water, as indicated by the ratio 0. The smaller the ratio, the higher the cost-corrected specific energy. The maximum specific energy at the c_d/c_1 ratio of 0.05 is 0.536 kWh/m^3, more than double that without cost correction. The maximum specific energy and the corresponding feed fraction at each p_d/p_f ratio are listed in Table 10.3.

Example 10.2

Calculate the cost correction indices (I_{CC}) for c_d/c_1 ratios of 0.05, 0.1, 0.25, 0.5, and 1.0 in the whole range of feed fraction.

SOLUTION

Calculate the indices at the feed fractions of 0.05, 0.1, 0.2, 0.3, 0.4, 0.5, 0.6, 0.7, 0.8, 0.9, and 1.0.

An example calculation for $c_d/c_3 = 0.05$ and $f = 0.05$, Using Eq. (10.6),

$$I_{CC} = 1 + \frac{1-f}{f}\left(\frac{c_d}{c_1}\right) = 1 + \frac{1-0.05}{0.05} \times 0.05 = 1.95$$

All results are given in the table below.

	c_d/c_1				
f	**0.05**	**0.1**	**0.25**	**0.5**	**1**
0.05	1.95	2.90	5.75	10.50	20.00
0.1	1.45	1.90	3.25	5.50	10.00
0.2	1.20	1.40	2.00	3.00	5.00
0.3	1.12	1.23	1.58	2.17	3.33
0.4	1.08	1.15	1.38	1.75	2.50
0.5	1.05	1.10	1.25	1.50	2.00
0.6	1.03	1.07	1.17	1.33	1.67
0.7	1.02	1.04	1.11	1.21	1.43
0.8	1.01	1.03	1.06	1.13	1.25
0.9	1.01	1.01	1.03	1.06	1.11
1.0	1.00	1.00	1.00	1.00	1.00

It can be seen from the table that the cost correction indices increase as the feed fraction decreases.

10.2.3 Economic Analysis

Rather than the cost-corrected specific energy, the feasibility for salinity energy harvest can be studied with the regular cost–benefit analysis. Similar to the analysis with the cost-corrected specific energy, cost–benefit analysis does not include membrane cost, either. Membrane-related cost will be discussed separately later. Therefore, the cost associated with salinity energy harvest is mainly the cost of pumping and pretreatment of feed water and seawater. The income is generated by selling the harvested salinity energy (electricity). Therefore, the net benefit is

$$B = c_e E - \left(c_1 V_f + c_d V_d\right) \tag{10.10}$$

where B is the benefit (\$), c_e is the price of electricity, and V_d is the volume of draw solution.

The total energy produced by processing a certain volume of feed water can be calculated from Eq. (10.5) as

$$E = -\theta \Delta\pi_0 V_f \left(\frac{1-f}{f}\right) \ln(1-f)$$

where θ $\left(= 2.778 \times 10^{-7}\ kWh/\left(Pa \cdot m^3\right)\right)$ is the conversion coefficient of energy units from Joule to kWh, $\Delta\pi_0$ is the osmotic pressure of the draw solution (Pa).

From Eq. (10.4), the volume of the draw solution involved can be calculated by

$$V_d = \frac{1-f}{f} V_f$$

Substituting the expressions for E and V_d into Eq. (10.10) results in

$$B = -V_d \left[c_e \theta \Delta\pi_0 \ln(1-f) + c_d \frac{f}{1-f} + c_1 \right] \tag{10.11}$$

The benefit per unit volume of feed water can be obtained by dividing Eq. (10.11) by V_f

$$\hat{B} = \frac{B}{V_f} = -\frac{1-f}{f}\left[c_e \theta \Delta\pi_0 \ln(1-f) + c_d\right] - c_1 \tag{10.12}$$

where $\hat{B}$ is the specific benefit. The second term on the right-hand side of Eq. (10.12) does not depend on the feed fraction. Therefore, it can be dropped from the analysis. Now Eq. (10.12) becomes,

$$\hat{B} = c_e \theta \Delta\pi_0 \left(\frac{1-f}{f}\right)\left[-\ln(1-f) - \frac{c_d}{c_e \theta \Delta\pi_0}\right] \tag{10.13}$$

The specific benefit $\hat{B}$ now does not include the pretreatment cost for feed water. A new parameter is introduced here, as defined by

$$\hat{B}_0 = c_e \theta \Delta\pi_0 \tag{10.14}$$

where $\hat{B}_0$ is the ultimate specific benefit. It is the maximum specific benefit that can be obtained when the pretreatment cost for the draw solution is zero, and the salinity energy is 100% harvested. For the pair of seawater and freshwater, the ultimate benefit is calculated with $c_e \times 0.71$ kWh/m^3.

Substituting Eq. (10.14) into Eq. (10.13) results in,

$$\hat{B} = \hat{B}_0 \left(\frac{1-f}{f}\right)\left[-\ln(1-f) - \frac{c_d}{\hat{B}_0}\right] \tag{10.15}$$

The specific benefit as expressed by Eq. (10.15) is only related to the ultimate specific benefit, the pretreatment price for draw solution, and feed water fraction. These terms are environmental and operational factors for salinity energy harvest. The cost associated with feed water is not included. The advantage of such a specific benefit will become clear later in the feasibility analysis.

Example 10.3

An investigation in a potential site for salinity energy harvest finds that the selling price to the power company is 10 ¢/kWh. The pretreatment price for seawater is 1 ¢/m^3. (1) Determine the ultimate specific benefit, and (2) calculate the specific benefit at two feed water fractions of 0.1 and 0.5.

SOLUTION

The osmotic pressure of seawater is

$$\Delta\pi_0 = 73.5\, Pa/(mg/L) \times 35{,}000\, mg/L = 2.57 \times 10^6\, Pa$$

The ultimate specific benefit is

$$\hat{B}_0 = c_e \theta \Delta\pi_0 = 10¢/kWh \times 2.778 \times 10^{-7}\, kWh/(Pa \cdot m^3) \times 2.57 \times 10^6\, Pa$$
$$= 7.15\, ¢/m^3$$

The specific benefits at a feed water fraction of 0.1 are

$$\hat{B} = \hat{B}_0 \left(\frac{1-f}{f}\right)\left[-\ln(1-f) - \frac{c_d}{\hat{B}_0}\right]$$
$$= 7.15\, ¢/m^3 \left(\frac{1-0.1}{0.1}\right)\left[-\ln(1-0.1) - \frac{1\, ¢/m^3}{7.15\, ¢/m^3}\right]$$
$$= -2.22\, ¢/m^3$$

The specific benefits at a feed water fraction of 0.5 are

$$\begin{aligned}\hat{B} &= \hat{B}_0\left(\frac{1-f}{f}\right)\left[-\ln(1-f)-\frac{c_d}{\hat{B}_0}\right]\\ &=7.15\,¢/m^3\left(\frac{1-0.5}{0.5}\right)\left[-\ln(1-0.5)-\frac{1\,¢/m^3}{7.15\,¢/m^3}\right]\\ &=3.95\,¢/m^3\end{aligned}$$

It can be seen from the calculations that the operation at a feed water fraction of 0.1 is unfeasible because of a negative benefit, even without considering the cost for feed water pretreatment. The specific benefits at a feed water fraction of 0.5 are $3.95\,¢/m^3$. It would be feasible if the pretreatment cost for feed water were smaller.

The benefits of different prices for draw solution pretreatment, represented by $c_d/\hat{B}_0$ ratios, are calculated in the whole domain of feed fraction. The same ultimate specific benefit $\hat{B}_0 = 7.15\,¢/m^3$ as in the above example is used in the calculations. The results are presented in Figure 10.4.

Figure 10.4 is fundamentally different from Figure 10.3. First, the curves in Figure 10.4 are the net economic benefit that excludes the pretreatment cost of freshwater. Therefore, it will be profitable if there exists a feed fraction region in which the benefit is greater than the pretreatment cost of freshwater. Second, the net benefit decreases to zero before the feed fraction approaches zero. A decrease in the feed fraction means that a larger volume of draw solution is needed to mix with a unit volume of freshwater. The cost for the draw solution pretreatment increases accordingly. The pretreatment cost of

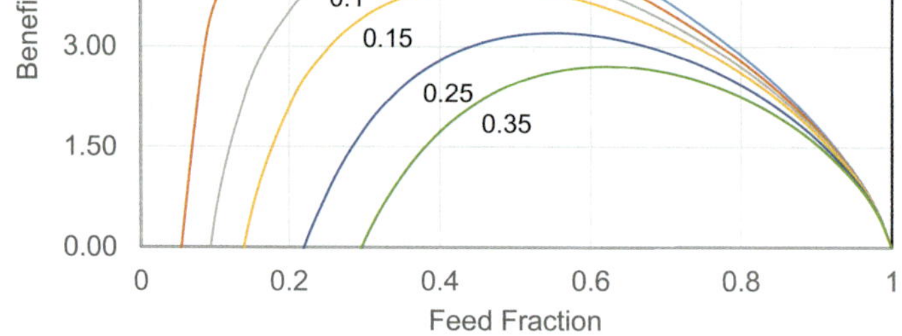

FIGURE 10.4
Benefits as functions of feed fraction for different ratios of draw solution pretreatment to the ultimate benefit, which are indicated on the curves.

the draw solution will equal the economic gain of salinity energy when the feed fraction reduces to a certain point.

10.2.4 Optimal Feed Water Fraction

The feed fraction that maximizes the profit can be found by taking the derivative of Eq. (10.15) and letting it equal 0, i.e.,

$$\frac{d\hat{B}}{df} = \hat{B}_0 \frac{1}{f^2}\left[\ln(1-f) + f + \frac{c_d}{\hat{B}_0}\right] = 0$$

Then

$$\ln(1-f) = -f - \frac{c_d}{\hat{B}_0}$$

Converting the logarithmic function to exponential form

$$f = 1 - e^{-f - \frac{c_d}{\hat{B}_0}}$$

The feed fraction f in the equation can be found with the successive substitution method for any given ratio $\frac{c_d}{\hat{B}_0}$ as below.

$$f^{(n+1)} = 1 - e^{-f^{(n)} - \frac{c_d}{\hat{B}_0}}$$

The optimal feed fraction thus calculated for various feed water pretreatment cost ratios is presented in Table 10.4. It can be seen that the optimal feed fraction increases with the increasing pretreatment price for feed water.

TABLE 10.4

The Optimal Feed Fractions at Various Pretreatment Costs of Draw Solution

$c_d/\hat{B}_0$	F
0.00	0.000
0.01	0.135
0.02	0.187
0.03	0.225
0.05	0.284
0.07	0.329
0.10	0.383
0.13	0.427
0.17	0.476
0.21	0.516
0.25	0.551

From the data in Table 10.4, a polynomial equation is developed to correlate the optimal feed water fraction to the parameter $c_d/\hat{B}_0$ in the Microsoft Excel spreadsheet with the trendline function.

$$f = 32.7\left(\frac{c_d}{\hat{B}_0}\right)^3 - 18.5\left(\frac{c_d}{\hat{B}_0}\right)^2 + 4.38\left(\frac{c_d}{\hat{B}_0}\right) + 0.10 \tag{10.16}$$

Eq. (10.16) is very convenient to use because it is universal for the cost–benefit analysis of all salinity energy harvest cases.

Example 10.4

There are two potential cases for salinity energy harvest with the pair of seawater and freshwater. The costs for the draw solution pretreatment are 10 and 25% of the ultimate specific benefit in case 1 and case 2, respectively. The cost for the freshwater pretreatment, as determined by pilot tests, is 3.3 ¢/m^3 for both cases. To answer (1) which case is economically feasible, (2) find the optimal feed water fraction and specific benefit for the feasible case(s) if the ultimate specific benefit is $\hat{B}_0 = 7.15$ ¢/m^3.

SOLUTION

1. Referring to Figure 10.4, case 1 is economically feasible, but case 2 is not.

   The curve for the ratio 0.1 has the benefit greater than 3.3 ¢/m^3 in the feed fraction range from about 0.2 to 0.7. It would be profitable to work in this range of feed fractions.

   Case 2 is not economically feasible because the peak value of benefit for the ratio of 0.25 is smaller than the pretreatment cost of 3.3 ¢/m^3 of the freshwater.
2. For Case 1, $c_d/\hat{B}_0 = 0.1$, using Eq. (10.16), the optimal feed water fraction is

$$f = 32.7(0.1)^3 - 18.5(0.1)^2 + 4.38(0.1) + 0.10 = 0.386$$

The specific benefit is

$$\hat{B} = \hat{B}_0\left(\frac{1-f}{f}\right)\left[-\ln(1-f) - \frac{c_d}{\hat{B}_0}\right]$$

$$= 7.15\ ¢/m^3\left(\frac{1-0.386}{0.386}\right)\left[-\ln(1-0.386) - 0.1\right]$$

$$= 4.41\ ¢/m^3$$

Even with the cost for pretreatment of feed water, there is a net benefit of 1.11 ¢/m^3.

10.3 Pressure-Retarded Osmosis

The most widely discussed idea for salinity energy harvest is the pressure-retarded osmosis process [19–24]. A schematic of the process is shown in Figure 10.5. The core of the process is an osmotic membrane that divides the membrane unit into two chambers. A draw solution is circulated over one side of the membrane under a hydraulic pressure, and feed water is provided on the other side of the membrane pressure-free. The hydraulic pressure on the draw solution should be smaller than its osmotic pressure so that the residual osmotic pressure can draw feed water through the membrane. Permeate of the pressure-free feed water acquires the hydraulic pressure applied on the draw solution after it enters the draw chamber. The mixture out of the draw chamber carries more energy than that which pumps draw solution into the membrane process because of increased volume and unchanged pressure. A portion of the mixture that equals the original volume of the draw solution goes to the pressure exchanger, where the energy it carries passes to the incoming draw solution. The energy carried by the increased volume of the mixture is converted to mechanical energy or electricity in a turbine. In this way, salinity energy is converted to mechanical or electrical energy. The technical feasibility of pressure-retarded osmosis has been experimentally proven.

10.3.1 Amount of Harvested Salinity Energy

As shown in Figure 10.5, the flow rate of feed water into the draw chamber is Q_f where it acquires the hydraulic pressure Δp. The total power generated in the process is therefore $Q_f \Delta p$. In the study of salinity energy harvest with membranes, it is more common to express the energy production rate per

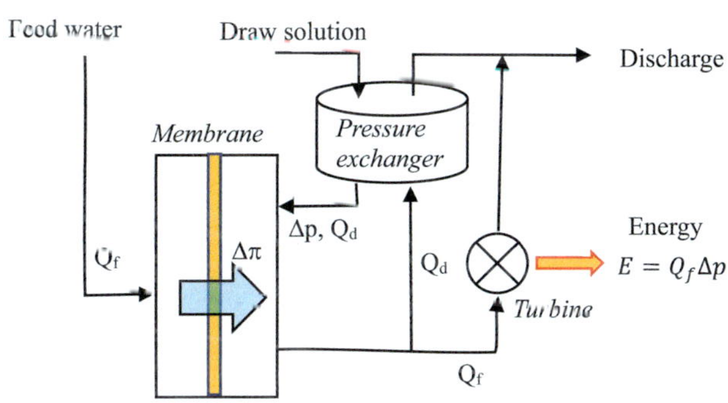

FIGURE 10.5
Schematic of a pressure retarded osmosis process. Permeate of the pressure-free feed water acquires the hydraulic pressure applied on the draw solution and converts salinity energy into mechanical or electrical energy.

unit membrane area, which is named *the membrane power density* and calculated by

$$W = v\Delta p \tag{10.17}$$

where W is the membrane power density, v is the permeate flux, and Δp is the hydraulic pressure applied on the draw solution, which is often referred to as the retarded pressure. Membrane power density is another important indicator for the efficiency of salinity power harvest. Sufficiently high membrane power density is required for the economic feasibility of salinity energy harvest and its competitiveness with other types of energy.

It seems from Eq. (10.17) that membrane power density increases with the retarded pressure. It is not always true because osmotic pressure serves two purposes in the pressure-retarded osmosis process. Part of the osmotic pressure is used to balance the retarded pressure that is used to increase the pressure of the permeate. The rest of the osmotic pressure is used to draw feed water through the membrane. There is an optimal retarded pressure that maximizes membrane power density. If permeate flux is linearly related to the net osmotic pressure as predicted by Eq. (9.24),

$$v = A_{os}\left(\Delta\pi - \Delta p\right)$$

Substituting it into Eq. (10.17) results in,

$$W = A_{os}\left(\Delta\pi - \Delta p\right)\Delta p \tag{10.18}$$

Taking the derivative of Eq. (10.18) with respect to Δp and letting it equal zero,

$$\frac{dW}{d\Delta p} = -A_{os}\Delta p + A_{os}\left(\Delta\pi - \Delta p\right) = 0$$

The retarded pressure that maximizes membrane power density can be determined as,

$$\Delta p = \frac{1}{2}\Delta\pi \tag{10.19}$$

Substituting Eq. (10.19) into Eq. (10.18), the maximum membrane power density is obtained as

$$W = \frac{1}{2}A_{os}\Delta\pi^2 \tag{10.20}$$

As discussed in Chapter 9, the linearity between permeate flux with net driving pressure as described by Eq. (9.24) does not exist under net osmotic pressure. Therefore, Eq. (10.20) cannot be practically used to calculate membrane power density. Eq. (10.20) might have only a theoretical value about the membrane power density. It indicates that membrane power density is related to the square of the osmotic pressure of the draw solution and maximizes when the retarded pressure equals half of the osmotic pressure.

10.3.2 Critical Membrane Power Density

The least requirement for salinity energy harvest is to pay back the cost of membranes. The critical membrane power density is named to satisfy this requirement. It can be estimated from the price of the membrane and the value of energy produced by the membrane in its entire service time, i.e.,

$$c_m = W \times c_e \tag{10.21}$$

where c_m is the price of the membrane, t is the service time of the membrane, and c_e is the selling price of the energy generated.

The critical membrane power density cannot be accurately determined because the membrane price and service time are not readily available, as the membranes for salinity energy harvest are not currently produced on an industrial scale. Reverse osmosis membranes have to be used as surrogates in the calculation. For example, if the reverse osmosis membrane is currently available at about \$18/m^2 with a warranted service time of five years, and the electricity price is \$0.08/kWh. The membrane price is the membrane cost divided by service time, i.e.,

$$c_m = \frac{18\ Dollar/m^2}{5y \times \frac{356d}{1y} \times \frac{24h}{1d}} = 4.11 \times 10^{-4}\ Dollar/\left(m^2 \cdot hr\right)$$

The critical membrane power density can be calculated as

$$W = \frac{c_m}{c_e} = \frac{4.11 \times 10^{-4}\ Dollar/\left(m^2 \cdot hr\right)}{0.08\ Dollar/kWh \times \frac{1\ kW}{1000\ W}} = 5.14\ W/m^2$$

Though the membranes and electricity prices vary with time and places, the commonly cited 5 W/m^2 can serve as a good benchmark to assess the economic feasibility of pressure-retarded osmosis in salinity energy harvest when only the membrane cost is considered.

Extensive studies have been carried out on salinity energy harvest since the 1970s. Although laboratory surprises of high power densities were occasionally reported in the literature, none of them showed sustainable membrane power density above the threshold of 5 W/m^2. For whatever reason, the fact is that the permeate flux under osmotic pressure is just too small to generate sufficient power density for profitable operation, even with extremely high osmotic pressure differences between the draw solution and the feed water. It is commonly accepted that the pressure-retarded osmosis process is unlikely to be economically feasible for salinity energy harvest [25–28].

10.4 Pressure Amplifier

In the conventional pressure-retarded osmosis, a fraction of osmotic pressure is used to drive feed water across the membrane, and the rest is used to raise the permeate to the retarded pressure on the draw side. In this way, the chemical potential of the draw solution is converted to mechanical potential. The failure of pressure-retarded osmosis to produce sufficient energy is largely due to the ineffectiveness of osmotic pressure to drive water across the membrane. Are there any better ways to utilize osmotic pressure (chemical potential of the draw solution) in salinity energy harvest? This interesting subject will be discussed in the rest of this chapter.

10.4.1 Unduly Accused Internal Concentration Polarization

Speculation of excessively strong internal concentration polarizations in the porous supporting layer is not supported by the basic physical characteristics or properties of the porous layer. Obviously, the role of internal concentration polarization on permeate flux is greatly exaggerated. According to the prevailing theory on internal concentration polarization, permeate flux or velocity is proportional to the effective osmotic pressure, i.e.,

$$v = A\Delta\pi_{eff} = A\Delta\pi_d e^{-\frac{v}{D}S} \tag{10.22}$$

where v is the permeate velocity, A is the hydraulic permeability, $\Delta\pi_{eff}$ is the effective osmotic pressure difference, $\Delta\pi_d$ is the osmotic pressure difference between draw solution and feed water, D is the salt diffusion coefficient, and S is the structural parameter. Hydraulic permeability is used here because the internal concentration polarization is explicitly considered. Initially proposed by Lee et al. [19], the structural parameter is a lamped parameter of physical properties of the porous supporting layer of the membrane, which is defined by

$$S = \frac{l\tau}{\varepsilon} \tag{10.23}$$

where l is the thickness, τ is the tortuosity, and ε is the porosity of the porous layer. Physically, the structural parameter is the length of the actual path that the salt moves across the porous layer. Because the development of internal concentration polarization is characterized by the structural parameter of the porous layer, it was intensively investigated for different membranes under various conditions. Eq. (10.22) differs from Eq. (9.19) by replacing the thickness l with the structure parameter S. With this replacement, an additional parameter, the tortuosity τ, is introduced into the equation for permeate velocity driven by net osmotic pressure.

The structural parameter in Eq. (10.22) or its variations is commonly treated as a fitting parameter to match the experimentally observed fluxes in forward

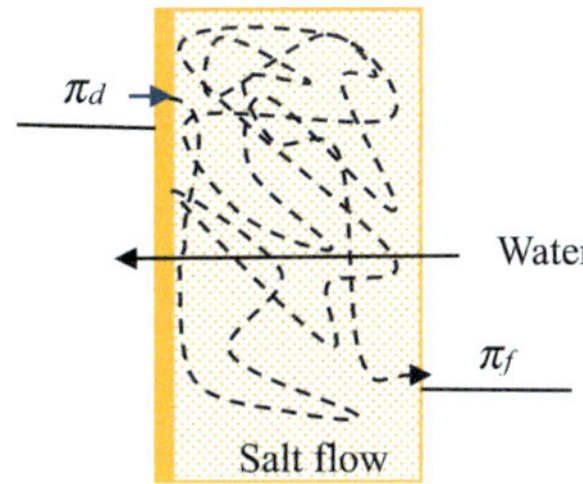

FIGURE 10.6
An improbable path of salt movement in the porous layer with high tortuosity.

osmosis and pressure-retarded osmosis tests. However, unreasonable structural parameters are often obtained with this method. For instance, there usually is a big discrepancy in structural parameter values of the same membrane at differential concentrations of draw solution or with different membrane orientations [29]. An even more serious problem is that excessively large structural parameters, which are more than ten times the thickness of the porous layer, are often needed to fit the experimentally observed fluxes. It means that a salt molecule is wandering along a path that is more than tenfold of the membrane thickness when it transports from one side to another side of the porous layer (see Figure 10.6). Considering porosity and pore size of the porous layer, such a large tortuosity does not make any sense physically, as discussed in Chapter 9. Direct measurement of membrane physical characteristics reveals that the tortuosity of the porous layer is only slightly greater than unity [30].

The need for such unrealistically large structural parameters in Eq. (10.22) to fit the experimental data itself just indicates that internal concentration polarization cannot be the main reason for the low permeate fluxes in the osmotically driven membrane processes. There must be other reasons for the low effectiveness of osmotic pressure as a driving force for water transport across membranes.

Example 10.5

The porous supporting layer of a forward osmosis membrane has a thickness of 152 μm and a porosity of 36%. The structural parameter determined from experimental fluxes is 21,000 μm. What is the tortuosity of the porous supporting layer?

SOLUTION

From Eq. (10.23), the tortuosity is

$$\tau = \frac{S\varepsilon}{l} = \frac{21,000\mu m \times 0.36}{152\mu m} = 49.7$$

The tortuosity of the porous supporting layer of the membrane is about 50, which is unrealistic.

10.4.2 Pressure Amplifier

Instead of internal concentration polarization, the breakup of water continuity inside the membrane was hypothesized in Chapter 9 to be one possible reason for low permeate flux under osmotic pressure. There might also be some other unknown mechanisms to hinder water transport under osmotic pressure. Though more studies are definitely needed to delineate the fundamental principles, there is no doubt about the fact that permeate flux is much smaller under osmotic pressure than under hydraulic pressure of the same magnitude. A membrane process of a particular configuration to harvest salinity energy with high efficiency is proposed based on this fact, without the need for knowing the underlying mechanisms.

The key idea of this membrane process is to keep the pressure positive inside the membrane and to drive water across the membrane with net hydraulic pressure [18]. In this way, water flux can be much higher than that in the conventional pressure-retarded osmosis, so that salinity energy can be harvested profitably.

A schematic of the membrane process is shown in Figure 10.7. To avoid negative pressure inside the membrane, the retarded pressure, which is the hydraulic pressure on the draw side, should be equal to or greater than the osmotic pressure of the draw solution. It is critically important that there is no net osmotic pressure left on the draw side to suck water through the membrane. A high-pressure pump is added on the feed side in an otherwise conventional pressure-retarded osmosis process. Feed water in this membrane process is driven by hydraulic pressure through the membrane. Though a circulation pump may be installed in the conventional pressure-retarded osmosis process, that pump is not intended to generate high pressure on feed water.

A pressure profile in the conventional pressure-retarded osmosis is portrayed in Figure 10.8(a). The draw solution is on the right side of the

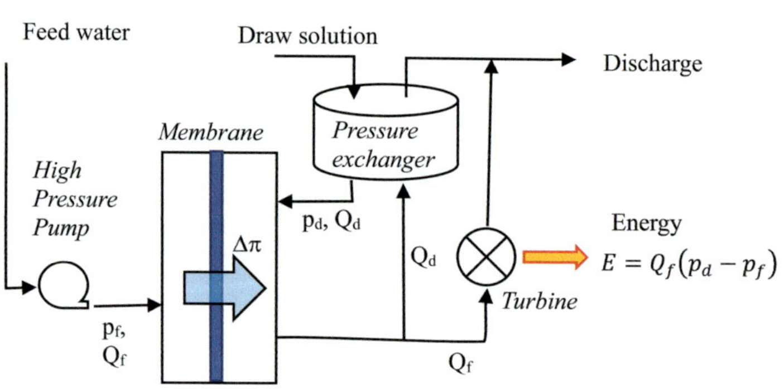

FIGURE 10.7
Schematic of a pressure amplifier (PA) process. Feed water under a hydraulic pressure enters the draw chamber, where its pressure is raised to a higher level by the chemical potential of the draw solution.

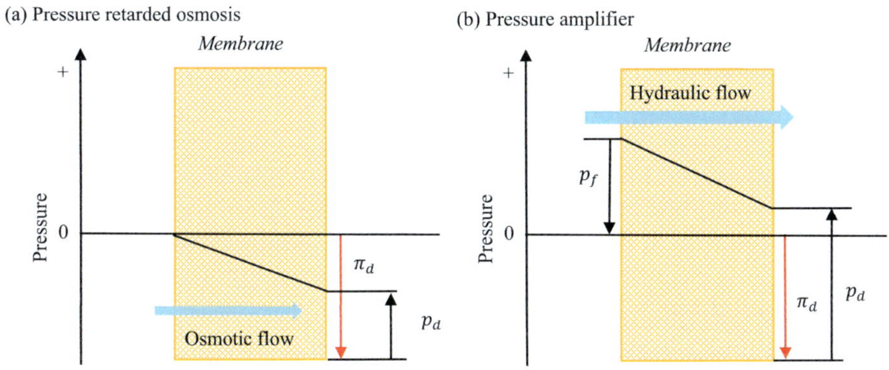

FIGURE 10.8
(a) Net osmotic pressure (with negative pressure inside membrane) and (b) net hydraulic pressure as driving force for water transport in pressure-retarded osmosis and pressure amplifier.

membrane. The osmotic pressure π_d is equivalent to a negative hydraulic pressure as indicated with a downward arrow. The retarded pressure p_d is indicated with an upward arrow. Because the hydraulic pressure in the pressure-retarded osmosis is smaller than the osmotic pressure of the draw solution, the pressure inside the membrane is negative. Therefore, water is driven by net osmotic pressure (osmotic flow) through the membrane.

On the contrary, as shown in Figure 10.8(b), the hydraulic pressure applied on the draw side is equal to or greater than the osmotic pressure of the draw solution in the membrane process, as shown in Figure 10.7. A hydraulic pressure is required on the feed side to drive water through the membrane. As can be seen in Figure 10.8(b), pressure remains positive across the whole membrane thickness. Water transport across the membrane is driven by net hydraulic pressure (hydraulic flow). Complications or negative impacts of net osmotic pressure on water transport will not occur in this membrane process.

A fundamental difference of this process from the conventional pressure-retarded osmosis is that the osmotic pressure (or more fundamentally chemical potential) of the draw solution is not used as the driving force for water transport. Instead, it is fully used to elevate the permeate from a hydraulic pressure on the feed side to a higher one on the draw side of the membrane. It is a more effective way to use the chemical potential of the draw solution than driving water across the membrane. On the other hand, hydraulic pressure, which is a much more effective driving force, is the only driving force for water transport across the membrane. It is not difficult to understand that orders of magnitude higher permeate flux can be obtained in the membrane process than in the conventional pressure-retarded osmosis process.

The membrane process in Figure 10.7 is no longer suitable to be named a "pressure retarded osmosis" process because there is no osmotic flow. The net effect of this membrane process is to raise the low pressure of the feed water at the entrance to the high hydraulic pressure at the outlet of the membrane device. Therefore, the membrane process can be named as "**pressure amplifier**" and abbreviated as PA. In conventional pressure-retarded osmosis, half of the chemical potential of the draw solution is used to draw water through the membrane, and the other half is used to raise the pressure of the permeate. In the PA, all chemical potential is used to raise the pressure of the permeate. Considering the low effectiveness of osmotic pressure in driving water through the membrane in pressure-retarded osmosis, the chemical potential of the draw solution is much more effectively utilized in the PA.

10.4.3 Maximum Membrane Power Density

Membrane power density of a PA is the difference between the energy input and energy output of this PA, which is

$$W = v\left(p_d - p_f\right) \tag{10.24}$$

where v is the permeate flux, and p_d and p_f are the hydraulic pressures added on the draw solution and feed water, respectively. Because the pressure is positive all throughout the membrane thickness, there would not be transmutations (breakup of water continuity). Permeate flux can be reliably predicted with the filtration equation,

$$v = A\left[\left(p_f - \pi_f\right) - \left(p_d - \pi_d\right)\right] \tag{10.25}$$

where π_f and π_d are the osmotic pressures of feed water and draw solution, respectively. The conditions for Eq. (10.25) to be valid are that pressure is positive throughout the whole thickness of the membrane, i.e.,

$$\begin{aligned} &p_f - \pi_f > 0, \\ &p_d - \pi_d \geq 0, \text{ and} \\ &p_f - \pi_f > p_d - \pi_d \end{aligned} \tag{10.26}$$

Combining Eqs. (10.24) and (10.25) results in

$$W = A\left[\left(p_f - \pi_f\right) - \left(p_d - \pi_d\right)\right]\left(p_d - p_f\right) \tag{10.27}$$

When the pressure on the draw side is equal to the osmotic pressure of the draw solution, Eq. (10.27) becomes,

$$W = A\left(p_f - \pi_f\right)\left(\pi_d - p_f\right) \tag{10.28}$$

Taking the derivative of Eq. (10.28) with respect to p_f and letting the derivative equal zero,

$$\frac{dW}{dp_f} = A\left[\left(\pi_d - p_f\right) - \left(p_f - \pi_f\right)\right] = A\left[\pi_d - 2p_f + \pi_f\right] = 0$$

It can be seen that the membrane power density is maximized when the hydraulic pressure on the feed water is half of the sum of the osmotic pressures of the draw solution and the feed water, i.e.,

$$p_f = \frac{\pi_d + \pi_f}{2} \tag{10.29}$$

Substituting Eq. (10.29) into Eq. (10.28), the maximum power density of the PA is

$$W = A\left(\frac{\pi_d + \pi_f}{2} - \pi_f\right)\left(\pi_d - \frac{\pi_d + \pi_f}{2}\right) = \frac{1}{4}A\Delta\pi^2 \tag{10.30}$$

where $\Delta\pi$ $(= \pi_d - \pi_f)$ is the osmotic pressure difference between the draw solution and the feed water.

The maximum power density Eq. (10.30) for the PA is the same as Eq. (10.20) for the conventional pressure-retarded osmosis except for the permeability. Osmotic permeability of the membrane has to be used in Eq. (10.20), while hydraulic permeability is used in Eq. (10.30). It is a proven fact that the hydraulic permeability can be orders of magnitude higher than the osmotic permeability of the same membrane. By the way, a primary objective in the past to increase membrane power density was to develop new osmotic membranes with sufficiently high permeability under osmotic pressure. The objective failed badly, even with substantial investments of time and resources over decades.

Because the driving force in the PA is hydraulic pressure, the osmotic membranes that are commonly used in reverse osmosis processes can be used without any negative concerns or reservations, as in the conventional pressure-retarded osmosis processes. There are generally three categories of reverse osmosis membranes, which are seawater reverse osmosis, brackish water reverse osmosis, and ultralow-pressure reverse osmosis (SWRO, BWRO, and ULPRO) membranes. Typical membrane permeabilities of the three categories are presented in Table 10.5.

Membrane power densities are calculated with Eq. (10.28) for the pair of seawater and freshwater, osmotic pressures of which are 25.7 bar and negligibly small, respectively. The power densities as functions of hydraulic pressure on feed water are presented in Figure 10.9. The horizontal broken line in the figure indicates the critical membrane power density 5 W/m^2 for profitable salinity energy harvest. It can be seen that SWRO membranes in general cannot produce power density above the critical value. The power densities of the other two types of reverse osmosis membranes (BWRO and ULPRO) are way above the critical value in most of the pressure range.

TABLE 10.5

Typical Permeabilities of Three Types of Reverse Osmosis Membranes

Membrane	SWRO	BWRO	ULPRO
Permeability, $m/s \cdot Pa$	3.0×10^{-12}	1.0×10^{-11}	3.0×10^{-11}

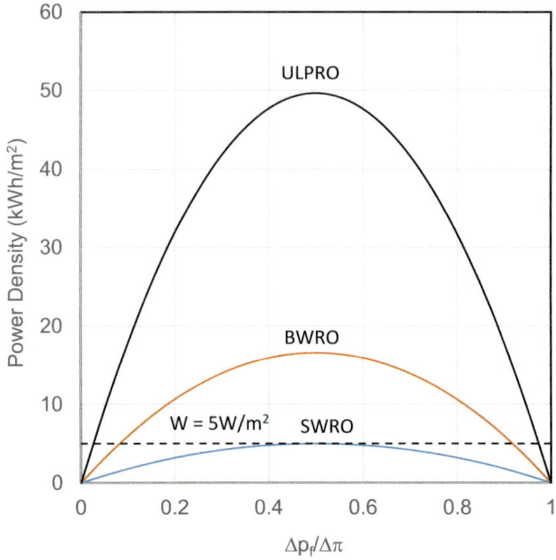

FIGURE 10.9
Membrane power densities as functions of pressure on feed water (presented as fraction of osmotic pressure of the draw solution). Seawater osmotic pressure of 25.7 bars is used for the draw solution. The osmotic pressure of feed water is assumed to be negligible.

10.4.4 Practical Membrane Power Density

In the calculation for membrane power density presented in Figure 10.9, the original osmotic pressure of the draw solution was used. They are actually the theoretical maximum power densities that can only be achieved at an infinitesimal feed fraction. The feed fraction has to be a finite value in real-world applications. Therefore, the power density is controlled by the osmotic pressure of the mixture leaving the membrane device, which is

$$\Delta\pi_{mix} = (1-f)\Delta\pi \tag{10.31}$$

using $\Delta\pi_{mix}$ for osmotic pressure in Eq. (10.30), the membrane power density with the optimal pressure on feed water is obtained as

$$W = \frac{1}{4}A(1-f)^2\,\Delta\pi^2 \tag{10.32}$$

Power densities of typical BWRO membranes and ULPRO membranes are calculated with Eq. (10.32) for different feed fractions, and the results are presented in Figure 10.10. SWRO membranes cannot produce the power density of 5 W/m^2 even at original full osmotic pressure of seawater and, therefore, are not included in the calculations. It is clear from the figure that both BWRO membranes and ULPRO membranes can have power densities above 5 W/m^2 in the ranges of feed fractions about 0–43% and 0–67%, respectively.

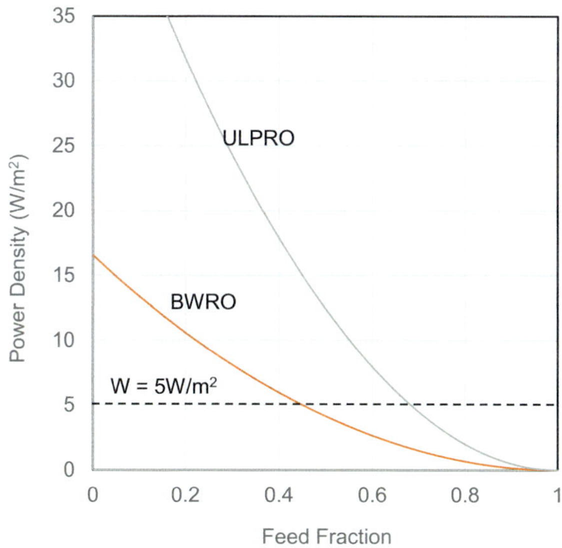

FIGURE 10.10
Maximum membrane power density as a function of feed fraction. Seawater osmotic pressure of 25.7 bars is used for the draw solution. The osmotic pressure of feed water is assumed to be negligible.

Example 10.6

A PA is designed to harvest salinity energy between seawater and freshwater from a river. The estimated costs for seawater and freshwater pretreatments are 1 ¢/m^3 and 2 ¢/m^3, respectively. The membrane permeability is 2×10^{-11} m/s·Pa. Is this process economically feasible?

SOLUTION

There are two criteria to assess the economic feasibility of a process for salinity energy harvest: the specific energy and membrane power density.

1. Specific energy
   Assuming the electricity price is 10 ¢/kWh, the ultimate benefit is

$$\hat{B}_0 = c_e \theta \Delta\pi_{d0} = 10\,¢/kWh \times 0.71\,kWh/m^3 = 7.1\,¢/m^3$$

The ratio of seawater pretreatment price to the ultimate benefit is

$$\frac{c_d}{\hat{B}_0} = \frac{1.0\,¢/m^3}{7.1\,¢/m^3} = 0.14$$

The benefit as a function of feed fraction is calculated with Eq. (10.15) for feed fractions from 0.1 to 1.0 with an increment of 0.10. The results are given in the table below.

f	0.1	0.2	0.3	0.4	0.5	0.6	0.7	0.8	0.9	1
$\hat{B}$(¢/m^3)	2.27	2.34	3.58	3.94	3.92	3.67	3.23	2.61	1.71	0

The results are graphically presented in the figure below.

The price for feed water pretreatment is also shown in the figure with a straight line. It can be seen that the benefit is higher than the pretreatment price of feed water in a wide range of feed fractions. Therefore, it is economically feasible according to the criterion of specific energy.

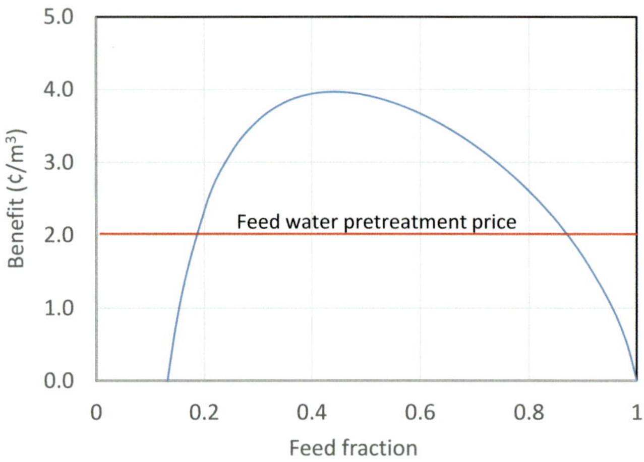

2. Membrane power density
   The osmotic pressure of seawater is

$$\Delta\pi = f_{0s}c_0 = 73.5\ Pa/(mg/L) \times 35{,}000\ mg/L$$
$$= 2.57 \times 10^6\ Pa$$

The practical membrane power density at any feed fraction is calculated with Eq. (10.32), and the results are presented in the table and figure below.

f	0.1	0.2	0.3	0.4	0.5	0.6	0.7	0.8	0.9	1
W (W/m^2)	26.8	21.2	16.2	11.9	8.27	5.29	2.98	1.32	0.33	0.00

It can be seen that the membrane power density is greater than 5 W/m^2 for a feed fraction smaller than 0.6.

From the two figures, it can be concluded that the PA is economically feasible for salinity energy harvest when the feed fraction is in the range of 0.2–0.6.

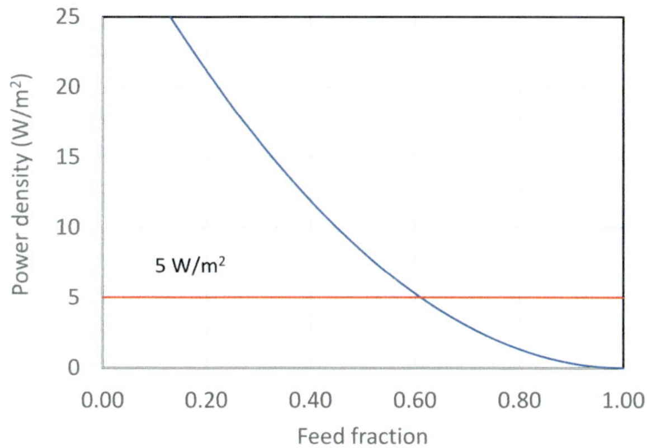

10.5 Conclusions

Salinity energy between seawater and river water or other pairs of waters represents a huge amount of clean, renewable energy, waiting for the development of appropriate technology to harness profitably. More appropriate economic analysis of specific salinity energy and membrane power density reveals the existence of potential windows for beneficial utilization of salinity energy from seawater and freshwater. Prospect to profitably harvest salinity energy is still alive.

A serious barrier to profitably harvesting salinity energy with the retarded osmosis processes, which is the most investigated technology in this regard, is the low membrane power density under pressure. The main reason is that the permeate flux under osmotic pressure is much lower than that under hydraulic pressure of the same magnitude. Internal concentration polarization in the porous supporting layer of the membrane was speculated to be responsible for the low permeate flux driven by osmotic pressure. However, this speculation cannot be validated because the effort in reducing or even eliminating the porous layer fails to elevate the membrane power density to the level anticipated.

A PA is proposed based on the fact that permeate flux can be orders of magnitude higher under hydraulic pressure than under osmotic pressure of the same magnitude. In a PA, a hydraulic pressure equal to or higher than its osmotic pressure is added to the draw solution to neutralize the osmotic driving force. A smaller hydraulic pressure is then added to the feed water to drive it through the membrane. The permeate acquires the higher hydraulic pressure on the draw side of the membrane and, therefore, carries higher energy than that drives it through the membrane. In this process, the chemical potential from the salinity difference is completely used to raise the hydraulic pressure of permeated water, which is a much smarter way to use it than to drive water through the membrane. Much higher flux and, consequently, much higher membrane power density can be obtained in the PA than in the conventional pressure-retarded osmosis.

10.6 Problems

1. What is salinity energy?
2. Solar energy, wind energy, and salinity energies. What are the advantages of salinity energy over the other two?
3. A river with a flow rate of 1000 m^3/s flows into a sea with a salinity of 35,000 mg/L. Calculate the salinity energy dispatched every day.
4. Calculate the specific salinity energies between seawater of osmotic pressure 25.7 bar and freshwater of negligible salinity at feed fractions of 0, 25, 50, 75%, respectively.
5. Derive the specific energy equation with respect to the total volume of draw solution and feed water processes.
6. Drive the specific energy equation with respect to the draw solution only.
7. Derive the equation for the salinity energy from two solutions with volumes V_1 and V_2 and salt concentrations c_1 and c_2, respectively, as water volume transport through an osmotic membrane. What is the maximum energy obtained? What is the water volume transported through the membrane for the maximum energy?
8. Two solutions of salinities of 10,000 mg/L and 2000 mg/L are separated by an osmotic membrane as shown in the figure below. The initial volumes are 1 m^3 and 2 m^3, respectively. Two freely movable pistons are sealed on the other sides of the chambers. (a) Determine the volumes of solution in the two chambers at equilibrium, and (b) Calculate the maximum possible energy obtained from this controlled mixing process.

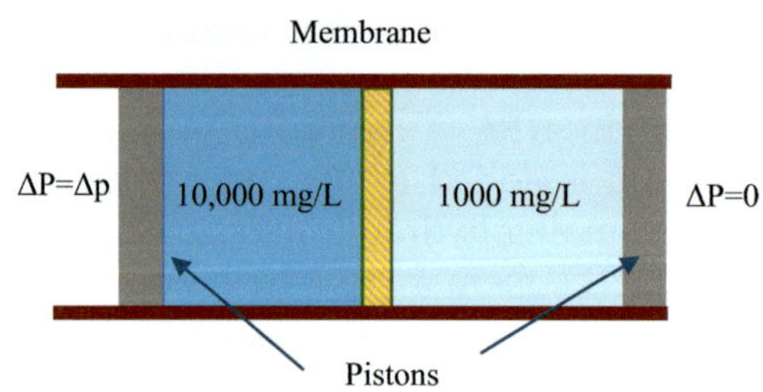

9. Calculate the pretreatment cost-corrected specific salinity energies between seawater of osmotic pressure 25.7 bar and freshwater of negligible salinity at a feed fraction of 25%. The pretreatment cost for freshwater is 2.0 ¢/m^3. The pretreatment costs for seawater at three different locations are 0.5, 1.0, 1.5, and 2.0 ¢/m^3, respectively.
10. Find the maximum cost-corrected specific salinity energy with draw solution of 70,000 mg/L and feed water of negligible salinity, and the feed fraction for it.
11. What is the ultimate benefit for salinity energy harvest with seawater and pure water when the selling price of electricity is 10 ¢/kWh?
12. Determine the benefit of salinity energy harvest with seawater and pure water when the feed fraction is 0.5. The selling price of electricity is 7 ¢/kWh and the pretreatment price of seawater is 0.5 ¢/m^3.
13. The Great Salt Lake in Utah state has a salinity of 140,000 mg/L. If the feed water has negligibly low salinity and the pretreatment price is 1 ¢/m^3, calculate
    a. the ultimate benefit,
    b. the benefits at feed fractions of 0.25, 0.50, and 0.75.
    c. the optimal feed fraction to maximize the benefit, and
    d. the maximum benefit.
14. Salinity energy harvest is considered between seawater of osmotic pressure 25.7 bar and freshwater of negligible salinity. A pilot test demonstrated that pretreatment of seawater at the cost of 1 ¢/m^3 is sufficient. Calculate the benefit of salinity energy harvest as a function of feed fraction without considering the cost for feed water pretreatment.
15. In Question 14, if the pretreatment price of feed water is 1 ¢/m^3, is there a net benefit from salinity energy harvest?
16. In a conventional pressure-retarded osmosis process with seawater and pure water, the permeability of an osmotic membrane is 10^{-11} m/(s·Pa) and independent of the net driving pressure. The pressure applied on the draw solution is 0, $0.25\Delta\pi_0$, $0.5\Delta\pi_0$, $0.75\Delta\pi_0$, and $1.0\Delta\pi_0$, where $\Delta\pi_0$ is the osmotic pressure of seawater. Calculate the membrane power density for each applied pressure.
17. The Salton Sea in California has a salinity of 44,000 mg/L. According to the theory for conventional pressure-retarded osmosis, calculate the maximum membrane power density. What is the retarded pressure on the draw solution for the maximum membrane power density?

18. The price of an osmotic membrane is $20/m² and the electricity price is 10 ¢/kWh. Determine the minimum membrane power density to pay back the membrane cost if the service time of the membrane is five years.
19. Critical membrane power density for potentially beneficial harvest of salinity energy is 5 W/m². Determine the critical permeate flux that is required for the critical power density with seawater and freshwater as draw solution and feed water, respectively.
20. What is the reason that the pressure-retarded osmosis processes cannot generate the required membrane power density >5 W/m²? Does the postulation of internal concentration polarization provide a convincing explanation for the low permeate flux under osmotic pressure? Give your opinion.
21. A PA is planned to harvest salinity energy between seawater and freshwater. If the hydraulic pressures added to the feed water and draw solution of the PA are $0.5\Delta\pi_0$ and $\Delta\pi_0$, respectively, what permeability of the osmotic membrane is required to produce membrane power density >5 W/m²?
22. For the same osmotic membrane, hydraulic pressure is much more effective (possibly an order of magnitude higher) than osmotic pressure to drive water. To take advantage of the effectiveness of hydraulic pressure, the hydraulic pressure that is equal to its osmotic pressure is added to the draw solution. In addition, a hydraulic pressure that is smaller than that on the draw side is added to the feed water. In this case, water flux through the membrane should be high because it is driven by net hydraulic pressure. Calculate membrane power density as a function of the hydraulic pressure on feed water.
23. The salinity energy between the concentrate discharged from a desalination plant and seawater is recovered to reduce energy consumption. The salinities of the desalination concentrate stream and seawater are 70,000 and 35,000 mg/L, respectively. The pressure added to the draw solution (the concentrate) equals its osmotic pressure. Determine (a) the optimal pressure added on the feed water (seawater), and (b) the maximum membrane power density if the membrane resistance is $1.2 \times 10^{-11}\ m/s \cdot Pa$.
24. A PA uses a very permeable osmotic membrane with permeability of $2.5 \times 10^{-11}\ m/s \cdot Pa$. Calculate the potential membrane power density of the PA using seawater and freshwater with a feed fraction of 0.3.
25. What is the reason that the membrane processes with a PA configuration can be more effective in recovering salinity energy than the conventional pressure-retarded osmosis?

References

1. Loeb, S. (1998). Energy production at the Dead Sea by pressure-retarded osmosis: Challenge or chimera? Desalination, 120, 247–262.
2. Nijmeijer, D.K., & Metz, S.J. (2010). Salinity gradient energy. Sustainability Science and Engineering, 95–139.
3. Panyor, L. (2006). Renewable energy from dilution of salt water with fresh water: Pressure retarded osmosis. Desalination, 199, 408–410.
4. Post, J.W., Veerman, J., Hamelers, H.V., Euverink, G.J., Metz, S.J., Nymeijer, K., & Buisman, C.J. (2007). Salinity-gradient power: Evaluation of pressure-retarded osmosis and reverse electrodialysis. Journal of Membrane Science, 288, 218–230.
5. Spiegler, K.S. (1977). Salt-Water Purification (2nd ed.). Plenum Press, New York, pp. 163–167.
6. Dai, A., & Trenberth, K.E. (2003, Feb. 9–13). New estimates of continental discharge and oceanic freshwater transport. AMS Symposium on Observing and Understanding the Variability of Water in Weather and Climate. Long Beach, CA.
7. Broadley, A., Stewart-Koster, B., Burford, M.A., & Brown, C.J. (2022). A global review of the critical link between river flows and productivity in marine fisheries. Reviews in Fish Biology and Fisheries. https://doi.org/10.1007/s11160-022-09711-0.
8. Trenberth, K.E., Fasullo, J.T., & Mackaro, J. (2011). Atmospheric moisture transports from ocean to land and global energy flows in reanalyses. Journal of Climate, 24, 4907–4924.
9. Emami, Y., Mehrangiz, S., Etemadi, A., Mostafazadeh, A., & Darvishi, S. (2013). A brief review about salinity gradient energy. International Journal of Smart Grid and Clean Energy, 2, 295–300.
10. Wick, G.L., & Schmitt, W.R. (1977). Prospects for renewable energy from sea. Marine Technology Society Journal, 11, 16–21.
11. IEA (2021). Electricity information: Overview, IEA, Paris. https://www.iea.org/reports/electricity-information-overview/electricity-consumption, accessed on 4/21/2026.
12. U.S. Energy Information Administration (2021). Electricity explained: Use of electricity. https://www.eia.gov/energyexplained/electricity/use-of-electricity.php, accessed on 4/21/2026.
13. Armstrong, C.A. (2008). New Orleans and the Mississippi River. https://www.hydro-international.com/content/article/new-orleans-and-the-mississippi-river, accessed on 4/21/2026.
14. Szalay, J. (2016). Facts about Great Salt Lake, Live Science. https://www.livescience.com/56941-great-salt-lake.html, accessed on 7/15/2022.
15. Stephens, D.W., & Gardner, J. (1999). Great Salt Lake, Utah. USGS Water-Resources Investigations Report 99–4189.
16. U. S. Department of Energy (2021). State of Utah Energy Sector Risk Profile. https://le.utah.gov/interim/2021/pdf/00002523.pdf, accessed on 7/15/2022.
17. Zhang, H., Yang, W., Rainwater, K.A., & Song, L. (2017). Limiting extractable energy from pressure retarded osmosis with different pretreatment costs for feed and draw solutions. Journal of Membrane Science, 544, 208–212.
18. Song, L. (2021). Modeling and optimization of membrane process for salinity gradient energy production. Separations, 8, 64.

19. Lee, K.L., Baker, R.W., & Lonsdale, H.K. (1981). Membranes for power generation by pressure-retarded osmosis. Journal of Membrane Science, 8, 141–171.
20. Seppälä, A., & Lampinen, M.J. (1999). Thermodynamic optimizing of pressure-retarded osmosis power generation systems. Journal of Membrane Science, 161, 115–138.
21. Sharma, M., Das, P.P., Chakraborty, A., & Purkait, M.K. (2022). Clean energy from salinity gradients using pressure retarded osmosis and reverse electrodialysis: A review. Sustainable Energy Technologies and Assessments, 49, 101687.
22. Boyu, L., Shaohong, L., & Gang, W. (2010). Progress in extracting power from salinity gradient. Renewable Energy Resources, 28, 141–144.
23. Altaee, A., Zhou, J., Alanezi, A.A., & Zaragoza, G. (2017). Pressure retarded osmosis process for power generation: Feasibility, energy balance and controlling parameters. Applied Energy, 206, 303–311.
24. O'Toole, G., Jones, L., & Coutinho, C. (2016). River-to-sea pressure retarded osmosis: Resource utilization in a full-scale facility. Desalination, 389, 39–51.
25. Patel, S. (2014). Statkraft shelves osmotic power project. https://www.powermag.com/statkraft-shelves-osmotic-power-project/, accessed on 7/20/2026.
26. Schaetzle, O., & Buisman, C.J. (2015). Salinity gradient energy: Current state and new trends. Engineering, 1, 164–166.
27. Tseng, S., Li, Y., Lin, C., & Hsu, J. (2016). Salinity gradient power: Optimization of nanopore size. Electrochimica Acta, 219, 790–797.
28. Straub, A.P., Deshmukh, A., & Elimelech, M. (2016). Pressure-retarded osmosis for power generation from salinity gradients: Is it viable? Energy & Environmental Science, 9, 31–48.
29. Manickam, S.S., & McCutcheon, J.R. (2015). Model thin film composite membranes for forward osmosis: Demonstrating the inaccuracy of existing structural parameter models. Journal of Membrane Science, 483, 70–74.
30. Manickam, S.S., Geld, J., & McCutcheon, J.R. (2014). Pore structure characterization of asymmetric membranes: Non-destructive characterization of porosity and tortuosity. Journal of Membrane Science, 483, 549–554.

11

Solute Transport Through Membrane

Real-world membranes may not be perfectly semipermeable. In general, most solutes can transport through a membrane at certain rates that are determined by properties of the membrane, the nature of the solutes, and the types of driving forces. Same as for water transport, convection and diffusion are the two major transport mechanisms for all solutes as long as there exist pressure and concentration gradients. For charged species, transport can be profoundly complicated by the involvement of electromigration under the applied or induced electric field [1–5]. Transport of uncharged solutes across the membrane is delineated and elaborated first in this chapter, followed by deep discussions about ion transport across the membrane with strong emphasis on accurate formulation of mathematical models for electromigration.

11.1 Transport of Uncharged Solutes

Transport of uncharged solutes across the membrane is accomplished in two forms. First, solutes move with water or other liquids that transfer across the membrane as a viscous flow (convective movement). When concentration gradients of solutes exist, there is a diffusive movement of solute superposing on the viscous flow.

11.1.1 Diffusive and Convective Transports

First, concentrations of solutes on the membrane surface are usually different from their concentrations in the bulk solution. Because of the special properties of the interface, there is usually a partition of solutes between the membrane surface and the solution [6]. A solute concentration on the membrane surface can be related to its bulk solution by

$$c_{m0} = Kc_0 \tag{11.1}$$

where c_{m0} and c_0 are the concentrations on the membrane surface and in the bulk solution, respectively, and K is the partition coefficient.

DOI. 10.1201/9781003713569-14

When there exists a concentration difference for a solute across a membrane, the solute can transport through the membrane by diffusion. According to Fick's law, the solute flux is

$$J_s = -D\frac{dc_m}{dx}$$

where J_s is the solute flux, D is the diffusion coefficient, and dc_m/dx is the concentration gradient of the solute.

When there is a viscous water flow in the membrane, a convective transport of solutes has to be included. The solute flux becomes the sum of convective and diffusive transports as

$$J_s = vc_m - D\frac{dc_m}{dx} \tag{11.2}$$

where v is the velocity of viscous water flow. Eq. (11.2) is the governing equation for solute flux.

11.1.2 Solution of Governing Equation

For a given solute flux, the concentration distribution along membrane thickness can be analytically determined. By rearranging the governing equation, the derivative of concentration can be expressed as

$$\frac{dc_m}{dx} = -\frac{v}{D}\left(\frac{J_s}{v} - c_m\right)$$

Separating variables and integrating,

$$\int_{c_{m0}}^{c_m} \frac{dc_m}{\left(\frac{J_s}{v} - c_m\right)} = -\frac{v}{D}\int_0^x dx$$

$$\ln\left[\left(\frac{J_s}{v} - c_m\right)\Big/\left(\frac{J_s}{v} - c_{m0}\right)\right] = \frac{V}{D}x$$

$$\left(\frac{J_s}{v} - c_m\right) = \left(\frac{J_s}{v} - c_{m0}\right) e^{\frac{v}{D}x}$$

The solute concentration along the membrane thickness is

$$c_m = \frac{J_s}{v} + \left(c_{m0} - \frac{J_s}{v}\right) e^{\frac{v}{D}x} \tag{11.3}$$

Taking derivative of Eq. (11.3) results in

$$\frac{dc_m}{dx} = \frac{v}{D}\left(c_{m0} - \frac{J_s}{v}\right) e^{\frac{v}{D}x} \tag{11.4}$$

11.1.3 Solute Passage Coefficient

The solute concentration at the membrane surface on the permeate side is determined with Eq. (11.3)

$$c_{ml} = \frac{J_s}{v} + \left(c_{m0} - \frac{J_s}{v} \right) e^{\frac{v}{D}l}$$

where l is the membrane thickness.

The solute flux is then determined as

$$J_s = v \frac{c_{ml} - c_{m0} e^{\frac{v}{D}l}}{1 - e^{\frac{v}{D}l}}$$

In membrane filtration process, the solute concentration in permeate can be calculated from the flux of solute divided by water flux. Because the permeate velocity v (m/s) is virtually the water flux J ($m^3/m^2 \cdot s$). The solute concentration in the permeate can be calculated by

$$c_p = \frac{J_s}{v} = K \frac{c_p - c_0 e^{\frac{v}{D}l}}{1 - e^{\frac{v}{D}l}}$$

where c_p is the solute concentration in the permeate. Eq. (11.1) is used to related c_{ml} to c_p.

Permeate solute concentration can be found explicitly from the above equation as follows,

$$c_p = \frac{Kc_0}{1 - (1 - K) e^{-\frac{v}{D}l}} \tag{11.5}$$

It can be seen from Eq. (11.5) that there would be no separation ($c_p = c_0$) if $K = 1$. It is mandatory to have $K < 1$ for a membrane to be a selective barrier for solute and water [7]. If $vl/D \ll 1$, the exponential item in Eq. (11.5) can be approximated with the first two items of its Tylor expansion, i.e.,

$$e^{-\frac{v}{D}l} = 1 - \frac{v}{D}l$$

When $K << 1$, Eq. (11.5) reduces to

$$c_p = \frac{Pc_0}{P + v} \tag{11.6}$$

where P is the solute passage coefficient of the membrane, defined as

$$P = \frac{KD}{l} \tag{11.7}$$

The rejection rate by the membrane to the solute can be determined from Eq. (11.6) as

$$r = 1 - \frac{c_p}{c_0} = \frac{v}{P+v} \tag{11.8}$$

It can be seen from Eqs. (11.6) and (11.8) that the solute passage coefficient B characterizes solely solute transport across the membrane, which is rigorously derived from the generic equation Eq. (11.2) for solute transport with both diffusion and convection.

11.1.4 Solute Fluxes

The total solute flux across the membrane, which is constant across the membrane thickness, can be readily calculated by

$$J_s = vc_p = \frac{vPc_0}{P+v} \tag{11.9}$$

Furthermore, both convective and diffusive solute fluxes can be determined analytically. By definition, the convective flux is

$$J_c = vc$$

where J_c is the convective flux of solute. Substituting Eq. (11.3) into the equation results in

$$J_c = v\left[\frac{J_s}{v} + \left(c_{m0} - \frac{J_s}{v}\right)e^{\frac{v}{D}x}\right]$$

In the equation, J_s/v is actually the permeate concentration c_p and can be substituted with Eq. (11.5). Furthermore, the term c_{m0} can be replaced by 11.1). The convective solute flux is then

$$J_c = \frac{Kvc_0}{1-(1-K)e^{-\frac{v}{D}l}}\left[1-(1-K)e^{-\frac{v}{D}(l-x)}\right] \tag{11.10}$$

On the other hand, the diffusive flux is defined by

$$J_d = -D\frac{dc_m}{dx}$$

where J_d is the diffusive flux of solute. Substituting Eq. (11.4) into the equation and carrying out the differentiation, we have

$$J_d = -v\left(c_{m0} - \frac{J_s}{v}\right)e^{\frac{v}{D}x}$$

Again, substituting Eqs. (11.5) and (11.1) in the equation result in

$$J_d = Kvc_0 \frac{(1-K)}{1-(1-K)e^{-\frac{v}{D}l}} e^{-\frac{v}{D}(l-x)} \tag{11.11}$$

By the way, solute concentration long the membrane thickness can be obtained by combining Eqs. (11.1), (11.3), and (11.5),

$$c_m = \frac{Kc_0}{1-(1-K)e^{-\frac{v}{D}l}}\left[1-(1-K)e^{-\frac{v}{D}(l-x)}\right] \tag{11.12}$$

Example 11.1

The thickness of a reverse osmosis membrane is 1.0 μm. The solute concentration in feed water is 2500 mg/L. The diffusion coefficient of solute is 1.29×10^{-9} m²/s, and the permeate velocity is 10^{-6} m/s. Calculate solute rejection rate and solute concentration in permeate for partition coefficient $K_d = 1.0 \times 10^{-4}$ and 5.0×10^{-5}, respectively.

SOLUTION

At $v = 10^{-6}$ m/s and $K_d = 1.0 \times 10^{-4}$, the solute passage coefficient

$$P = \frac{KD}{l} = \frac{10^{-4} \times 1.29 \times 10^{-9}\ m^2/s}{10^{-6}\ m} = 1.29 \times 10^{-7}\ m/s$$

The solute rejection is

$$r = \frac{v}{P+v} = \frac{10^{-6}\ m/s}{1.29 \times 10^{-7}\ m/s + 10^{-6}\ m/s} = 88.6\%$$

$$c_p = \frac{Pc_0}{P+v} = \frac{1.29 \times 10^{-7}\ m/s \times 2500\ mg/L}{1.29 \times 10^{-7}\ m/s + 10^{-6}\ m/s} = 286\ mg/L$$

At $v = 10^{-6}$ m/s and $K = 5.0 \times 10^{-5}$, the solute passage coefficient

$$P = \frac{KD}{l} = \frac{5 \times 10^{-5} \times 1.29 \times 10^{-9}\ m^2/s}{10^{-6}\ m} = 6.45 \times 10^{-8}\ m/s$$

The solute rejection is

$$r = \frac{v}{P+v} = \frac{10^{-6}\ m/s}{6.45 \times 10^{-8}\ m/s + 10^{-6}\ m/s} = 93.9\%$$

$$c_p = \frac{Pc_0}{P+v} = \frac{6.45 \times 10^{-8}\ m/s \times 2500\ mg/L}{6.45 \times 10^{-8}\ m/s + 10^{-6}\ m/s} = 151\ mg/L$$

The results show that solute rejection rate and solute concentration in permeate are solute passage coefficient.

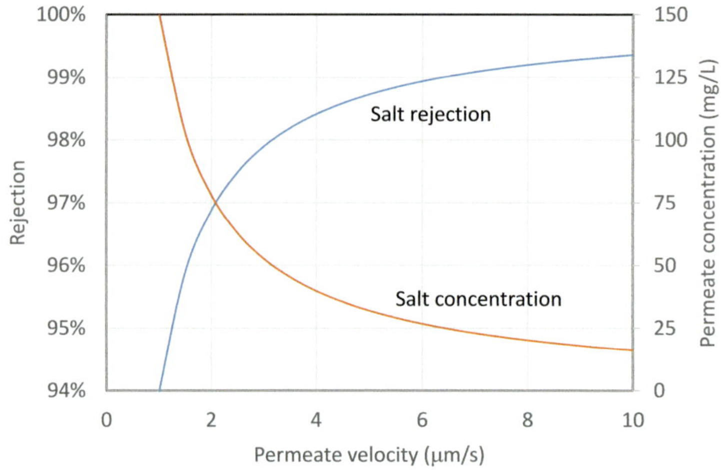

FIGURE 11.1
Solute concentration in permeate and rejection rate as functions of permeate velocity. Parameters are $D = 1.29 \times 10^{-9}$ m²/s, $l = 1.0$ μm, $K = 5\times10^{-5}$, and $c_0 = 2500$ mg/L.

Solute concentration in the permeate and rejection rate by the membrane for the same parameters as used in Example 11.1 are calculated with Eqs. (11.6) and (11.8) for permeate velocities in the range of 10^{-6}–10^{-5} m/s. Results are presented graphically in Figure 11.1. Solute concentration in the permeate declines and solute rejection rate increases with increasing permeate velocity. This is the dilution effect of permeate velocity on solute rejection rate by membranes, which is experimentally observed and confirmed in reverse osmosis membrane desalination.

Solute concentrations along membrane thickness are calculated with Eq. (11.12) for different permeate velocities. Profiles of solute concentrations are presented in Figure 11.2. The numbers in the figure indicate different permeate velocities from small to large magnitudes. The permeate velocities are given in the caption of the figure. The shape of concentration profiles can reveal the mechanisms for solute transport through the membrane. For instance, the profile labeled number "1" is for a permeate velocity of 1 μm/s. This profile is almost a straight line, indicating a constant diffusive flux across the membrane thickness. Therefore, solute transport is primarily accomplished by diffusion only. The curvature of the concentration line increases with permeate velocity, which means that convection plays a more important role in solute transport in a certain part of the membrane thickness. Because the total solute flux is constant across membrane thickness, convective transport must make up the difference in the regions where the concentration gradient is small, such as in the initial section of the membrane thickness on the curve indicated by number "6."

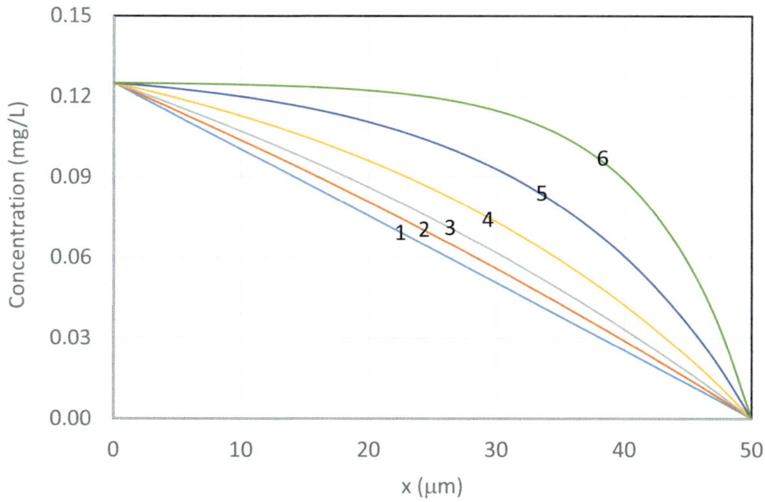

FIGURE 11.2
Solute concentration profiles for different permeate velocities: (1) 1 μm/s, (2) 10 μm/s, (3) 20 μm/s, (4) 40 μm/s, (5) 80 μm/s, and (6) 160 μm/s. Other parameters are $D = 1.29 \times 10^{-9}$ m²/s, L = 50 μm, $K = 5 \times 10^{-5}$, $c_0 = 2500$ mg/L.

Convective and diffusive solute fluxes along the membrane thickness are calculated with Eqs. (11.10) and (11.11) and presented in Figure 11.3 together with the total solute flux. The parameters used in calculations are the same as for the concentration profile number "4" in Figure 11.2. This concentration profile is chosen because both convection and diffusion play similar important roles for solute transport. Convective transport contributes more to the total solute flux in the first part of the membrane thickness, while diffusive transport takes over in the last part of the membrane thickness. More simulations on the contribution of convective and diffusive solute transports reveal an interesting phenomenon. At the surface of the membrane on the permeate side, solute transport is solely accomplished by diffusion, and convective flux drops to zero. On the other hand, both convection and diffusion contribute to solute transport at the membrane surface on the feed side.

11.2 Electrostatic Force on Ion Transport

Electrostatic interaction plays an extremely important role in the transport of ions across the membrane. It is not only a critical driving force for ions to move across the membrane but also can be a consequence of ion transport [8–11]. Because ion transport interplays with the electric potential developed

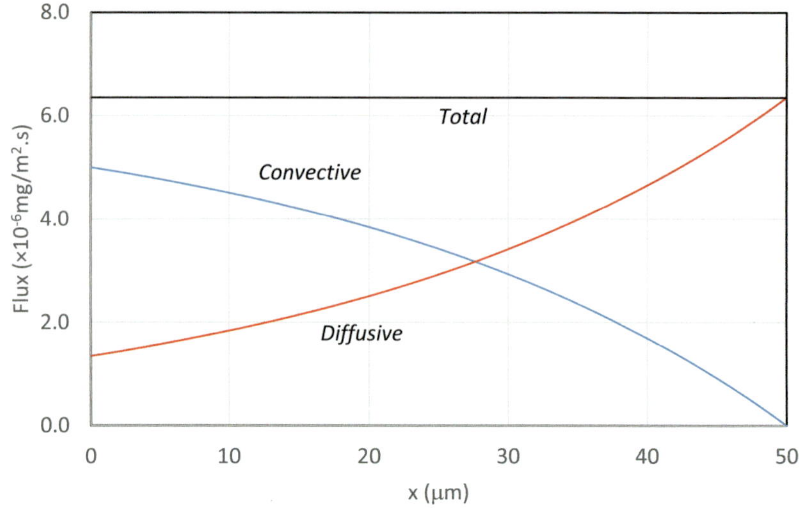

FIGURE 11.3
Total solute flux and its diffusive and convective components as functions of permeate velocity. Parameters are $D = 1.29 \times 10^{-9}$ m²/s, L = 50 μm, $K = 5\times10^{-5}$, $c_0 = 2500$ mg/L, and $v = 4\times10^{-5}$m/s.

on the membrane in a complicated manner, a mathematical description of ion transport across the membrane itself is a serious challenge [12–17]. The correctly formulated models for this interplay are essential in the study of ion transport across the membrane. Before touching on ion transport, a brief review of basic knowledge of electricity and electrochemistry that is useful to understand ion transport is conducted.

11.2.1 Basic Electronics

Electrostatic force Neutral molecules can gain or lose one or several electrons to become positively or negatively charged ions. Large particles or objects can become charged by natural or engineering processes. The electrostatic force developed between two charged particles (point charges) can be quantified by Coulomb's law,

$$F_e = \frac{1}{4\pi\varepsilon\varepsilon_0}\frac{q_1 q_2}{r^2}$$

where F_e is the electrostatic force ε_0 ($= 8.854\times10^{-12}\,F/m$ with F being the Farad, a unit of capacitance), is the vacuum permittivity, ε is the relative permittivity of the medium, q_1 and q_2 are the charges carried on the two particles, respectively, and r is the distance between the two particles. The basic unit for electrical charge is a coulomb, denoted by the symbol C. The force is attractive for oppositely charged particles but repulsive for similarly charged particles.

Electric field Electrostatic interactions of a charged particle can exist with many charged objects. It would be very tedious to deal with the interactions one by one. This case can be more conveniently handled with the concept of the electric field. An electric field is created in the space by one or more charged objects. An electrostatic force on a charge particle develops when it is situated in the electric field. The strength of an electric field at a spatial point is defined as the force on a test charge, which carries a unit positive charge. Therefore, the force experienced by a particle in an electric field is proportional to the charge that the particle carries, i.e.,

$$F_e = qE$$

where q is the charge on the particle (it can take the positive sign or negative sign), and E is the electric field strength. An electric field is a vector. Positive charge moves along the electric field direction while negative charge moves against the electric field direction.

The electric field of a single charge can be determined directly from Coulomb's law by letting one charge be the unit charge,

$$E = \frac{1}{4\pi\varepsilon\varepsilon_0}\frac{q}{r^2}$$

The direction of the electric field is along the radial direction.

The electric field generated by multiple point charges or charged objects of various shapes can be calculated with the superposition of all charges. For an electric field generated with charged objects of certain symmetry, Gauss's law is a very useful method to calculate the electric field strength. Gauss's law states that the integration of the electric field perpendicular to a closed surface (electric flux) is proportional to the total electric charge enclosed by this surface, i.e.,

$$\oint E\cos\theta\, ds = \frac{q}{\varepsilon\varepsilon_0}$$

where θ is the angle between the electric field and the normal direction out of the surface.

For example, the electric field strength of a uniformly charged plane surface can be conveniently determined with Gauss's law. When a positively charged surface can be viewed as infinitely large (without boundary effects), the direction of the electric field is perpendicular to the surface and points outward. Draw a cylinder of arbitrary length with a wall parallel with the field direction, as shown in Figure 11.4. The side wall and the two circles on the two ends of the cylinder form a closed surface. The two circles are perpendicular to the field direction, or in other words, the angle between the direction of the electric field and the normal direction of the outer surface of the circles is zero. The total charge is the charge density multiplied by the

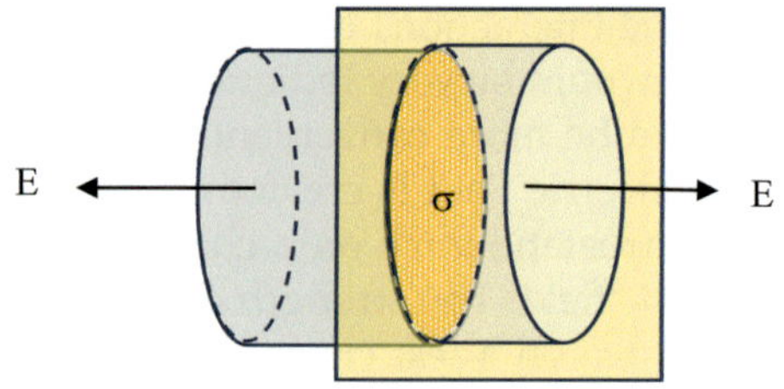

FIGURE 11.4
Application of Gauss's law to a charged plane surface.

surface enclosed. Integration of electric flux on the side wall of the cylinder is zero because the angle between the field direction and the surface normal direction is 90°.

Then the field strength on the circles of the cylinder can be calculated according to Gauss's law

$$2SE = \frac{\sigma S}{\varepsilon \varepsilon_0}$$

where S is the area of a circle on the end of the cylinder, and σ is the charge density of the surface. The electric field strength is

$$E = \frac{\sigma}{2\varepsilon \varepsilon_0}$$

Electric potential When a test charge is moved in an electric field from one point to another, an amount of work is needed from the external force. An electric potential at a location is defined as the work done by the external force to move a unit positive charge from a reference point to the location. The amount of work does not depend on the path of movement but only on the positions. Usually, the reference point is taken at an infinitely far away from the point of concern where the potential is defined as zero. In other words, the potential is the work needed to move a unit positive charge to the point of concern.

$$\varphi = -\int_{\infty}^{x} E(x')\,dx'$$

where φ is the potential, x is the coordinate of the point of concern, and x' is the integration variable. Because potential is a scalar quantity and conservative, it is very convenient to use potential to deal with many practical problems rather than the use of electric force or field directly. The electric field is readily obtainable by taking the derivative of the above potential equation, i.e.,

$$E(x) = -\frac{d\varphi}{dx}$$

11.2.2 Electric Migration of Ions in Solution

The electrostatic force exerted on an ion in solution by an electric field is

$$F_e = qE = -ze\frac{d\varphi}{dx}$$

where z is the valence, and e $(=1.60\times10^{-19}\,C)$ is the elementary charge. The electrostatic force is equal to the frictional force on the ion when it migrates in solution at constant velocity, i.e.,

$$F_e = \frac{kT}{D}v_m$$

where v_m is the electromigration velocity of the ion. Combining the above two equations, the migration velocity of the ion is determined as

$$v_m = -\frac{Dze}{kT}\frac{d\varphi}{dx} = -\frac{DzF}{RT}\frac{d\varphi}{dx} \tag{11.13}$$

where F is the Faraday constant, which is

$$F = N_a e = 96485\ C/mol$$

where N_a $\left(=6.022\times10^{22}/mol\right)$ is the Avogadro constant.

With the knowledge of electromigration of ions, the flux of ions under the influence of a concentration gradient and an electric field can be expressed as

$$J = -D\frac{dc}{dx} + cv_m = -D\frac{dc}{dx} - c\frac{DzF}{RT}\frac{d\varphi}{dx}$$

This equation is the well-known Nernst–Planck equation.

11.2.3 Linearized Poisson–Boltzmann Equation

As shown in Figure 11.5, a charged membrane is in contact with a solution of multiple ions of bulk concentrations c_{i0} (for clarity, only a cation and an anion are presented in the figure). Counterions are attracted to the surface, and coins are repelled away from the surface. A transition layer of deviated concentrations of cations and anions forms between the membrane surface and the bulk solution. Because of an unbalanced charge, an electrical potential develops in the transition layer. This potential is affected not only by the charge on the membrane but also by the distributions of the ions in the transition layer. The electric potential and ion concentrations in the transition layer are well described by the Poisson–Boltzmann model. For the membrane that can be regarded as extended infinitely in two dimensions, ion concentrations only change in the direction perpendicular to the membrane surface.

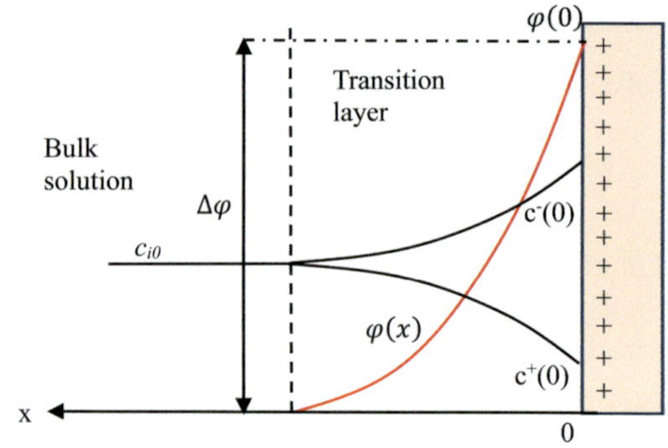

FIGURE 11.5
Distributions of the cation concentration c^+, anion concentration c^-, and potential in the transition layer adjacent to a charged surface.

Potential in solution induced by a charged surface is governed by the Poisson equation,

$$\frac{d^2\varphi}{dx^2} = -\frac{q(x)}{\varepsilon\varepsilon_0} \tag{11.14}$$

Because of the volumetric charge density, Eq. (11.14) is actually written for a unit membrane area. The most convenient vacuum permittivity to be used in Eq. (11.4) is given in $\varepsilon_0 = 8.854\times10^{-12}C/(V\cdot m)$, where the symbol V is volts, a unit for potential difference.

The net charge density in solution is calculated by

$$q(x) = \sum_i z_i\, Fc_i(x) \tag{11.15}$$

When ions in the solution are at equilibrium, i.e., the electrochemical potential of ions is constant,

$$\mu_i(x) = \mu_i^0 + RT\ln c_i(x) + z_iF\varphi(x) = constant \tag{11.16}$$

Far away from the membrane surface into the bulk solution, the ion concentrations become c_{i0} and an electrical potential diminishes, i.e., $c_i(x) \to c_{i0}$ and $\varphi(x) \to 0$ as $x \to \infty$, then

$$\mu_i(x) = \mu_i(\infty) = \mu_i^0 + RT\ln c_{i0} \tag{11.17}$$

Combining Eqs. (11.16) and (11.17) and rearranging, the distribution of ion concentrations in the transition layer is obtained as

$$c_i(x) = c_{i0}e^{-\frac{z_iF\varphi(x)}{RT}} \tag{11.18}$$

Eq. (11.18) is the well-known Boltzmann distribution of ions.

Substituting Eq. (11.18) into Eq. (11.14) results in the Poisson–Boltzmann equation,

$$\frac{d^2\varphi}{dx^2} = -\frac{1}{\varepsilon\varepsilon_0}\sum_i z_i F c_{i0} e^{-\frac{z_i F\varphi(x)}{RT}} \tag{11.19}$$

For $\left|\frac{z_i FV(x)}{RT}\right| \ll 1$, the exponential term can be approximated by the first two terms of the Taylor expansion, i.e.,

$$\frac{d^2\varphi}{dx^2} = -\frac{1}{\varepsilon\varepsilon_0}\left[\sum_i z_i F c_{i0} - \sum_i z_i F c_{i0} \frac{z_i F\varphi(x)}{RT}\right]$$

Because of electroneutrality in the bulk solution, the first term in the square brackets is zero, i.e.,

$$\sum_i z_i F c_{i0} = 0$$

Eq. (11.19) becomes

$$\frac{d^2\varphi}{dx^2} = \frac{F^2 \sum_i c_{i0} z_i^2}{\varepsilon\varepsilon_0 RT}\varphi(x) \tag{11.20}$$

Let us introduce a parameter

$$\lambda = \sqrt{\frac{\varepsilon\varepsilon_0 RT}{F^2 \sum_i c_{i0} z_i^2}} \tag{11.21}$$

where λ the Debye length. Eq. (11.20) can be rewritten as

$$\frac{d^2\varphi}{dx^2} = \frac{1}{\lambda^2}\varphi(x) \tag{11.22}$$

Eq. (11.22) is the linearized Poisson–Boltzmann equation.

11.2.4 Potential on a Charged Surface in Solution

Electrical potential on a charged surface can be obtained from the linearized Poisson–Boltzmann equation. The general solution of Eq. (11.22) is

$$\varphi(x) = a_1 e^{x/\lambda} + a_2 e^{-x/\lambda} \tag{11.23}$$

where a_1 and a_2 are coefficients to be determined. Because φ is 0 in the bulk (far away from the surface, i.e., $x \to \infty$), $a_1 = 0$. Then

$$\varphi(x) = a_2 e^{-x/\lambda} \tag{11.24}$$

The total charge in the solution per unit area of membrane is

$$\sigma = \int_0^\infty q(x)\,dx = \int_0^\infty \sum_i z_i F c_i(x)\,dx \tag{11.25}$$

where σ is the total net charge in solution over per unit area of membrane. Substituting Eq. (11.18) into Eq. (11.25) results in

$$\sigma = \int_0^\infty \sum_i z_i F c_{i0} e^{-\frac{z_i F \varphi(x)}{RT}}\,dx \tag{11.26}$$

Substituting Eq. (11.24) into Eq. (11.26) results in

$$\sigma = \int_0^\infty \sum_i z_i F c_{i0} e^{-\frac{z_i F}{RT} B e^{-x/\lambda}}\,dx \tag{11.27}$$

Keeping the first two terms of the Taylor expansion of the exponential function in Eq. (13.27),

$$\sigma = \int_0^\infty \sum_i z_i F c_{i0}\,dx - a_2 \sum_i \frac{z_i^2 F^2 c_{i0}}{RT} \int_0^\infty e^{-x/\lambda}\,dx$$

The first integration is zero because of electroneutrality in the bulk solution. Then

$$\sigma = -a_2 \left(\sum_i \frac{z_i^2 F^2 c_{i0}}{RT} \right) \int_0^\infty e^{-x/\lambda}\,dx = -a_2 \left(\frac{\varepsilon\varepsilon_0}{\lambda^2} \right) \lambda = -B \frac{\varepsilon\varepsilon_0}{\lambda}$$

Then, we have

$$a_2 = -\frac{\sigma\lambda}{\varepsilon\varepsilon_0}$$

Finally,

$$\varphi(x) = -\frac{\sigma\lambda}{\varepsilon\varepsilon_0} e^{-x/\lambda} \tag{11.28}$$

And let $x = 0$ in Eq. (11.28), the potential at the membrane surface is obtained as

$$\varphi(0) = -\frac{\sigma\lambda}{\varepsilon\varepsilon_0} \tag{11.29}$$

Eq. (11.29) shows that the potential on the membrane surface is linearly related to the net charge in the transition layer adjacent to the membrane surface. This important relationship will be used later to set up boundary conditions for the differential equations governing ion transport across the membrane. Since the potential in the bulk is set as zero, the potential on the membrane surface is also the potential difference across the transition layer, as indicated by $\Delta\varphi$ in Figure 11.5.

Example 11.2

Calculate Debye length for solutions at 20°C that contain 0.1 and 0.01 mol/L NaCl, respectively.

SOLUTION

Using Eq. (11.21) for Debye length. The permittivity in units of C/J·m is most convenient for this question.

At concentration 0.1 mol/L, $c_{Na} = c_{Cl} = 0.1\ mol/L = 100\ mol/m^3$.

$$\lambda_0 = \sqrt{\frac{\varepsilon\varepsilon_0 RT}{F^2 \sum_i c_{i0} z_i^2}} = \sqrt{\frac{8.85\times10^{-12}\ C/J\cdot m \times 80 \times 8.31\ J/K\cdot mol \times 293\ K}{(96485\ C/mol)^2 \times \left[100\times1^2 + 100\times(-1)^2\right]\ mol/m^3}}$$

$$= 9.62\times10^{-10}\ m = 0.962\ nm$$

At concentration 0.01 mol/L, $c_{Na} = c_{Cl} = 0.01\ mol/L = 10\ mol/m^3$,

$$\lambda_L = \sqrt{\frac{\varepsilon\varepsilon_0 RT}{F^2 \sum_i c_{i0} z_i^2}} = \sqrt{\frac{8.85\times10^{-12}\ C/J\cdot m \times 80 \times 8.31\ J/K\cdot mol \times 293\ K}{(96485\ C/mol)^2 \times \left[10\times1^2 + 10\times(-1)^2\right]\ mol/m^3}}$$

$$= 3.04\times10^{-9}\ m = 3.04\ nm$$

11.3 Membrane Potentials

Electric potential develops on a membrane when there is an unbalanced charge distribution along the membrane thickness or between the membrane surface and the solution in contact with it [18–20]. The unbalanced charge distribution is caused by ion transport between the membrane and the solutions in contact with it. In turn, the potential developed on the membrane strongly affects or regulates the transport of ions across the membrane. Below is a brief review of membrane potentials developed in some simple cases that can be presented with analytical equations.

11.3.1 Nernst Potential

Nernst potential is the membrane voltage at which the electrical force acting on an ion equals the force from its concentration gradient, leading to zero net ion flow. It represents the potential for a single ion species at equilibrium, where the forces from the concentration gradient and the electrical gradient are in balance.

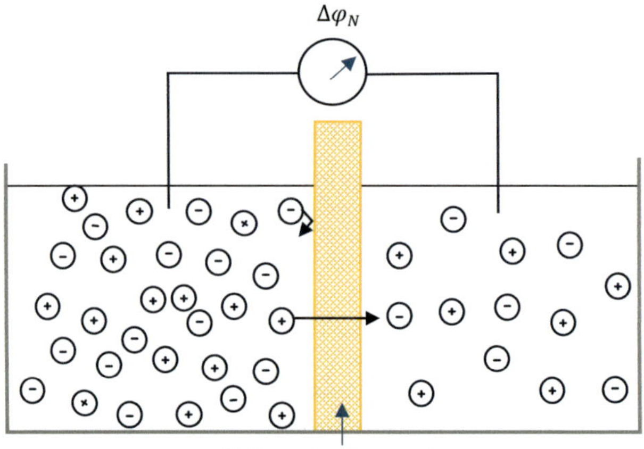

FIGURE 11.6
Nernst potential develops across a cation-selective membrane that separates two solutions of different cation concentrations.

Because the Nernst potential is the potential at equilibrium, it can be determined relatively simply and straightforwardly with thermodynamic principles. Considering ion transport through an ion-selective membrane between two salt solutions with different concentrations. As shown in Figure 11.6, a cation-selective membrane separates two salt solutions with the cation concentration on the left side being higher than on the right side.

Initially, cations transported from the left side to the right side of the membrane under a concentration gradient. Because the counterions, which are anions in this case, cannot transport across the membrane, an extra positive charge accumulates on the right side of the membrane. As a result of this unbalanced charge distribution, an electric potential develops across the membrane. The electric potential tends to slow down the movement of cations. Transport of cations eventually stops when the potential increases to a certain value. At that moment, the electrochemical potentials of cations on two sides of the membrane are equal to each other. According to Eq. (9.9), we have

$$\mu^0 + RT\ln c_1 = \mu^0 + RT\ln c_2 + zF\Delta\varphi_N \tag{11.30}$$

where $\Delta\varphi_N$ is the Nernst potential, c_1 and c_2 are the cation concentrations on left side and right side of the membrane, and z is the valence of cations. From Eq. (11.30), the Nernst potential is determined as

$$\Delta\varphi_N = \frac{RT}{zF}\ln\frac{c_1}{c_2} \tag{11.31}$$

Example 11.3

A cation-selective membrane separates two solutions of sodium chloride with a concentration ratio $c_2/c_1 = 0.1$. Determine the potential difference across the membrane at equilibrium at 20°C.

SOLUTION

Using the Nernst equation

$$\Delta\varphi_N = \frac{RT}{zF}\ln\frac{c_1}{c_2} = \frac{8.31\,J/K\cdot mole\times 293\,K}{1\times 96485\,J/mol\cdot volt}\times ln10 = 0.0581\,V = 58.1\,mV$$

11.3.2 Donnan Potential

Ion exchange membranes contain immobile charges that can cause uneven distributions of mobile ions when they are in contact with a solution. The electrical potential difference thus induced at the interface of membrane and solution is the Donnan potential [21–23]. As schematically presented in Figure 11.7, there are two phases with the membrane phase on the right side and the solution phase on the left side. Both phases contain mobile cations and anions. In the membrane phase, however, there are immobile positive charges in addition to the mobile ions. The immobile charges that are fixed with the membrane matrix are indicated with the "+" signs. The mobile cations and anions that can move freely across the interface of the membrane with the solution are indicated by the circles.

Because the concentration of mobile cations in the membrane phase is small due to the existence of the immobile positive charges, cations in the solution phase diffuse into the membrane phase under a concentration gradient when the two phases are brought into contact. The extra positive charge induces a higher potential in the membrane phase than in the solution phase. This potential difference tends to oppose further movement of cations into the membrane phase but attracts anions into the membrane phase to lower the

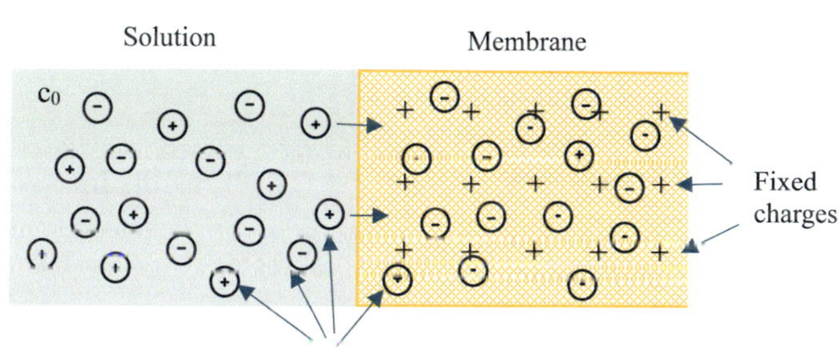

FIGURE 11.7
Donnan potential develops at the interface of a solution with a membrane of fixed charge in it, which is indicated with "+" signs.

potential, which causes more cations to transfer into the membrane phase. The movement of ions from the solution phase into the membrane phase stops when an equilibrium is established between the chemical (concentration) driving force and the electrical (potential) driving force.

At equilibrium, the electrochemical potentials of cations in the solution phase and the membrane phase are equal to each other, and so are anions. Therefore, for the solution of a univalent 1–1 salt, the potential difference across the interface is

$$\Delta\varphi_D = \frac{RT}{F}\ln\frac{c_0}{c_+{}^m} = \frac{RT}{F}\ln\frac{c_-{}^m}{c_0} \tag{11.32}$$

where $\Delta\varphi_D$ is the Donnan potential, c_0 is the concentration of cations and anions in the solution phase (they are equal for a univalent 1–1 salt), $c_+{}^m$ and $c_-{}^m$ are the cation concentration and anion concentration in the membrane phase.

From Eq. (11.32), we have

$$c_+{}^m c_-{}^m = c_0{}^2 \tag{11.33}$$

Applying electroneutrality to the membrane phase, we have

$$c_+{}^m - c_-{}^m + X = 0 \tag{11.34}$$

where X is the concentration of immobile elementary charges of the membrane. The parameter X takes a positive or negative value depending on the immobile charge.

Using Eq. (11.32), the concentrations of cations and anions in the membrane phase are, respectively

$$c_+{}^m = c_0 e^{-\frac{F\Delta\varphi_D}{RT}} \tag{11.35}$$

and

$$c_-{}^m = c_0 e^{\frac{F\Delta\varphi_D}{RT}} \tag{11.36}$$

Substituting Eqs. (11.35) and (11.36) into Eq. (11.34) results in

$$sinh\left(\frac{F\Delta\varphi_D}{RT}\right) = \frac{X}{2c_0} \tag{11.37}$$

The Donnan potential is then determined as

$$\Delta\varphi_D = \frac{RT}{F} arsinh\left(\frac{X}{2c_0}\right) = \frac{RT}{F}\ln\left[\frac{1}{2c_0}\left(X + \sqrt{X^2 + 4c_0{}^2}\right)\right] \tag{11.38}$$

Substituting Eq. (11.38) into Eq. (11.35), the cation concentration in the membrane phase is determined as,

$$
\begin{aligned}
c_+^m &= c_0 e^{-\ln\left[\frac{1}{2c_0}\left(X+\sqrt{X^2+4c_0^2}\right)\right]} \\
&= c_0 \frac{1}{e^{\ln\left[\frac{1}{2c_0}\left(X+\sqrt{X^2+4c_0^2}\right)\right]}} \\
&= c_0 \frac{1}{\frac{1}{2c_0}\left(X+\sqrt{X^2+4c_0^2}\right)} \\
&= c_0 \frac{\left(X-\sqrt{X^2+4c_0^2}\right)}{\frac{1}{2c_0}\left(X+\sqrt{X^2+4c_0^2}\right)\left(X-\sqrt{X^2+4c_0^2}\right)} \\
&= \frac{1}{2}\left(\sqrt{X^2+4c_0^2}-X\right)
\end{aligned}
\tag{11.39}
$$

Substituting Eq. (11.38) into Eq. (11.36), the anion concentration in the membrane phases is determined as

$$c_-^m = \frac{1}{2}\left(\sqrt{X^2+4c_0^2}+X\right) \tag{11.40}$$

Example 11.4

A cation-selective membrane carries a positive elementary charge $X = 5$ mol/m³. Calculate the Donnan potential for a range of monovalent electrolyte concentrations from 0.01 to 100 mol/m³. Determine cation and anion concentrations at the same time.

SOLUTION

Sample calculations of the Donnan potential, cation concentration, and anion concentration for $c_0 = 0.01$ with Eqs. (11.38)–(11.40).

$$
\begin{aligned}
\Delta\varphi_D &= \frac{RT}{F} arsinh\left(\frac{X}{2c_0}\right) = \frac{8.31\,J/K\cdot mol \times 293\,K}{96485\,J/V\cdot mol} arsinh\left(\frac{5\,mol/m^3}{2\times 0.01\,mol/m^3}\right) \\
&- 0.156\,V = 156\,mV
\end{aligned}
$$

$$
\begin{aligned}
c_+^m &- \frac{1}{2}\left(\sqrt{X^2+4c_0^2}-X\right) \\
&= \frac{1}{2}\left(\sqrt{\left(5\,mol/m^3\right)^2+4\left(0.01\,mol/m^3\right)^2}-5\,mol/m^3\right) \\
&= 2\times 10^{-5}\,mol/m^3
\end{aligned}
$$

$$c_-^m = \frac{1}{2}\left(\sqrt{X^2 + 4c_0^2} + X\right)$$
$$= \frac{1}{2}\left(\sqrt{\left(5\ mol/m^3\right)^2 + 4\left(0.01\ mol/m^3\right)^2} + 5\ mol/m^3\right)$$
$$= 5.00\ mol/m^3$$

The results of all calculations are listed in the table below.

c_0 (mol/m³)	$\Delta\varphi$(mV)	c_+ (mol/m³)	c_- (mol/m³)
0.01	156.83	2.00×10^{-5}	5.00
0.03	129.10	1.80×10^{-4}	5.00
0.1	98.73	2.00×10^{-3}	5.00
0.3	71.09	1.79×10^{-2}	5.02
1	41.57	1.93×10^{-1}	5.19
3	19.14	1.41	6.41
10	6.24	7.81	12.81
30	2.10	27.60	32.60
100	0.63	97.53	102.53

11.3.3 Diffusion Potential

Besides the Nernst potential and the Donnan potential at equilibrium with zero ion fluxes, the membrane potential can be generated by moving cations and anions of different mobilities driven initially by concentration gradient [24–27]. When a membrane separates two solutions of a single electrolyte of different concentrations, the ions initially move from the high concentration side to the low concentration side by diffusion. If cations and anions have the same mobility in the membrane, no electric potential will develop. However, if cations and anions have different mobilities, the ions with larger mobility move faster across the membrane than the ions with smaller mobility. As a result, an excess amount of the charge carried by the fast ions accumulates on the low concentration side and the opposite charge leaves behind in the high concentration side, as schematically shown in Figure 11.8. The shadowed ions in the figure represent the net charges accumulated on two sides of the membrane.

The electric potential thus generated slows down the fast ions but accelerates the slow ions. Eventually, the charge fluxes carried by cations and anions are equal to each other, and ion movement reaches a steady state. This potential at steady state is the diffusion potential across the membrane. Diffusion potential is of essential importance to maintain the electroneutrality of solutions on both sides of the membrane for sustainable ion transport.

The calculation of diffusion potential in general cases with arbitrary types of cations and anions cannot be done analytically. However, the diffusion potential associated with a binary monovalent electrolyte is manageable analytically.

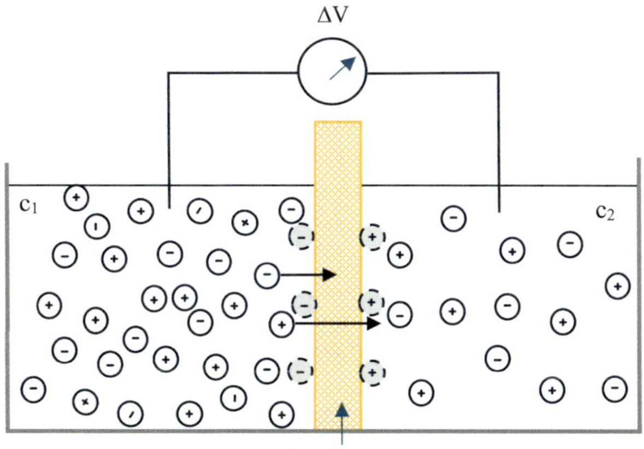

FIGURE 11.8
Diffusion potential develops as a result of transport of cations and anions with different mobilities.

According to the Nernst–Planck equation, the fluxes of monovalent cations and anions, respectively, are,

$$J_+ = -D_+ \frac{dc_+}{dx} - c_+ \frac{D_+ F}{RT} \frac{d\varphi}{dx} \tag{11.41}$$

and

$$J_- = -D_- \frac{dc_-}{dx} + c_- \frac{D_- F}{RT} \frac{d\varphi}{dx} \tag{11.42}$$

With the electroneutrality assumption, concentrations of cations and anions are equal to each other at every point across the membrane thickness, i.e.,

$$c_-(x) = c_+(x) = c(x) \tag{11.43}$$

At steady state, the charge fluxes carried by cations and anions are equal to each other (zero current), i.e.,

$$J_+ = J_-$$

Substituting Eqs. (11.41) and (11.42) into the above equation, we have

$$-D_+ \frac{dc}{dx} - c \frac{D_+ F}{RT} \frac{d\varphi}{dx} = -D_- \frac{dc}{dx} + c \frac{D_- F}{RT} \frac{d\varphi}{dx} \tag{11.44}$$

where D_+ and D_- are the diffusion coefficients of cations and anions, respectively.

Rearranging and integrating,

$$\int_{\varphi_1}^{\varphi_2} d\varphi = -\frac{RT}{F}\frac{D_+ - D_-}{D_+ + D_-}\int_{c_1}^{c_2}\frac{1}{c}dc$$

we obtain

$$\Delta\varphi_{dif} = \varphi_2 - \varphi_1 = \frac{RT}{F}U\ln\frac{c_1}{c_2} \qquad (11.45)$$

where U is the mobility coefficient defined as:

$$U = \frac{D_+ - D_-}{D_+ + D_-} \qquad (11.46)$$

It is clear from Eqs. (11.45) and (11.46) that diffusion potential develops when cations and anions have different diffusion coefficients, and there is a concentration difference across the membrane.

Example 11.5

The concentration ratio across a membrane is $c_1/c_2 = 10$. Calculate diffusion potentials for the following combinations of diffusion coefficients for cations and anions. (1) $D_+ = 3 \times 10^{-9}$ m²/s and $D_- = 1 \times 10^{-9}$ m²/s; (2) $D_+ = 2.5 \times 10^{-9}$ m²/s and $D_- = 1.5 \times 10^{-9}$ m²/s; (3) $D_+ = 1 \times 10^{-9}$ m²/s and $D_- = 1 \times 10^{-9}$ m²/s; (4) $D_+ = 1.5 \times 10^{-9}$ m²/s and $D_- = 2.5 \times 10^{-9}$ m²/s; (5) $D_+ = 1 \times 10^{-9}$ m²/s and $D_- = 3 \times 10^{-9}$ m²/s.

SOLUTION

Sample calculations of the diffusion potential for $D_+ = 3 \times 10^{-9}$ m²/s and $D_- = 1 \times 10^{-9}$ m²/s with Eqs. (13.46) and (13.45).

$$U = \frac{D_+ - D_-}{D_+ + D_-} = \frac{3\times10^{-9}\,m^2/s - 1\times10^{-9}\,m^2/s}{3\times10^{-9}\,m^2/s + 1\times10^{-9}\,m^2/s} = 0.5$$

$$\Delta\varphi_{dif} = \frac{RT}{F}U\ln\frac{c_1}{c_2} = \frac{8.31\,J/K\cdot mol \times 293\,K}{96485\,J/mol\cdot volt}\times 0.5\times\ln 10 = 0.029\,V = 29.05\,mV$$

The complete results are shown in the table below.

Group	(1)	(2)	(3)	(4)	(5)
D_+ $(10^{-9}\,m^2/s)$	3.00	2.50	1.00	1.50	1.00
D_- $(10^{-9}\,m^2/s)$	1.00	1.50	1.00	2.50	3.00
U	0.50	0.25	0.00	−0.25	−0.50
$\Delta\varphi$ (mV)	29.05	14.53	0.00	−14.53	−29.05

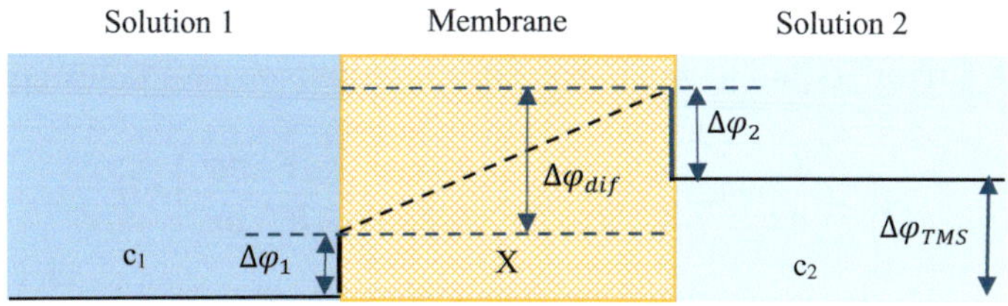

FIGURE 11.9
Teorell–Meyer–Sievers potential develops across an ion-selective membrane as a combination of the Donnan potential and the diffusion potential.

11.3.4 Teorell–Meyer–Sievers (TMS) Potential

When a membrane with a fixed charge separates two solutions of different ion concentrations. Donnan potentials develop at the interfaces of the membrane with solutions. At the same time, a diffusion potential also develops across the membrane as a result of the transports of cations and anions with different mobilities. Therefore, the total membrane potential consists of two Donnan potentials and one diffusion potential, which is the Teorell–Meyer–Sievers (TMS) potential [8,28], as illustrated in Figure 11.9. Electrolyte concentrations are c_1 and c_2 on the left side and right side of the membrane, respectively, while the density of fixed charge in the membrane is X.

The total membrane potential for cases with the same single electrolyte solution on both sides of the membrane can be calculated with the knowledge of the Donnan potential and the diffusion potential discussed above. The total membrane potential in this case is

$$\Delta\varphi_{TMS} = \Delta\varphi_1 + \Delta\varphi_{dif} + \Delta\varphi_2 \tag{11.47}$$

where $\Delta\varphi_{TMS}$ is the total membrane potential, $\Delta\varphi_{dif}$ is the diffusion potential, and $\Delta\varphi_1$ and $\Delta\varphi_2$ are two Donnan potentials at the two interfaces of the membrane with solutions, respectively. According to Eq. (11.38), the Donnan potential components are

$$\begin{aligned}\Delta\varphi_1 + \Delta\varphi_2 &= \frac{RT}{F}\ln\left[\frac{1}{2c_1}\left(X+\sqrt{X^2+4c_1^{\,2}}\right)\right] - \frac{RT}{F}\ln\left[\frac{1}{2c_2}\left(X+\sqrt{X^2+4c_2^{\,2}}\right)\right] \\ &- \frac{RT}{F}\ln\left[\frac{c_2}{c_1}\frac{\left(X+\sqrt{X^2+4c_1^{\,2}}\right)}{\left(X+\sqrt{X^2+4c_2^{\,2}}\right)}\right]\end{aligned} \tag{11.48}$$

That the negative sign is added before the expression for the Donnan potential on the right side of the membrane, $\Delta\varphi_2$, is because the orientation of the solution to the membrane is opposite from that on the left side.

The diffusion potential component is the potential across the membrane that makes the current equal to zero. According to the Nernst–Planck equation,

$$-D_+\left(\frac{dc_+}{dx}+c_+\frac{F}{RT}\frac{d\psi}{dx}\right)+D_-\left(\frac{dc_-}{dx}-c_-\frac{F}{RT}\frac{d\psi}{dx}\right)=0$$

Assuming electroneutrality on the membrane, i.e.,

$$c_- = c_+ + X$$

Rearranging the equation results in

$$\frac{d\psi}{dx}=-\frac{RT}{F}\frac{(D_+-D_-)}{(D_++D_-)c_++D_-X}\frac{dc_+}{dx}$$

Integrating

$$\Delta\varphi_{dif}=-\frac{RT}{F}\frac{D_+-D_-}{D_++D_-}\ln\frac{(D_++D_-)c_2^{\,m}+D_-X}{(D_++D_-)c_1^{\,m}+D_-X} \tag{11.49}$$

where $c_1^{\,m}$ and $c_2^{\,m}$ are the cation concentrations in the membrane at the interfaces with solutions.

According to Eq. (11.39),

$$c_1^{\,m}=\frac{1}{2}\sqrt{X^2+4c_1^{\,2}}-\frac{1}{2}X$$

and

$$c_2^{\,m}=\frac{1}{2}\sqrt{X^2+4c_2^{\,2}}-\frac{1}{2}X$$

Substituting the two expressions into Eq. (13.49) and rearranging results in

$$\Delta\varphi_{dif}=-U\frac{RT}{F}\ln\frac{\sqrt{X^2+4c_2^{\,2}}-UX}{\sqrt{X^2+4c_1^{\,2}}-UX} \tag{11.50}$$

Finally, substituting Eqs. (11.48) and (11.50) into Eq. (11.47), the total membrane potential, the TMS potential, is then obtained as,

$$\Delta\varphi_{TMS}=\frac{RT}{F}\ln\left[\frac{c_2}{c_1}\frac{\left(X+\sqrt{X^2+4c_1^{\,2}}\right)}{\left(X+\sqrt{X^2+4c_2^{\,2}}\right)}\right]-U\frac{RT}{F}\ln\frac{\sqrt{X^2+4c_2^{\,2}}-UX}{\sqrt{X^2+4c_1^{\,2}}-UX} \tag{11.51}$$

The total membrane potentials of a cation exchange membrane between solutions of one monovalent electrolyte are calculated for a concentration ratio $c_2/c_1 = 0.1$ at a temperature of 20°C. The results are graphically presented in Figure 11.10. The mobility coefficients are indicated under the potential

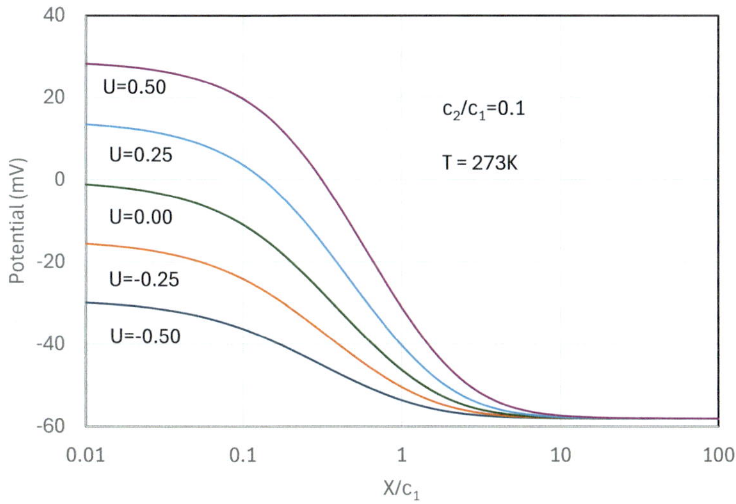

FIGURE 11.10
Membrane potential calculated with the Teorell–Meyer–Sievers model as a function of x/c_1 for various mobility coefficients as indicated in the figure.

curves. The abscissa is the ratio of immobile charge of the membrane to electrolyte concentration on the left side of the membrane, X/c_1. Positive immobile charge X is used in the calculations.

The potential lines decrease smoothly as the immobile charge increases. It can be seen from the lines that the contributions of the Donnan potentials and diffusion potential to the total potential change with the ratio of X/c_1. When the immobile charge is much greater than the electrolyte concentration in solution $X \gg c_1$, the total membrane potential is a constant because it is predominantly controlled by the Donnan potentials, which are fixed with the c_2/c_1 ratio. On the other hand, when the immobile charge is much smaller than the electrolyte concentration in solution, the total membrane potential is mainly controlled by diffusion potential, which varies with the mobility coefficient.

11.4 Governing Equations for Ion Transport

The concise analytical equations for the membrane potentials are only available for simple cases. Most of them are obtained with the use of local electroneutrality (or Nernst–Planck electroneutrality) assumption for solutions of one electrolyte of a single monovalent cation and anion [29–32]. However, there would not be any membrane potential if electroneutrality were strictly held locally on any point in a membrane system. Furthermore, real-world membrane separation systems usually involve more ions, including many multivalent ones. Development of the membrane potential, or in general, ion transport across the membranes is a complicated phenomenon and process. However, the general cases are still manageable, although they are quite challenging.

Let us start with the uncharged solute. Transport of uncharged conservative particles in the membrane is governed by the common convection–diffusion equation

$$\frac{dc_i}{dt} = D_i \frac{d^2 c_i}{dx^2} - v \frac{dc_i}{dx}$$

The first term and the second term on the right-hand side are for diffusion and convection, respectively. The derivation of the equation is a commonplace that can be easily found in the literature. We are not going to rederive it here.

When the particles are charged, electromigration has to be added to the governing equation. The electromigration term can be formulated with the mass balance principle. One-dimensional electromigration in the x direction is shown in Figure 11.11, in which the migration velocity of ion i is denoted by $v_{i,m}$.

Considering a volume with length dx and unit cross-sectional area perpendicular to the migration direction, as shown in Figure 11.11, the concentration change over time due to electromigration is

$$\frac{dc_i}{dt} = \frac{c_i v_{im} - \left[c_i v_{i,m} + d\left(c_i v_{i,m}\right)\right]}{dx} = -\frac{d\left(c_i v_{i,m}\right)}{dx}$$

Substituting Eq. (11.13) for $v_{i,m}$ in the equation results in

$$\frac{dc_i}{dt} = \frac{D_i F z_i}{RT} \frac{d}{dx}\left(c_i \frac{d\varphi}{dx}\right)$$

Adding this electromigration term to the convection–diffusion equation results in

$$\frac{dc_i}{dt} = D_i \frac{d^2 c_i}{dx^2} + \frac{DFz_i}{RT} \frac{d}{dx}\left(c_i \frac{d\varphi}{dx}\right) - v \frac{dc_i}{dx} \tag{11.52}$$

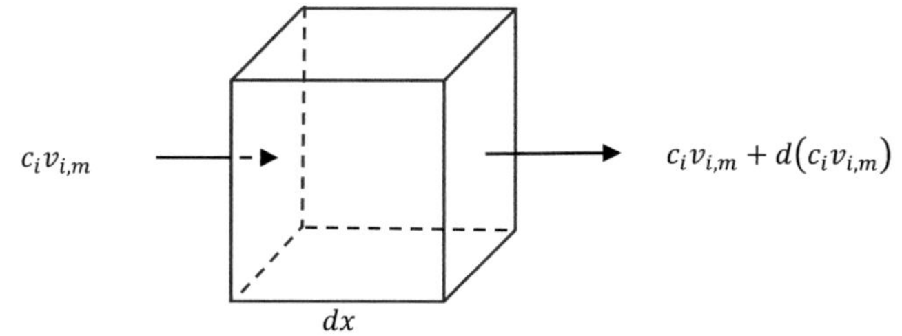

FIGURE 11.11
Analysis of electromigration of ions in the x direction.

Eq. (11.52) is the extended Nernst–Planck equation. It reduces to the Nernst–Planck equation when the convection term is zero. At steady state, Eq. (11.52) becomes

$$D_i \frac{d^2 c_i}{dx^2} + \frac{DFz_i}{RT} \frac{d}{dx}\left(c_i \frac{d\varphi}{dx} \right) - v \frac{dc_i}{dx} = 0 \tag{11.53}$$

The potential φ in the equations is related to the distributions of ions and fixed charge on the membrane with the Poisson equation,

$$\frac{d^2 \varphi}{dx^2} = -\frac{F}{\varepsilon \varepsilon_0} \left(X + \sum_i z_i c_i \right) \tag{11.54}$$

where X is the density of fixed charge (C/m^3) on the membrane.

Eqs. (11.53) and (11.54) are the governing equations for ion transport across the membrane at steady state, which is commonly referred to as Nernst–Planck–Poisson (NPP) equations [33–37].

When the solution of the NPP equations is obtained, ion fluxes across the membrane can be calculated with

$$J_i = vc_i - D_i \frac{dc_i}{dx} - \frac{D_i F z_i}{RT} c_i \frac{d\varphi}{dx} \tag{11.55}$$

In addition, the electrical potential across the membrane is also obtained with ion fluxes at the same time when the NPP equation is solved.

11.5 Boundary Conditions

Development of governing differential equations is usually an easy task for a problem. Problems controlled by the same principles or mechanisms have the same differential equations as governing equations. The setup of appropriate boundary conditions for the problem is a more complex and sometimes challenging task. Incorrectly imposed boundary conditions would represent completely different problems than the intended ones. For example, completely different flow fields of a vortex in an open tank and a Poiseuille flow in a pipe are governed by the same Navier–Stokes equation coupled with the continuity equation. It is the boundary conditions that virtually determine whether the flow field is a vortex or a Poiseuille flow. A pertinent solution of differential equations for a particular problem cannot be obtained without the appropriately specified boundary conditions.

Some differential equations may have meaningful solutions with different boundary conditions. However, some other differential equations have certain constraints on the selection of boundary conditions. Otherwise, solutions

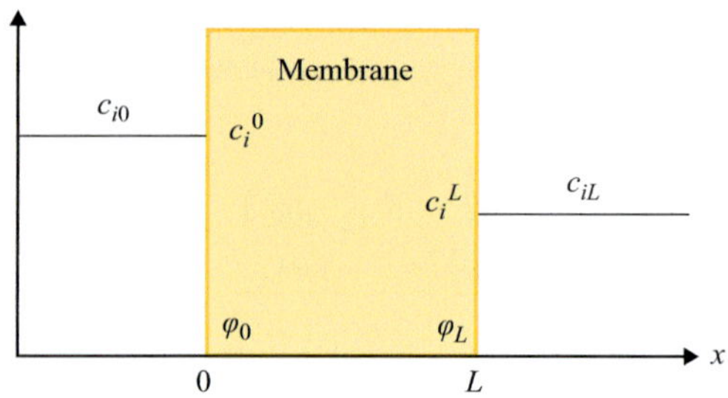

FIGURE 11.12
Boundary conditions of Nernst–Planck–Poisson equations for an uncharged membrane with two surfaces at location 0 and L.

of the differential equations may not exist or may not make any sense physically. Nernst–Planck equation coupled with Poisson equation belongs to the latter. Ion concentrations and electrical potential cannot be independently specified on the boundaries of the equations because they are required to be consistent with each other. This critical issue that is overlooked in many studies is discussed below.

11.5.1 Unspecifiable Boundary Conditions at Steady State

First, let us consider a trivial case of ion transport as shown in Figure 11.12. In this case, the membrane is uncharged with two surfaces of the membrane at locations of *0* and *L*, separating solutions of different ion concentrations. It is further assumed that there is no any partition mechanism (e.g., preferential adsorption) for either cations or anions. Because the membrane is uncharged, there are no electrical interactions between ions in the solution with the membrane. Therefore, the electrical potential differences between the bulk solutions and the membrane surfaces are zero. Ion concentrations on the membrane surfaces are equal to bulk concentrations. The boundary conditions on the membrane surfaces can be naturally specified as

$$c_i^{\,0} = c_{i0},\; c_i^{\,L} = c_{iL} \tag{11.56}$$

$$\varphi_0 = 0,\; \varphi_L = 0 \tag{11.57}$$

where $c_i^{\,0}$ and $c_i^{\,L}$ are the ion concentrations on the membrane surfaces, c_{i0} and c_{iL} are the ion concentrations in the bulk solutions, and φ_0 and φ_L are the electrical potentials on the membrane surface.

Boundary conditions as in Eqs. (11.56) and (11.57) looks appropriate for this case. But such boundary conditions cannot be applied to general ion transport problems simply because the membrane cannot be assumed uncharged when there is ongoing ion transport. Even with a membrane that is initially uncharged, an electrical potential would develop on it as a consequence of ion transport. When the membrane is charged, because of the repulsive interaction with the co-ions and attractive interaction with the counter-ions, ion concentrations on the membrane surfaces are no longer equal to their bulk concentrations. Furthermore, potentials are no longer equal to zero on both surfaces of the membrane.

There is a deadlock in applying NPP equations to study ion transport at steady state. Boundary conditions on ion concentrations are required to solve the differential equations to find the electric potentials on the membrane surfaces. But the electric potentials on the membrane surfaces are needed to specify the appropriate boundary conditions on ion concentrations for the differential equations. At this moment, there is no method that is convincingly valid to break the deadlock to specify the appropriate boundary conditions for NPP equations at steady state.

11.5.2 Potential Boundary Conditions

Because the boundary conditions for ion transport are not readily available yet, the solution of NPP equations at steady state cannot be pursued directly. However, the problem can be rigorously handled with the unsteady state Nernst–Planck equation (Eq. 11.52). When the time domain is included, calculation can start with an initial situation, for which the boundary conditions can be easily specified. For example, the solution process can always start with an uncharged membrane. Then Eqs. (11.56) and (11.57) are the boundary conditions at time $t = 0$. After ion transport starts, an unbalanced charge may develop in the transition layers adjacent to the membrane surfaces and on the membrane itself. There are known physical principles to modify the boundary conditions according to charge development.

First, boundary conditions on potential for the Poisson equation can be rigorously related to the charge in the transition layer. As shown in Figure 11.13,

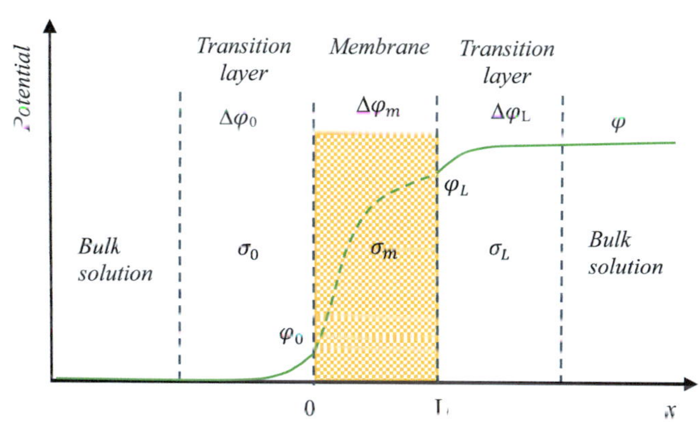

FIGURE 11.13
Schematic for boundary conditions on potential for the Poisson equation. Potential differences between the membrane surfaces and the bulk solutions can be rigorously related to the net charges in two adjacent transition layers.

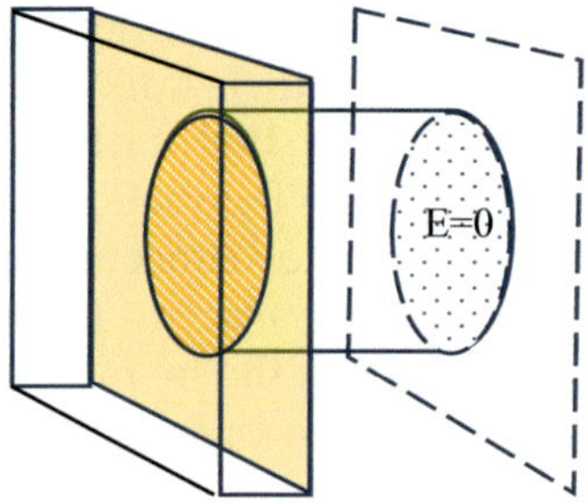

FIGURE 11.14
A closed surface for the transition layer on the right side of the membrane.

a membrane is sandwiched by two transition layers that contact the membrane surfaces at *0* and *L*, respectively. Net charges per unit membrane area in the two transition layers are σ_0 and σ_L, respectively. When the potential in the bulk solution on the left side of the membrane is arbitrarily taken as zero, a first type (Dirichlet) boundary condition can be specified for the left membrane surface at $x = 0$ by directly applying Eq. (11.29) as

$$\varphi_0 = \Delta\varphi_0 = -\frac{\lambda}{\varepsilon\varepsilon_0}\sigma_0 \tag{11.58}$$

The Dirichlet boundary condition cannot be specified on the right membrane surface because the potential difference across the membrane is unknown yet. Actually, to find out the membrane potential is one of the purposes of solving the differential equation. However, a second type (Neumann) boundary condition can be set up on the right membrane surface. Draw a closed surface as shown in Figure 11.14 to enclose the whole transition layer on the right-hand side of the membrane. The electric field on the circle on the right-hand side of the transition layer is zero because there is no electrostatic interaction in the bulk solution. The electric field on the side wall of the cylinder is perpendicular to the normal direction of the wall surface. Therefore, the electric flux across the side wall of the cylinder is zero, too. Therefore, the electric field on the right-hand surface of the membrane can be determined by applying Gauss's law

$$E = -\frac{\sigma_L}{\varepsilon\varepsilon_0} \tag{11.59}$$

The negative sign is added in Eq. (11.59) because the outward direction of the membrane surface as a part of the closed surface is opposite to the positive direction of this problem. Because the electric field is the negative gradient of potential, the boundary condition on the right-hand surface of the membrane can be specified as

$$\left.\frac{d\varphi}{dx}\right|_L = \frac{\sigma_L}{\varepsilon\varepsilon_0} \tag{11.60}$$

By the way, the potential difference across the transition layer on the right surface of the membrane can be determined by applying Eq. (11.29) as

$$\Delta\varphi_L = \frac{\lambda_L}{\varepsilon\varepsilon_0}\sigma_L \tag{11.61}$$

where $\Delta\varphi_L$ is the potential difference between the bulk solution and the right membrane surface, σ_L is the charge density of the transition layer on the right surface of the membrane, and λ_L is the Debye length of the solution on the right side of the membrane. The drop of the negative sign in Eq. (11.61) is because the positive direction of potential is from the membrane to the transition layer.

11.5.3 Poisson–Boltzmann Constraint

Concentrations of ions on the membrane surfaces cannot be specified independently of membrane potential. They have to satisfy the Poisson–Boltzmann constraint of Eq. (11.18). Otherwise, the boundary conditions for ion transport across the membrane would be ill-imposed and unphysical.

With known potential differences across the transition layers, ion concentrations on the membrane surfaces are readily determined with Eq. (11.18)

$$c_i^{\,0} = c_{i0}e^{-\frac{z_i F\Delta\varphi_0}{RT}} \tag{11.62}$$

$$c_i^{\,L} = c_{iL}e^{-\frac{z_i F\Delta\varphi_L}{RT}} \tag{11.63}$$

Now all four boundary conditions, which are Eqs. (11.58), (11.60), (11.62), and (11.63), for the NPP equations can be specified at any moment when net charges in the transition layers are known. Therefore, ion transport across the membrane at that moment is completely defined, and the solution can be pursued.

Example 11.6

Determine the concentration boundary conditions for a membrane with 0.1 mol/L and 0.01 mol/L of NaCl solution on the left and right sides of the membrane. The net charges in the transition layers are −0.01 C/m^2 and 0.005 C/m^2, respectively.

SOLUTION

From Example 11.1, the Debye lengths are

$$\lambda_0 = 9.62\times10^{-10}\,m$$

$$\lambda_L = 3.04\times10^{-9}\,m$$

The potential difference on the left membrane surface over the bulk solution is calculated with Eq. (11.29), i.e.,

$$\Delta\varphi_0 = -\frac{\lambda_0}{\varepsilon\varepsilon_0}\sigma_0 = -\frac{9.62\times10^{-10}\,m}{8.85\times10^{-12}\,C/V\cdot m\times 80}\times\left(-0.01\,C/m^2\right)$$
$$= 1.36\times10^{-2}\,V = 13.6\,mV$$

The sodium concentration and chloride concentration are

$$c_{Na}{}^{0} = c_{Na0}e^{-\frac{z_iF\Delta\varphi_0}{RT}} = 0.1\,mol/L\times\exp\left(-\frac{96485C/mol\times1.36\times10^{-2}\,V}{8.31\,J/K\cdot mol\times293\,K}\right) = 0.058\,mol/L$$

$$c_{Cl}{}^{0} = c_{Cl0}e^{-\frac{z_iF\Delta\varphi_0}{RT}} = 0.1\,mol/L\times\exp\left(\frac{96485\,C/mol\times1.36\times10^{-2}\,V}{8.31\,J/K\cdot mol\times293\,K}\right) = 0.171\,mol/L$$

The potential difference on the right membrane surface over the bulk solution is calculated with the same Eq. (11.29), i.e.,

$$\Delta\varphi_L = -\frac{\lambda_L}{\varepsilon\varepsilon_0}\sigma_L = -\frac{3.04\times10^{-9}\,m}{8.85\times10^{-12}\,C/V\cdot m\times 80}\times\left(0.005\,C/m^2\right)$$
$$= -2.15\times10^{-2}\,V = -21.5\,mV$$

The sodium concentration and chloride concentration are

$$c_{Na}{}^{L} = c_{NaL}e^{-\frac{z_iF\Delta\varphi_L}{RT}} = 0.01\,mol/L\times\exp\left(-\frac{96485C/mol\times\left(-2.15\times10^{-2}\,V\right)}{8.31\,J/K\cdot mol\times293\,K}\right)$$
$$= 0.023\,mol/L$$

$$c_{Cl}{}^{L} = c_{ClL}e^{-\frac{z_iF\Delta\varphi_L}{RT}} = 0.01\,mol/L\times\exp\left(\frac{96485C/mol\times\left(-2.15\times10^{-2}\,V\right)}{8.31\,J/K\cdot mol\times293\,K}\right)$$
$$= 4.27\times10^{-3}\,mol/L$$

11.5.4 Net Charges in the Transition Layers

From the solution of the NPP equations, ion fluxes can be calculated at the membrane surfaces and any point inside the membrane. Then the net charges in the transition layers at any time can be tracked with

$$\sigma_0 = \int_0^t \sum_i z_i J_{i0}\, d \tag{11.64}$$

$$\sigma_L = \int_0^t \sum_i z_i J_{iL}\, dt \tag{11.65}$$

where J_{i0} and J_{iL} are ion fluxes at the membrane surfaces at x = 0, and L, respectively.

Because of the electroneutrality of a membrane system, the net charge on the membrane can be determined by

$$\sigma_m = -\sigma_0 - \sigma_L \tag{11.66}$$

where σ_m is the charge on the membrane per area of the membrane. A steady state of ion transport is reached when the electric current across the membrane becomes zero (the "null current" condition), which is mathematically expressed as

$$\sum_i z_i J_i = 0$$

Up to now, the mathematical model for general ion transport across the membrane is fully constructed [38,39]. Most importantly, the boundary conditions for governing equations are consistent with each other and pertinent to ion transport across the membrane. Therefore, it can be a reliable tool for studies on issues related to ion transport across the membrane, such as the development of the membrane potential.

11.6 Numerical Solution Method

With completion of the mathematical model, it is time to seek a solution of the extended NPP equations. Analytical solutions are generally unavailable or virtually nonexistent for such complicated equations. Therefore, a numerical solution procedure will be introduced in this section that will guarantee a solution of the problem [40–42].

For convenience, the unsteady state Nernst–Planck equation is rewritten as,

$$\frac{dc_i}{dt} = A\frac{d^2c_i}{dx^2} + B\frac{dc_i}{dx} + Cc_i \tag{11.67}$$

where *A*, *B*, and *C* are the coefficients as given below

$$A = D_i$$

$$B = \frac{D_i z_i F}{RT}\frac{d\varphi}{dx} - v$$

$$C - \frac{D_i z_i F}{RT}\frac{d^2\varphi}{dx^2}$$

The derivatives of the potential in the coefficient will be calculated numerically with the numerical solution of the Poisson equation. All the coefficients A, B, and C are known at any step of the numerical solution procedure of the Nernst–Planck equation.

Eq. (11.67) can be readily solved numerically with the Crank–Nicolson method. In this method, concentration and its derivatives over time and space are discretized as

$$\frac{dc_i}{dt} = \frac{c_i^{j+1,k} - c_i^{j,k}}{\Delta t}$$

$$\frac{d^2c_i}{dx^2} = \frac{1}{2}\left(\frac{c_i^{j+1,k+1} - 2c_i^{j+1,k} + c_i^{j+1,k-1}}{\Delta x^2} + \frac{c_i^{j,k+1} - 2c_i^{j,k} + c_i^{j,k-1}}{\Delta x^2}\right)$$

$$\frac{dc_i}{dx} = \frac{1}{2}\left(\frac{c_i^{j+1,k+1} - c_i^{j+1,k-1}}{2\Delta x} + \frac{c_i^{j,k+1} - c_i^{j,k-1}}{2\Delta x}\right)$$

$$c_i = \frac{1}{2}\left(c_i^{j+1,k} + c_i^{j,k}\right)$$

where Δt and Δx are the time and space step sizes used in discretization, j is the time step (j = 0, 1, 2, …, n), and k is the spatial step (k = 0, 1, 2, …, m).

Substituting the discretized concentration and derivatives in Eq. (11.68) and rearranging results in

$$\alpha c_i^{j+1,k-1} + \beta c_i^{j+1,k} + \gamma c_i^{j+1,k+1} = d^k \qquad (11.68)$$

where

$$\alpha = \frac{A}{2\Delta x^2} + \frac{bB}{4\Delta x}$$

$$\beta = \frac{1}{\Delta t} + \frac{A}{\Delta x^2} + \frac{C}{2}$$

$$\gamma = -\frac{A}{2\Delta x^2} - \frac{B}{4\Delta x}$$

The values of d^k are calculated from the known concentrations at the current time step by

$$d^k = d_1 c_i^{j,k-1} + d_2 c_i^{j,k} + d_3 c_i^{j,k+1}$$

where

$$d_1 = \frac{A}{2\Delta x^2} - \frac{B}{4\Delta x}$$

$$d_2 = \frac{1}{\Delta t} - \frac{A}{\Delta x^2} + \frac{C}{2}$$

$$d_3 = \frac{A}{2\Delta x^2} + \frac{B}{4\Delta x}$$

Because of the boundary conditions, the first and the last constants need to modified. The Dirichlet boundary conditions are specified on both the membrane surfaces for the Nernst–Planck equation. The modifications are as follows,

$$d^1 = d_1 c_i^{j,0} + d_2 c_i^{j,1} + d_3 c_i^{j,2} - \alpha c_i^{j+1,0}$$

and

$$d^m = d_1 c_i^{j,m-1} + d_2 c_i^{j,m} + d_3 c_i^{j,m+1} - \gamma c_i^{j+1,m}$$

At any time level j, the concentration of ion *i* for the next time level j+1 can be found by solving the following matrix equation,

$$\begin{pmatrix} \beta^1 & \gamma^1 & 0 & 0 & 0 & 0 & 0 & 0 & 0 \\ \alpha^2 & \beta^2 & \gamma^2 & 0 & 0 & 0 & 0 & 0 & 0 \\ 0 & \alpha^3 & \beta^3 & \gamma^3 & 0 & 0 & 0 & 0 & 0 \\ \cdot & \cdot & \cdot & \cdot & \cdot & \cdot & \cdot & \cdot & \cdot \\ \cdot & \cdot & \cdot & \cdot & \cdot & \cdot & \cdot & \cdot & \cdot \\ \cdot & \cdot & \cdot & \cdot & \cdot & \cdot & \cdot & \cdot & \cdot \\ 0 & 0 & 0 & 0 & 0 & 0 & \alpha^{m-1} & \beta^{m-1} & \gamma^{m-1} \\ 0 & 0 & 0 & 0 & 0 & 0 & 0 & \alpha^m & \beta^m \end{pmatrix} \begin{pmatrix} c_i^{j+1,1} \\ c_i^{j+1,2} \\ c_i^{j+1,3} \\ \cdot \\ \cdot \\ \cdot \\ c_i^{j+1,m-1} \\ c_i^{j+1,m} \end{pmatrix} = \begin{pmatrix} d^1 \\ d^2 \\ d^3 \\ \cdot \\ \cdot \\ \cdot \\ d^{m-1} \\ d^m \end{pmatrix} \tag{11.69}$$

The equation is readily solvable by Thomas elimination algorithm.

The Poisson equation can be readily solved using the central difference method at any time step. The central finite difference equation for the Poisson equation

$$\alpha \varphi_i^{k+1} - \beta \varphi_i^k + \gamma \varphi_i^{k-1} = d^k \tag{11.70}$$

where

$$\alpha = \frac{1}{\Delta x^2}$$

$$\beta = -\frac{2}{\Delta x^2}$$

$$\gamma - \frac{1}{\Delta x^2}$$

For $k = 1, 2, \ldots, m\text{-}1$.
and

$$d^k = -\frac{F}{\varepsilon\varepsilon_0}\left(X + \sum_i z_i c_i^k\right)$$

The Dirichlet boundary condition on the left membrane surface is incorporated in the first linear equation as

$$d^1 = -\frac{F}{\varepsilon\varepsilon_0}\left(X + \sum_i z_i c_i^1\right) - \alpha\psi^{j,0}$$

The Neumann boundary on the right membrane surface is used to set up the linear equation for the last linear equation ($k = m$) as

$$\alpha = -\frac{1}{\Delta x}$$

$$\beta = \frac{1}{\Delta x}$$

$$\gamma = 0$$

$$d^m = \frac{\lambda\sigma_L}{\varepsilon\varepsilon_0}$$

A matrix equation similar to that for the Nernst–Planck equation is obtained, which can be solved with the same Thomas algorithm. The only difference is that the number of unknowns increases from m–1 to m due to the Neumann boundary condition on the right membrane surface.

11.7 Boundary Updating Scheme

The numerical solutions at time step j = 0 are first sought for the initially uncharged membrane, i.e., with the boundary conditions

$$\varphi^0 = 0,\ \varphi^m = \varphi^{m-1} \tag{11.71}$$

Because the membrane is uncharged, the initial ion concentrations on the membrane surfaces are equal to their bulk concentrations,

$$c_i^0 = c_{i0} \ and \ c_i^m = c_{iL} \tag{11.72}$$

With boundary conditions Eqs. (11.71) and (11.72), the numerical solutions of NPP equations can be obtained.

With the numerical solution, ion fluxes at the membrane surfaces $x = 0$, and L can be calculated with

$$J_{i0}^{\ 0} = vc_i^{\ 0} - D_i \frac{c_i^{\ 1} - c_i^{\ 0}}{\Delta x} - \frac{DFz_i}{RT} c_i^{\ 0} \frac{\varphi^1 - \varphi^0}{\Delta x}$$

$$J_{iL}^{\ 0} = vc_i^{\ m} - D_i \frac{c_i^{\ m} - c_i^{\ m-1}}{\Delta x} - \frac{DFz_i}{RT} c_i^{\ m} \frac{\varphi^m - \varphi^{m-1}}{\Delta x}$$

where J_{i0} and J_{iL} are fluxes at left and right membrane surfaces, respectively, the subscripts *0, 1, m–1, m* are the spatial nodes in the finite difference method with node *0* and node *m* on the left surface and right surface of the membrane, respectively.

For time step j > 0, the cumulative charge in the two transition layers can be updated at the beginning of every time step with

$$q_0^{\ j} = q_0^{\ j-1} - F\left(\sum_i z_i J_{i0}^{\ j-1}\right)\Delta t$$

$$q_L^{\ j} = q_L^{\ j-1} + F\left(\sum_i z_i J_{iL}^{\ j-1}\right)\Delta t$$

For the initially uncharged membrane, $q_0^{\ 0} = q_L^{\ 0} = 0$. Now, the first type of (Dirichlet) boundary condition for the Poisson equation on the left membrane surface can be specified as

$$\varphi_0^{\ j} = -\frac{\lambda}{\varepsilon\varepsilon_0}\sigma_0^{\ j} \tag{11.73}$$

The second type of (Neumann) boundary condition for the Poisson equation on the right membrane surface can be specified as

$$\varphi_L^{\ j,m} - \varphi_L^{\ j,m-1} = \frac{\sigma_L^{\ j}}{\varepsilon\varepsilon_0}\Delta x \tag{11.74}$$

The ion concentrations on the left and right membrane surfaces are readily determined as

$$c_i^{\ j,0} = c_{i0} e^{\frac{z_i F}{RT}\frac{\lambda}{\varepsilon\varepsilon_0}\sigma_0^{\ j}} \tag{11.75}$$

$$c_i^{\ j,m} = c_{iL} e^{-\frac{z_i F}{RT}\frac{\lambda_L}{\varepsilon\varepsilon_0}\sigma_L^{\ j}} \tag{11.76}$$

Eqs. (11.73)–(11.76) are the updated boundary conditions at the current time step. Now the solutions for the NPP equations can be sought at this time level. When the new solutions are obtained, the boundary conditions will be updated for the next time step with the same procedure.

A flowchart of the numerical procedure for the coupled NPP equations is presented in Figure 11.15. Because the solutions for the Nernst–Planck

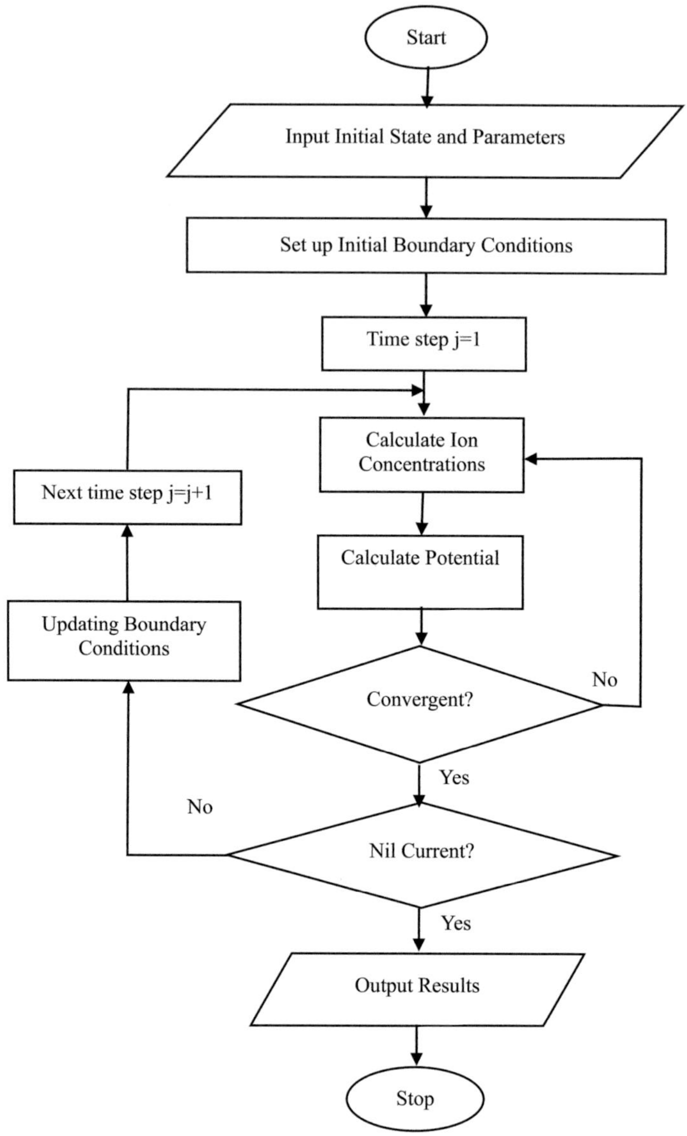

FIGURE 11.15
Flowchart for numerical solution of the coupled Nernst–Planck and Poisson equations with boundary condition updating scheme.

equation and the Poisson equation are coupled, iterations between numerical solutions of the two equations are required for convergence. The computation is executed along the time steps until the preset period is reached, or a steady state is obtained.

11.8 A Case Study

A case study for ion transport across a membrane is presented to demonstrate the numerical method for NPP equations with the boundary updating scheme. In this case, solutions of hydrochloric acid (HCl) of different concentrations are placed on two sides of a membrane. The membrane is treated as a porous layer that allows free passage of ions by diffusion and migration but acts as a barrier to prevent mechanical mixing (no convection) between two solutions. No viscous flow through the membrane can be practically achieved by application of hydraulic pressure to balance the osmotic pressure. The membrane is further assumed to have no fixed charge on it and to have no specific interactions with the ions (no partition between the membrane surface and solution). Protons and chlorides move across the membrane initially under the concentration difference. Because of the difference between the mobilities of protons and chlorides, a diffusive electric potential develops on the membrane. The membrane potential accelerates the slow chlorides but retards the fast protons until a "null current" condition is reached. This case will be investigated with the numerical procedure.

A program is developed in the C++ language for the numerical the solution of NPP equations (Attached to the end of this chapter). Parameters used in the simulation are listed in Table 11.1. Calculation was completed on a personal computer in a fraction of a second. The results are discussed below.

TABLE 11.1

Default Parameter Values Used in the Case Study

Parameter	Symbol	Unit	Value
Membrane permittivity	$\varepsilon\varepsilon_0$	F/m	6.92×10^{-10}
Membrane thickness	L	m	0.2×10^{-6}
Temperature	T	K	293.15
Time step	Δt	s	10^{-8}
Number of spatial steps	N		2000
Number of time steps	M		10000
Number of ions			2
Valence of H^+	z_+		+1
Valence of Cl^-	z_-		−1
Diffusivity of H^+	D_+	m^2/s	9.30×10^{-9}
Diffusivity of Cl^-	D_-	m^2/s	2.03×10^{-9}
Concentrations on the left side	c_{i0}	mol/m^3	10
Concentrations on the right side	c_{iL}	mol/m^3	1

TABLE 11.2

Boundary Conditions on the Nernst–Planck–Poisson Equations

Parameter	Left Side	Right Side
Charge (C/m^2)	-2.527×10^{-5}	1.979×10^{-4}
Potential Difference (mV)	1.10	2.72
H^+ concentration (mol/m^3)	9.957	1.114
Cl^- concentration (mol/m^3)	10.04	0.898
H^+ flux ($mol/m^2{\cdot}s$)	0.150	0.150
Cl^- flux ($mol/m^2{\cdot}s$)	0.150	0.150

11.8.1 A True Solution of Nernst–Planck–Poisson Equations

For solution of NPP equations to be valid, boundary conditions on ion concentrations for the Nernst–Planck equation have to be consistent with boundary conditions on potentials for the Poisson equation. Charges in transition layers, potential differences across transition layers, ion concentrations on the membrane surfaces, and ion fluxes across the membrane surfaces from the numerical solution are listed in Table 11.2. Because proton H^+ has a bigger mobility than chloride Cl^-, positive charge accumulates in the right transition layer, and the negative charge is left in the left transition layer. Potential differences across the transition layers are 1.10 mV (the left membrane surface over the bulk solution) and 2.72 mV (the bulk solution over the right membrane surface), respectively. Proton concentration is lower than chloride concentration on the left membrane surface, but it is higher than chloride concentration on the right membrane surface. It can be further verified that the ion concentrations on the membrane surfaces satisfy the Poisson–Boltzmann constraint. It means that they are related to the potential differences of transition layers with Eqs. (11.62) and (11.63). Therefore, this numerical solution is the true solution of the differential equations obtained with consistent boundary conditions.

It can also be seen from Table 11.2 that fluxes of protons and chlorides entering the membrane at the left surface and leaving the membrane at the right surface are all equal. It indicates that a steady state of ion transport has been achieved. Therefore, the solution of Eq. (11.53) at steady state, which cannot be pursued directly due to the difficulty in setting up boundary conditions, can be obtained naturally with the numerical procedure for the unsteady state ion transport Eq. (11.52).

11.8.2 Features of Ion Transport

Fig. 11.16 shows distributions of H^+ and Cl^- ions across the membrane thickness and transition layers adjacent to the membrane surfaces. The membrane is located from $x = 0$ to 0.2 μm. The left transition layer lies on $x < 0$, and the right transition layer lies on $x > 0.2$ μm. Ion concentrations along the

TABLE 11.3

Concentrations of Ions Along the Membrane Thickness

x (μm)	H^+ (mol/m³)	Cl^- (mol/m³)
0.00	9.957	10.040
0.02	9.099	9.102
0.04	8.199	8.202
0.06	7.299	7.303
0.08	6.398	6.404
0.10	5.497	5.505
0.12	4.596	4.607
0.14	3.694	3.712
0.16	2.789	2.821
0.18	1.889	1.931
0.20	1.114	0.898

membrane thickness are taken from the numerical solutions of the Nernst–Planck equation, while the ion concentrations in the transition layers are calculated using Eqs. (11.18) and (11.28) with the charges in transition layers obtained from the numerical solution as listed in Table 11.2.

Both H^+ and Cl^- concentrations are approximated well with straight lines almost across the entire membrane thickness. Though it looks as through the two lines coincide with each other in the figure, except for a small region near the right surface, concentrations of the two ions are noticeably different in numerical values in the entire membrane thickness. Ion concentrations at a few points along the membrane thickness are shown in Table 13.3. The difference between concentrations of cations and anions is of critical importance for ion transport across the membrane because it is the source of electromotive force for ion migration.

Net electric charge density is a result of the unbalanced distribution of cations and anions, which is calculated with

$$q(x) = F\left[c_H(x) - c_{Cl}(x)\right] \tag{11.77}$$

where q is the volumetric charge density and F is the Faraday's constant.

Charge density is also presented in Figure 11.16. There are a few important observations about the charge density. First and most importantly, the local electroneutrality that charge is balanced at every point along the membrane thickness is not true. There are net charges all over the membrane thickness except at one point, across which the charge changes sign from negative to positive. Second, there are two charge peaks near the membrane surfaces, which are a small negative one on the left membrane surface and a big

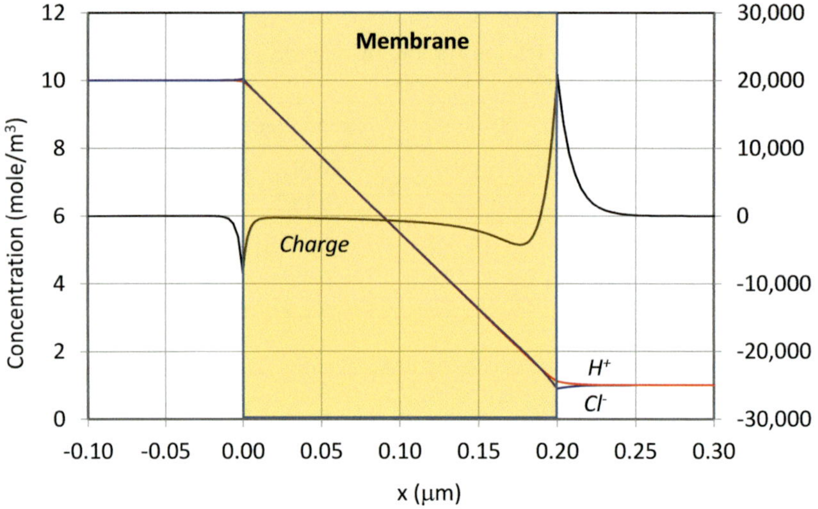

FIGURE 11.16
Distributions of ion concentrations and net charge across the membrane thickness and transition layers.

positive one on the right membrane surface. Furthermore, charge changes nonlinearly along the membrane thickness with a secondary minimum before it shoots up near the right membrane surface.

Profiles of potential and electric field along the membrane thickness from the numerical solutions are presented in Figure 11.17. Also presented in the

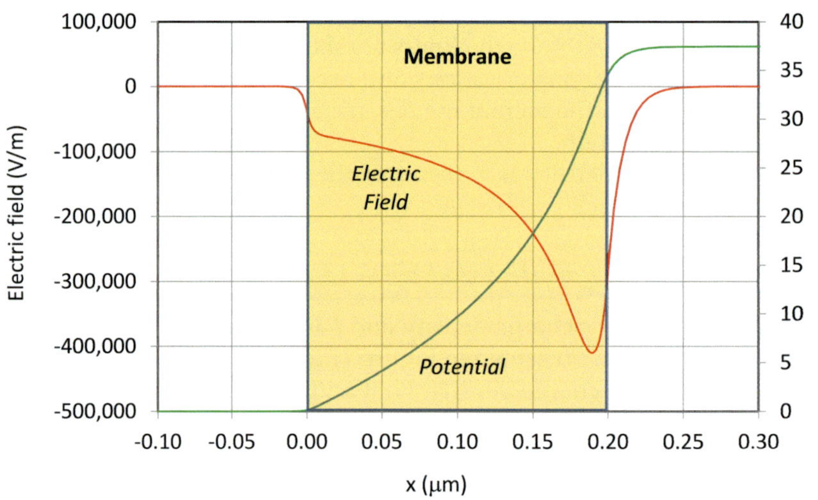

FIGURE 11.17
Distributions of potential and electric field across the membrane and transition layers.

figure are the potential and electric field in transition layers, which are calculated using Eq. (11.28) with the charge on the membrane surfaces from Table 11.2. Potential starts from zero in the bulk solution on the left side and increases through the membrane and transition layers monotonically to the bulk solution on the right side of the membrane. The electric field behaves more complicatedly and drastically than the potential across the membrane and transition layers. It can be seen that the electric field remains negative over the entire membrane thickness and transition layers. It means that the electromotive force always to retard cations but accelerates anions to move across the membrane system. The electric field varies significantly inside the membrane, with the strongest negative electric field occurring before the right membrane surface.

It can be seen from Figures 11.10 and 11.11, a relatively narrow transition layer occurs on the left membrane surface with a thickness of about 0.01 μm. The transition layer on the right side is much wider, with a thickness of about 0.05 μm. The difference in thickness of transition layers reflects the impact of ion concentrations on the range of electric interactions in solution.

11.8.3 A Powerful Tool for Studies of Ion Transport Across the Membrane

Because of the complexity involving electrical migration, ion transport across the membrane represents a serious challenge in the field of membrane science and technology. The membrane potential is a natural consequence induced by the exchange of ions between the membrane and solutions and the transports of ions with different mobilities. Unfortunately, quantification of the membrane potentials in general cannot be done, even though they are of profound importance to understand the chemical or biological processes involved. At this moment, the membrane potentials can only be calculated analytically for a few limited simple cases, such as those involve a solution of two monovalent ions at steady state. Assumptions of either electroneutrality or a constant electric field on the membrane have to be used to derive the analytical expressions. However, as shown in this case study, neither electroneutrality nor a constant electric field is valid.

Numerical methods have to be used to determine the membrane potential when multivalent ions are involved in the membrane transport. It is well established that ion transport across a membrane is governed by NPP equations. Numerical methods for differential equations are also well developed. However, as discussed in this chapter, a particular challenge to having valid numerical solutions for ion transport across the membrane is to set up appropriate boundary conditions for governing equations. Solutions of differential equations are heavily dependent on the boundary conditions. Completely different solutions can be obtained from the same differential equations with different boundary conditions.

Particularly, the boundary conditions for the Nernst–Planck equation and the Poisson equation have to be consistent with each other because of the underlying physical principles. Because independently specified boundary conditions for each of the equations may represent irrational cases that do not exist physically, precautions have to be taken in the pursuit of numerical solutions for NPP equations. The updating scheme introduced can ensure that the boundary conditions of the two equations are consistent during the solution process.

The case study presented here is one of the simplest cases for ion transport across the membrane. In this case, there are only two monovalent ions involved without a fixed charge, an external electric field, and convection. The C++ program is attached at the end of this chapter that can be easily modified to include a greater number of ions with different (multi)valence, the membrane with fixed charges, viscous flow, and the existence of an external electric field, as in real-world applications in reverse osmosis and electrodialysis processes. The same ideas and numerical procedure may also be used in the study of ion transport across biological membranes.

11.9 Problems

1. Calculate the force between two elementary charges of the same sign separated by a distance of 0.1 nm in vacuum, air, and water.
2. Determine the electric field strength at a point on the symmetry line of a charged rod as shown in the figure below in a vacuum. The charge density on the rod is σ C/m. The distance of the point to the rod is y.

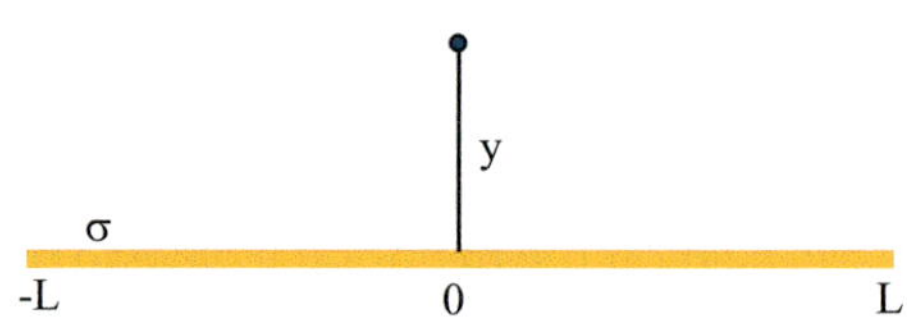

3. A large plate in vacuum has a charge density of 1 C/m^2; calculate the electric field strength of a point over the plate at a distance much smaller than the plate size.
4. A membrane with a charge density of -2.0×10^{-4} C/m^2 is submerged in a NaCl solution of 1 mol/m^3. (1) Calculate the Debye length of the solution. (2) Calculate the surface potential at the membrane surface. (3) Calculate the concentrations of sodium and chloride ions in the solution at the membrane surface.
5. The electric potential difference across a membrane is 4.0×10^{-2} V, and the membrane thickness is 0.1 μm. Calculate (a) the electrostatic

forces on the sodium ion and the chloride ion, and (b) the electromigration velocity of the sodium ion and the chloride ion across the membrane. Assume the diffusion coefficients of sodium ions and chloride ions inside the membrane are the same as in water, which are 1.33×10^{-9} and 1.78×10^{-9} m^2/s at 20°C, respectively.

6. A solution contains 0.1 M Na^+, 0.05 M Ca^{2+}, and 0.2 M Cl^-. Calculate the Debye length in the solution at 20°C.
7. A membrane with a surface charge of 0.03 C/m^2 is in contact with a solution of 0.05 M sodium chloride. Calculate the electrical potential on the membrane surface.
8. In Question 7, calculate the potentials in solution at the distances of 0.25, 0.5, 1.0, and 2.0 times of the Debye length.
9. Determine the equilibrium (Nernst) potential at 20°C of an ion-selective membrane that separates two solutions of the same salt, $CaCl_2$, but with different concentrations of 0.1 M and 0.05 M, respectively. Cl^- ions cannot pass through the membrane, but Ca^{2+} ions can.
10. The Donnan potential develops at the interface when a membrane with fixed charge (ion-selective or exchange membrane) is in contact with an electrolyte solution. Calculate the Donnan potential for the membrane of fixed charge density of 0.1 mol/m^3 and the monovalent electrolyte concentrations of 0.5 mol/m^3.
11. In Question 10, determine the cation and anion concentrations inside the membrane at equilibrium.
12. A porous membrane separates the two solutions of monovalent salt of different concentrations. There is no bulk flow across the membrane, but both cations and anions can transport through the membrane by diffusion. The salt concentrations in the solutions are 0.1 M and 0.02 M, respectively. The diffusion coefficients of cations and anions are 3×10^{-9} m^2/s and 1×10^{-9} m^2/s, respectively. Calculate the diffusion potentials when the steady state is reached.
13. In Question 12, if the membrane is negatively charged with a density of 0.05 M (mol/L). Determine the TMS potential across the membrane.
14. The potential increases from left to right across a membrane 1 μm thick with a difference of 30 mV. The diffusion coefficient of Na^+, Ca^{2+}, Cl^-, and SO_4^{2-} are 1.33×10^{-9}, 0.79×10^{-9}, 2.03×10^{-9}, and 1.07×10^{-9} m^2/s, respectively.
    a. Calculate the electric field strength inside the membrane, assuming a linear distribution of membrane potential across the membrane.
    b. Determine electromigration velocities of Na^+, Ca^{2+}, Cl^-, and SO_4^{2-} ions inside the membrane.

15. In Question 14, if the viscous flow velocity is 0.1 m/s and the Na^+ and Cl^- concentrations are 0.1 M on the left side and 0.01 M on the right side of the membrane, calculate the fluxes of Na^+ and Cl^- ions across the membrane. Assume a linear distribution of ion concentrations across the membrane thickness.
16. Determine the electric potential as a function of the thickness of a membrane. The membrane has a constant charge density of 5 mole/m^3 and a thickness of 100 μm. The electric strength on one side of the membrane is zero.
17. A membrane 50 μm thick has a constant charge across the membrane thickness. The potential on the two sides of the membrane is 10 and 50 mV, respectively. Determine the membrane charge density.
18. The ion concentrations and electric potential on the membrane surface are not independent but related to each other with the Poisson–Boltzmann equation. In one case, the potential difference across the transition layer between a membrane surface and a bulk solution is 35 mV. The solution contains NaCl of 0.1 mol/m^3. Calculate the concentrations of sodium ions and chloride ions on the membrane surface (Temperature is 20°C).
19. At steady state, the net charge in the transition layers on the left and right sides of a membrane is $\sigma_0 = -2.6 \times 10^{-5}\,\text{C/m}^2$ and $\sigma_L = 1.50 \times 10^{-4}\,\text{C/m}^2$ as shown in the figure below.
    a. What is the net charge on the membrane?
    b. What is the potential difference across the left transition layer $\Delta\varphi_0$? and
    c. What is the potential difference across the right transition layer $\Delta\varphi_L$?

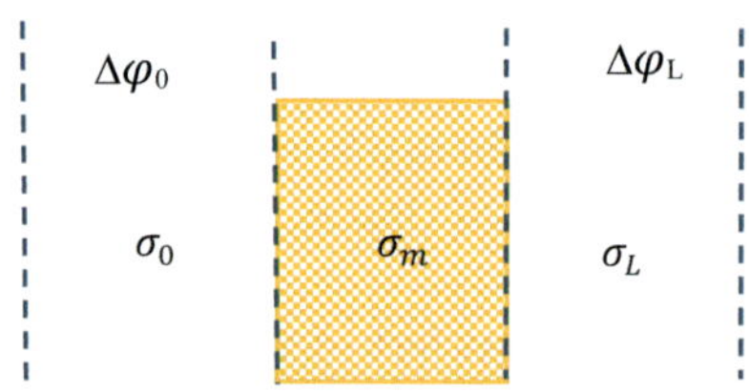

20. Construct the equation for potential in the right-side transition layer. Show that the Neumann boundary condition on the right membrane surface can be obtained simply by taking the derivative of the potential equation at the membrane surface.
21. Why is it difficult to set the consistent boundary conditions for the NPP equation at steady state?

22. Numerical studies of NPP equations are frequently reported in the literature without sufficient attention to the consistency of potential and ion concentrations on membrane surfaces (the conditions). What are your opinions about these numerical methods for the study of ion transport across the membrane?
23. Using the provided C++ program, carry out numerical solutions of the NPP equation with parameters different from Table 13.1.
24. The assumption of electroneutrality across the membrane thickness is commonly used in the studies of ion transport across the membrane and potential development on the membrane. What is the validity of this assumption according to the case study reported in this chapter?
25. A constant electric field is another commonly used assumption for membrane potential and ion transport across the membrane. How about its validity from the results of the case study in this chapter?

References

1. Bard, A.J., & Faulkner, L.F. (1980). Electrochemical Methods – Fundamentals and Applications. Wiley, New York.
2. Baschuk, J.J., & Li, X. (2010). Modeling of ion and water transport in the polymer electrolyte membrane of PEM fuel cells. International Journal of Hydrogen Energy, 35, 5095–5103.
3. Tong, X., Zhang, B., Fan, Y., & Chen, Y. (2017). Mechanism exploration of ion transport in nanocomposite cation exchange membranes. ACS Applied Materials & Interfaces, 9(15), 13491–13499.
4. Nonner, W., & Eisenberg, B. (2000). Electrodiffusion in Ionic Channels of Biological Membranes. Journal of Molecular Liquids, 87, 149–162.
5. Shi, L., Rossi, R., Son, M., Hall, D.M., Hickner, M.A., Gorski, C.A., & Logan, B.E. (2020). Using reverse osmosis membranes to control ion transport during water electrolysis. Energy and Environmental Science, 13, 3138–3148.
6. Kwon, Y. (2001). Partition and Distribution Coefficients. Handbook of Essential Pharmacokinetics, Pharmacodynamics and Drug Metabolism for Industrial Scientists. Kluwer Academic/Plenum Publishers, New York, p. 44.
7. Bellona, C., Drewes, J.E., Xu, P., & Amy, G. (2004). Factors affecting the rejection of organic solutes during NF/RO treatment—A literature review. Water Research, 38, 2795–2809.
8. Galama, A.H., Post, J.W., Hamelers, H.V.M., Nikonenko, V.V., & Biesheuvel, P.M. (2016). On the origin of the membrane potential arising across densely charged ion exchange membranes: How well does the Teorell-Meyer-Sievers theory work? Journal of Membrane Science and Research, 2, 128–140.

9. Benavente, J., & Jonsson, G. (2000). Electrokinetic characterization of composite membranes: Estimation of different electrical contributions in pressure induced potential measured across reverse osmosis membranes. Journal of Membrane Science, 172, 189–197.
10. Dickinson, E.J.F., Freitag, L., & Compton, R.G. (2010). Dynamic theory of liquid junction potentials. The Journal of Physical Chemistry B, 114, 187–197.
11. Delemotte, L., Dehez, F., Treptow, W., & Tarek, M. (2008). Modeling membranes under a transmembrane potential. The Journal of Physical Chemistry B, 112(18), 5547–5550.
12. Goldman, D.E. (1942). Potential, impedance, and rectification in membranes. Journal of General Physiology, 27, 37–60.
13. Blank, M. (1987). The surface compartment model: A theory of ion transport focused on ionic processes in the electrical double layers at membrane protein surfaces. Biochimica et Biophysica Acta, 906, 277–294.
14. Kimani, E.M., Kemperman, A.J., van der Meer, W., & Biesheuvel, P.M. (2021). Multicomponent mass transport modeling of water desalination by reverse osmosis including ion pair formation. The Journal of Chemical Physics, 154(12), 124501.
15. Rohman, F.S., & Aziz, N. (2008). Mathematical model of ion transport in electrodialysis process. Bulletin of Chemical Reaction Engineering & Catalysis, 3, 3–8.
16. Nikonenko, V., Zabolotsky, V., Larchet, C., Auclair, B., & Pourcelly, G. (2002). Mathematical description of ion transport in membrane systems. Desalination, 147, 369–374.
17. Miękisz, J., Gomułkiewicz, J., & Miekisz, S. (2014). Mathematical models of ion transport through cell membrane channels. Mathematica Applicanda, 42, 39–62.
18. Galama, A.H., Post, J.W., Cohen Stuart, M.A., & Biesheuvel, P.M. (2013). Validity of the Boltzmann equation to describe Donnan equilibrium at the membrane–solution interface. Journal of Membrane Science, 442, 131–139.
19. Cımen, A., Ersöz, M., & Yıldız, S. (2012). The Donnan potentials of polysulfonated cation exchange membranes. Gazi University Journal of Science, 25(2), 355–362.
20. Benevante, J., Munoz, A., Heredia, A., & Canas, A. (1999). Fixed charge and transport numbers in isolated pepper fruit cuticles from membrane potential measurements: Donnan and diffusion potential contributions. Colloids and Surfaces A: Physicochemical and Engineering Aspects, 159, 423.
21. Hodgkin, A.L., & Huxley, A.F. (1952). A quantitative description of membrane current and its application to conduction and excitation in nerve. The Journal of Physiology, 117, 500–544.
22. Hodgkin, A.L., & Katz, B. (1949). The effect of sodium ions on the electrical activity of the giant axon of the squid. The Journal of Physiology, 108, 37–77.
23. Ibanez, J.A., & Tejerina, F. (1982). Surface charge density, membrane potential and salt flux. Physics Letters A, 88, 262–264.
24. Pellicer, J., Aguilella, V.M., & Mafé, S. (1986). Validity of the electroneutrality and Goldman constant-field assumptions in describing the diffusion potential for ternary electrolyte systems in simple, porous membranes. Journal of Membrane Science, 29, 117–126.
25. Hirvonen, J., Murtomaki, L., & Kontturi, K. (1998). Effect of diffusion potential, osmosis and ion-exchange on transdermal drug delivery: Theory and experiments. Journal of Controlled Release, 56, 33–39.

26. Patlak, C.S. (1960). Derivation of an equation for the diffusion potential. Nature, 188, 944–945.
27. Kernan, R.P. (1962). Membrane potential changes during sodium transport in frog sartorius muscle. Nature, 193, 986–987.
28. Teorell, T. (1935). An attempt to formulate a quantitative theory of membrane permeability. Proceedings of the Society for Experimental Biology and Medicine, 33, 282–285.
29. Sten-Knudsen Ove (2002). Biological Membranes. Cambridge University Press, Cambridge, p. 277.
30. Feldberg, S.W. (2000). On the dilemma of the use of the electroneutrality constraint in electrochemical calculations. Electrochemistry Communications, 2, 453–456.
31. Syganow, A., & von Kitzing, E. (1999). (In)validity of the constant field and constant currents assumptions in theories of ion transport. Biophysical Journal, 76, 768–781.
32. Smith, C.P., & White, H.S. (1991). Theory of the voltammetric response of electrodes of submicron dimensions. Violation of electroneutrality in the presence of excess supporting electrolyte. Analytical Chemistry, 65, 3343–3353.
33. Park, J.H., & Jerome, J.W. (1997). Qualitative properties of steady-state Poisson-Nernst-Planck systems: Mathematical study. SIAM Journal on Applied Mathematics, 57, 609–630.
34. Buck, R.P. (1984). Kinetics of bulk and interfacial ionic motion: Microscopic bases and limits for the Nernst-Planck equation applied to membrane systems. Journal of Membrane Science, 17, 1–62.
35. Tedesco, M., Hamelers, H.V.M., & Biesheuvel, P.M. (2016). Nernst-Planck transport theory for (reverse) electrodialysis: I. Effect of co-ion transport through the membranes. Journal of Membrane Science, 510, 370–381.
36. Zhang, M. (2018). Boundary layer effects on ionic flows via classical Poisson-Nernst-Planck systems. Computational and Mathematical Biophysics, 6, 14–27.
37. Aitbayev, R., Bates, P.W., Lu, H., Zhang, L., & Zhang, M. (2019) Mathematical studies of Poisson-Nernst-Planck systems: Dynamics of ionic flows without electroneutrality conditions. Journal of Computational and Applied Mathematics, 362, 510–527.
38. Sun, Y., & Song, L. (2019). On rigorous definition of ion transport process and accurate determination of membrane potential at steady state. AIChE Journal, 65, e16715.
39. Zhou, W.W. (2004). Study on solute transport through RO/NF membranes. PhD thesis, National University of Singapore.
40. Brumleve, T.R., & Buck, R.P. (1978). Numerical solution of the Nernst-Planck and Poisson equation system with applications to membrane electrochemistry and solid state physics. Journal of Electroanalytical Chemistry, 90, 1–31.
41. Masayuki, K. (1995). Numerical analysis of the Nernst-Planck-Poisson system. Journal of Theoretical Biology, 177, 299–304.
42. Sokalski, T., Lingenfelter, P., & Lewenstam, A. (2003). Numerical solution of the coupled Nernst-Planck and Poisson equations for liquid junction and ion selective membrane potentials. The Journal of Physical Chemistry B, 107, 2443–2452

A Computer Program for Case Study

```
#include <iostream>
#include <fstream>
#include <iomanip>
#include <cmath>
#include <string>
using namespace std;

void tald(double A[], double B[], double C[], double D[],
            double Conc[], int M1, int M2);

int main ()
{

//Steady State Ion Transport
//Variables and constant declaration

//Constants
     double F=96485, R=8.31;

//eps is the water dielectric constant at 20 degree Celsius.
Unit is F/m.

      double thickness, T, eps, dx, dt, Q0, QL, dQ0, dQL, Debye0;
      double DebyeL, dU0, dUm, dUL, Convflux, Allowance;
      double C[2][3001][2],U[3001], X[3001], Charge[3001],
Field[3001];        double Flux[6][3001];
      double Cf[2], Ce[2], Z[2], Flux0[2][2], FluxL[2][2],
Diff[2];
      double a[3001], b[3001], c[3001], solution[3001],
d[3001];
      double pa, pb, pc, d1, d2, d3, conv, conv0, rate;

      int N, Iteration, Ncount;
      bool linear;
      string OutputFileName;

// Open input file and read parameter values

     ifstream input_file("Input.txt");
     input_file
          >> linear
          >> thickness >> T  >>  eps
          >> Cf[0] >> Cf[1]
          >> Ce[0] >> Ce[1]
          >> Z[0] >> Z[1]
          >> Diff[0] >> Diff[1]
```

```
          >> dt
          >> Allowance
          >> N
          >> Iteration
          >> OutputFileName;

//Information about the software
       if (linear) {cout << "Initial Linear Concentration
Distribution"
                            << "\n";}
       else {cout << "Initial Zero Concentration Distribution"
<< "\n";}

     cout << "The Name of Output File = " << OutputFileName <<
"\n"
                            << "\n";

//Preparation

     dx = thickness/double(N);
     X[0]=0.0;
     for (int i=1; i<=N; i++)

     {X[i]=X[i-1]+dx;}

     Debye0 = sqrt(eps*R*T/(F*F*(Z[0]*Z[0]*Cf[0] +
Z[1]*Z[1]*Cf[1])));
     DebyeL = sqrt(eps*R*T/(F*F*(Z[0]*Z[0]*Ce[0] +
Z[1]*Z[1]*Ce[1])));

//Initial conditions

    for (int j=0; j<2; j++)
    {
    rate = (Ce[j]-Cf[j])/double(N);
    if (linear)
        {
        for (int i=0; i<=N; i++)
            {
            C[j][i][0]= Cf[j] + rate*i;
            U[i]=0.0;
            }
        }
        else
        {
        for (int i=0; i<=N; i++)
            {
            C[j][i][0]= 0.0;
            U[i]=0.0;
               }
```

```
        }
        Flux0[0][j] = 0.0;
        FluxL[0][j] = 0.0;
    }
    for (int i=0; i<=N; i++)
    {U[i]=0.0;}

    Q0 = 0.0;
    QL = 0.0;
    dU0 = 0.0;
    dUm = 0.0;
    dUL = 0.0;
    Ncount = 0;

//Start Calculation

//Iteration for iteration in time domain

      while (Ncount <= Iteration)
      {

//Iteration to convergent

       conv0 = 1.0;
       while (conv0 > Allowance)

       {
       dQ0 = 0.0;
       dQL = 0.0;

//Numerical Soultion of Nernst-Planck Equation

       for (int j=0; j<2; j++)
       {
//Set up boundary conditions
       C[j][0][1] = Cf[j]*exp(-Z[j]*F*U[0]/(R*T));
       C[j][N][1] = Ce[j]*exp(Z[j]*F*dUL/(R*T));

//Prepare coefficient of the linear equation
       pa = Diff[j];
       for (int i=1; i<N; i++)
       {
       pb = Diff[j]*F*Z[j]*(U[i+1]-U[i-1])/(2.0*dx*R*T);
       pc = Diff[j]*F*Z[j]*(U[i+1]-2.0*U[i]+U[i-1])/
(dx*dx*R*T);

       a[i] = -pa/(2.0*dx*dx) + pb/(4.0*dx);
       b[i] = 1.0/dt + pa/(dx*dx) - pc/2.0;
       c[i] = -pa/(2.0*dx*dx) - pb/(4.0*dx);
```

```
      d1 = pa/(2.0*dx*dx) - pb/(4.0*dx);
      d2 = 1.0/dt - pa/(dx*dx) + pc/2.0;
      d3 = pa/(2.0*dx*dx) + pb/(4.0*dx);

      d[i] = d1*C[j][i-1][0] + d2*C[j][i][0] + d3*C[j][i+1][0];
      }

      d[1] = d[1]-a[1]*C[j][0][1];
      d[N-1] = d[N-1] - c[N-1]*C[j][N][1];

//Solving linear equations for concentration of ion j
      tald(a, b, c, d, solution, N-1, 1);

      for (int i=1; i<N; i++)
      {
      C[j][i][1] = solution[i];
      }
//Calculate the fluxes for ion j at x = 0 and x = L

     Flux0[1][j] = -Diff[j]*((C[j][2][1]-C[j][0][1])/(2.0*dx)
               +F*Z[j]*(U[2]-U[0])/(2.0*dx*R*T)*C[j][1][1]);
     FluxL[1][j] = -Diff[j]*((C[j][N][1]-C[j][N-2][1])/(2.0*dx)
               +F*Z[j]*(U[N]-U[N-2])/(2.0*dx*R*T)*C[j][N-1][1]);

//Calculate the net charge passing two sides of the membrane
     dQ0 = dQ0 -F*Z[j]*0.5*(Flux0[0][j] + Flux0[1][j])*dt;
     dQL = dQL +F*Z[j]*0.5*(FluxL[0][j] + FluxL[1][j])*dt;
     }

//Numerical solution of Poisson Equation

     U[0] = -Debye0*(Q0 + dQ0)/eps;
     dUL = DebyeL*(QL + dQL)/eps;

     for (int i=1; i<N; i++)
     {
     a[i] = 1.0;
     b[i] = -2.0;
     c[i] = 1.0;
     d[i] =  - dx*dx*F*(Z[0]*C[0][i][1]+Z[1]*C[1][i][1])/eps;
         }
     d[1] = d[1] - a[1]*U[0];
     a[N] = -1.0;
     b[N] = 1.0;
     d[N] = (QL + dQL)*dx/eps;

     tald(a, b, c, d, solution, N, 1);
```

```
//Check if U convergent

      conv = 0.0;
      for (int i=1; i<=N; i++)
      {
      conv = conv + (solution[i]-U[i])*(solution[i]-U[i]);
      U[i] = solution[i];
      }
      conv0 = conv;

      }

//Close loop of while for convergency

//Calculate the charge out two sides of the membrane

      Q0 += dQ0;
      QL += dQL;

      for (int j=0; j<2; j++)
      {
      Flux0[0][j] = Flux0[1][j];
      FluxL[0][j] = FluxL[1][j];
      }
      dU0 = -Debye0*Q0/eps;
      U[0] = dU0;
      dUm = U[N] - U[0];
      dUL = DebyeL*QL/eps;

      Convflux = pow((Z[0]*Flux0[0][0]+Z[1]*Flux0[0][1]), 2)
               + pow((Z[0]*FluxL[0][0]+Z[1]*FluxL[0][1]), 2);

      for (int j=0; j<2; j++)
      {
      for (int i=0; i<=N; i++)
          {
          C[j][i][0] = C[j][i][1];
            }
      }

      Ncount++;
      if(Ncount%1000 == 0)
      {cout << "  Time Steps =    "  << Ncount << "\n";}
      }

      cout << "\n";

for (int i=0; i<=N; i++)
      {
      Charge[i]= F*(C[0][i][1]*Z[0] + C[1][i][1]*Z[1]);
```

```
      for (int j=0;j<2; j++)
      {
      if(i>0 && i<N)
      {
      Flux[3*j][i] = -Diff[j]*(C[j][i+1][1]-C[j][i-1][1])/
(2.0*dx);
      Flux[3*j+1][i] = -Diff[j]*F*Z[j]*(U[i+1]-U[i-1])/(2.0
                *dx*R*T)*C[j][i][1];
      Flux[3*j+2][i] = Flux[3*j][i]+ Flux[3*j+1][i];
      Field[i] = -(U[i + 1] - U[i - 1]) / (2.0 * dx);
      }
      Flux[3*j][0] = -Diff[j]*(C[j][2][1]-C[j][0][1])/(2.0*dx);
      Flux[3*j+1][0]=-Diff[j]*F*Z[j]*(U[2]-U[0])/(2.0*dx*R*T)
                *C[j][1][1];
      Flux[3*j+2][0] = Flux[3*j][0] + Flux[3*j+1][0];
      Flux[3*j][N] = -Diff[j]*(C[j][N][1]-C[j][N-2][1])/(2.0*dx);
      Flux[3*j+1][N] = -Diff[j]*F*Z[j]*(U[N]-U[N-2])/(2.0*dx*R*T)
                *C[j][N-1][1];
      Flux[3*j+2][N] = Flux[3*j][N] + Flux[3*j+1][N];
      Field[0] = -(U[2] - U[0]) / (2.0 * dx);
      Field[N] = -(U[N] - U[N - 2]) / (2.0 * dx);
      }
      }

//Output results
// Open output file
// the output of x in micron and potential in mV

       ofstream output_file(OutputFileName);
       if (!output_file.is_open())
       {
       cout << "Output file could not be opened! Terminating!"
<< endl;
           return 1;
       }

       output_file << "CONDITIONS" << "\n" ;
       output_file << setw(13) << "Linear Init"  << setw(13)
                << boolalpha << linear << "\n";
       output_file << fixed;
       output_file.precision(0);
       output_file << setw(13) << "Space-Step="  << setw(13) <<
N
                << "\n";
       output_file.precision(2);
       output_file << setw(13) << "Temperature"  << setw(13)
<< T
                << "\n";
```

```
      output_file << scientific;
      output_file.precision(4);
      output_file  << setw(13)  << "Thickness ="  << setw(13)
                                << thickness << "\n";
      output_file  << setw(13) << "Dielectric="  << setw(13) << eps
                                << "\n";
      output_file  << setw(13) << "Cin - Ion1="  << setw(13)
<< Cf[0];
              output_file  << setw(13) << "Cin - Ion2="  <<
setw(13)
                       << Cf[1] << "\n";
      output_file  << setw(13) << "Cout -Ion1="  << setw(13)
<< Ce[0];
      output_file  << setw(13) << "Cout -Ion2="  << setw(13)
<< Ce[1]
                       << "\n";
      output_file  << setw(13) << "Val - Ion1="  << setw(13)
<< Z[0];
      output_file  << setw(13) << "Val - Ion2="  << setw(13)
<< Z[1]
                       << "\n";
      output_file  << setw(13) << "Diff -Ion1="  << setw(13)
<< Diff[0];
      output_file  << setw(13) << "Diff -Ion2="  << setw(13)
<< Diff[1]
                    << "\n" << "\n";
      output_file  << setw(13) << "Time  Int ="  << setw(13)
<< dt
                  << "\n" << "\n";

      output_file << fixed << "RESULTS  " << "\n";
      output_file.precision(0);
      output_file  << setw(13) << "Iteration ="  << setw(11)
                   << Iteration << "\n";

      output_file << scientific;
      output_file.precision(3);
      output_file  << setw(13) << "Flux-Squar="  << setw(13)
<< Convflux
                    << "\n";
      output_file  << setw(13) << "Conv-Crite="  << setw(13)
                  << Allowance << "\n";
      output_file << setw(13) << "Flux-Ion1 ="<< setw(13) <<
Flux0[0][0]
                   << setw(13) << FluxL[0][0] << "\n";
      output_file << setw(13) << "Flux-Ion2 ="<< setw(13) <<
Flux0[0][1]
                   << setw(13) << FluxL[0][1] << "\n";
```

```
      output_file   << setw(13) << "Q0 ="  << setw(13) << Q0
<< setw(13)
                    << "Qm =" << setw(13) << -Q0 - QL  <<
setw(13)
                    << "QL ="  << setw(13) << QL << "\n";
      output_file   << setw(13) << "dU0 ="  << setw(13) <<
dU0*1000
                    << setw(13) << "dUm =" << setw(13) <<
dUm*1000
                    << setw(13) << "dUL ="  << setw(13) <<
dUL*1000
                    << setw(13) << "Total U ="  << setw(13)
                            << dU0 +dUm+dUL << "\n" << "\n";

      output_file  << setw(13) << "X (miron)"  << setw(13)
<< "Conc1"
                 << setw(13) <<  "Conc2" << setw(13) <<
"Potent (mV)"
                 << setw(13) <<  "Charge"  << setw(13)
                 <<  "Flux1-Diff" << setw(13) <<  "Flux1-Drif"
                 << setw(13) <<  "Flux1-Totl"   << setw(13)
                 <<  "Flux2-Diff" << setw(13) <<  "Flux2-Drif"
                 << setw(13) <<  "Flux2-Totl" << setw(13)
                       << "Field" << "\n";

      for (int i=0; i<=19; i++)
      {
      output_file << setw(13) << X[i]*1e6  << setw(13)  <<
C[0][i][0]
                  << setw(13) <<  C[1][i][0] << setw(13) <<
U[i]*
                  1000  << setw(13) <<  Charge[i]  << setw(13)
                  <<  Flux[0][i] << setw(13) <<  Flux[1][i]
                  << setw(13) <<  Flux[2][i]  << setw(13)
                  <<  Flux[3][i] << setw(13) <<  Flux[4][i]
                  << setw(13) <<  Flux[5][i] << setw(13) <<
Field[i]
                  << "\n";
      }
      for (int i=20; i<=N-20; i+=10)
      {
      output_file << setw(13) << X[i]*1e6  << setw(13)  <<
C[0][i][0]
                 << setw(13) <<  C[1][i][0] << setw(13) <<
U[i]*
                 1000  << setw(13) <<  Charge[i]  << setw(13)
                 <<  Flux[0][i] << setw(13) <<  Flux[1][i]
                 << setw(13) <<  Flux[2][i]  << setw(13)
                 <<  Flux[3][i] << setw(13) <<  Flux[4][i]
```

```
                << setw(13) <<  Flux[5][i] << setw(13) <<
Field[i]
                 << "\n";
      }
      for (int i=N-21; i<=N; i++)
      {
      output_file << setw(13) << X[i]*1e6  << setw(13)  <<
C[0][i][0]
                 << setw(13) <<  C[1][i][0] << setw(13) <<
U[i]*
                 1000  << setw(13) <<  Charge[i]  << setw(13)
                 <<  Flux[0][i] << setw(13) <<  Flux[1][i]
                 << setw(13) <<  Flux[2][i]  << setw(13)
                 <<  Flux[3][i] << setw(13) <<  Flux[4][i]
                 << setw(13) <<  Flux[5][i] << setw(13) <<
Field[i]
                 << "\n";
      }
      cout << "DONE. Check the results please" << "\n";

      system("PAUSE");
      return 0;
}

//Function that solve the linear equation group

void tald(double A[], double B[], double C[], double D[],
         double Conc[], int M1, int M2)
{
      D[M2] = D[M2]/B[M2];
      for (int i=M2+1; i<=M1; i++)
      {
      B[i] = B[i] - A[i]*C[i-1]/B[i-1];
      D[i] = (D[i] - A[i]*D[i-1])/B[i];
      }
      Conc[M1] = D[M1];
      for (int i=1; i<=M1-M2; i++)
      {
      int j = M1 - i;
      Conc[j] = D[j] - C[j]*Conc[j+1]/B[j];
      }
}
```

Input file

```
  1
  0.20e-06  293.15  6.92e-10
  1.0e+00  1.0e+00
  1.0e-01  1.0e-01
  1.0      -1.0
  9.30e-09  2.03e-09
```

```
  1.0e-09
  1.0e-18
  2000
  10000
  Results.txt
     input_file       >> linear
                    >> thickness >> T  >>  eps
                       >> Cf[0] >> Cf[1]
                       >> Ce[0] >> Ce[1]
                       >> Z[0] >> Z[1]
                       >> Diff[0] >> Diff[1]
                       >> dt
                       >> Allowance
                       >> N
                       >> Iteration;
Note: 1. Concentration unit has to be mole/m3
      2. linear: 0-not; 1-yes
```

Appendices

Appendix I

TABLE A.1

SI Units

Quantity	SI Units	Symbol	Unit
Base units			
Length	Meter	m	
Mass	Kilogram	kg	
Time	Second	s	
Electrical current	Ampere	A	
Temperature	Kelvin	K	
Amount of substance	Mole	mol	
Luminous intensity	Candela	cd	
Some derived units			
Area		S	m^2
Volume		V	m^3
Density		ρ	kg/m^3
Velocity		v	m/s
Pressure	Pascal	Pa	$kg/(m \cdot s^2)$
Force	Newton	N	$kg \cdot m/s^2$
Energy	Joule	J	$kg \cdot m^2/s^2$
Power	Watt	W	$kg \cdot m^2/s^3$
Electrical potential	Volt	V	$kg \cdot m^2/(s^3 \cdot A)$
Electrical field		E	volt/m
Charge	Coulomb	C	A·s
Diffusion coefficient		D	m^2/s
Viscosity		μ	Pa·s

TABLE A.2

Physical Constants

Quantity	Symbol	Value	Unit
Acceleration of gravity	g	9.81	m/s^2
Avogadro number	N_{avo}	6.02×10^{23}	1/mol
Boltzmann constant	k	1.38×10^{-23}	J/K
Elementary charge	e	1.60×10^{-19}	C
Faraday constant	F	9.65×10^{4}	C/mol
Gas constant	R	8.31	J/K·mol
Vacuum permittivity	ε_0	8.85×10^{-12}	F/m

Appendix II

Commonly Used Mathematical Equation and Formulas

1. Solution of algebraic equations
   a. Quadratic equation

$$ax^2+bx+c=0$$

Solutions

$$x=\frac{-b\pm\sqrt{b^2-4ac}}{2a}$$

   b. Cubic equation

$$x^3+ax^2+bx+c=0$$

Let

$$Q=\frac{3b-a^2}{9}$$

$$R=\frac{9ab-27c-2a^2}{54}$$

$$S=\sqrt[3]{R+\sqrt{Q^3+R^2}}$$

$$T=\sqrt[3]{R-\sqrt{Q^3+R^2}}$$

Solutions

$$\begin{cases} x_1=S+T-\frac{1}{3}a \\ x_2=-\frac{1}{2}(S+T)-\frac{1}{3}a+\frac{1}{2}i(S-T) \\ x_3=-\frac{1}{2}(S+T)-\frac{1}{3}a-\frac{1}{2}i(S-T) \end{cases}$$

2. Taylor series

$$f(x)=f(a)+f'(a)(x-a)++\frac{f''(a)}{2!}(x-a)^2\cdots$$

$$\frac{1}{1+x}=1-x+x^2-x^3+x^4+\cdots\quad -1<x<1$$

$$e^x=1+x+\frac{x^2}{2!}+\frac{x^3}{3!}+\cdots\quad -\infty<x<\infty$$

$$\ln(1+x)=x-\frac{x^2}{2}+\frac{x^3}{3}-\frac{x^4}{4}+\cdots\quad -1<x\le 1$$

3. Derivatives

$$\frac{d}{dx}\left(e^x\right)=e^x$$

$$\frac{d}{dx}(uv)=u\frac{dv}{dx}+v\frac{du}{dx}$$

$$\frac{d}{dx}\left(\frac{u}{v}\right)=\frac{1}{v^2}\left(v\frac{du}{dx}-u\frac{dv}{dx}\right)$$

$$\frac{d}{dx}\left(u^n\right)=nu^{n-1}\frac{du}{dx}$$

$$\frac{dy}{dx}=\frac{dy}{du}\frac{du}{dx}$$

4. Integrations

$$\int u\,dv=uv-\int v\,du$$

$$\int u^n\,du=\frac{u^{n+1}}{n+1}$$

$$\int\frac{du}{u}=\ln|u|$$

$$\int e^u\,du-e^u$$

$$\int e^{au}\,du=\frac{e^{au}}{a}$$

5. Differential equations

   a. Separation of variables

$$f_1(x)g_1(y)dx+f_2(x)g_2(y)dy=0$$

   Solution

$$\int\frac{f_1(x)}{f_2(x)}dx+\int\frac{g_2(y)}{g_1(y)}dy=c$$

   b. Linear first-order

$$\frac{dy}{dx}+P(x)y=Q(x)$$

   Solution

$$ye^{\int P dx}=\int Qe^{\int P dx}\,dx+c$$

   b. Linear homogeneous second-order equation

$$\frac{d^2y}{dx^2}+a\frac{dy}{dx}+by=0$$

   Solution

   Let m_1 and m_2 be the roots of the characteristic equation

$$m^2+am+b=0$$

There are three cases.

Case 1. m_1 and m_2 are real and distinct:

$$y = c_1 e^{m_1 x} + c_2 e^{m_2 x}$$

Case 2. m_1 and m_2 are real and equal:

$$y = c_1 e^{m_1 x} + c_2 x e^{m_1 x}$$

Case 3. $m_1 = p + qi$, $m_2 = p - qi$:

$$y = e^{px}\left(c_1 \cos qx + c_2 \sin qx\right)$$

where $p = -a/2$, $q = \sqrt{b - a^2/4}$

Bibliography

Spiegel, M.R. (1969). Mathematical Handbook of Formulas and Tables. Schaum's Outline Series. McGraw-Hill, New York.

Appendix III

TABLE A.3

Physical Properties of Water

Temperature (°C)	Density (kg/m³)	Dynamic Viscosity (×10⁻³ Pa·s)	Molar Volume (L/mol)	Vapor Pressure (Pa)
0	999.8	1.781	18.00	610
5	1000.0	1.518	18.00	870
10	999.7	1.307	18.01	1230
15	999.1	1.139	18.02	1700
20	998.2	1.002	18.03	2340
25	995.7	0.890	18.08	3170
30	995.7	0.798	18.08	4240
40	992.2	0.653	18.14	7380
50	988.0	0.547	18.22	12,330
60	983.2	0.466	18.31	19,920
70	977.8	0.404	18.41	31,160
80	971.8	0.354	18.52	47,340
90	965.3	0.315	18.65	70,100
100	958.4	0.282	18.78	101,330

Source: Adapted from Venard, J.K., & Street, R.L. (1975). Elementary Fluid Mechanics, 5th ed. Wiley, New York.

Appendix IV

TABLE A.4

Diffusion Coefficients of Major Ions in Water

Ion	D (×10⁻⁹ m²/s)	Ion	D (×10⁻⁹ m²/s)
H^+	9.31	OH^-[a]	5.17
K^+[a]	1.92	Cl^-	2.03
Na^+	1.33	NO_3^-	1.90
Li^+	1.03	SO_4^{2-}	1.07
NH_4^+	1.96	HCO_3^-[a]	1.16
Ca^{2+}	0.79		
Mg^{2+}	0.71		

Appendix V

TABLE A.5

Diffusion Coefficients of Common Salt in Water[a]

Salt	D ($\times10^{-9}$ m²/s)	Ion	D ($\times10^{-9}$ m²/s)
HCl	3.34	$CaCl_2$	1.34
NH_4Cl	1.99	$MgCl_2$	0.25
NaCl	1.61	Na_2SO4	1.23
$NaNO_3$	1.57	$MgSO_4$[b]	0.85
LiCl	1.37		

Source: From Nielen et al. (2020). Ion specific effects on aqueous phase separation of responsive copolymers for sustainable membranes. Journal of Colloid and Interface Science, 576, 186–194.

[a] Calculated with the data from MacInnes, D.A. (1961). The Principles of Electrochemistry. Dover, New York, p. 342.

[b] From Applin, K.R., & Lasaga, A.C. (1984). The determination of SO_4^{2-}, $NaSO_4^-$, and $MgSO_4^\circ$ tracer diffusion coefficients and their application to diagenetic flux calculations. Geochimica et Cosmochimica Acta, 48, 2151–2162.

Index

For Product Safety Concerns and Information please contact our
EU representative GPSR@taylorandfrancis.com Taylor & Francis
Verlag GmbH, Kaufingerstraße 24, 80331 München, Germany